Published by NY Research Press,
23 West, 55th Street, Suite 816,
New York, NY 10019, USA
www.nyresearchpress.com

Metallurgy and Materials
Edited by Ricky Peyret

International Standard Book Number: 978-1-63238-320-4 (Hardback)

Printed in China.

Contents

Preface

Metallurgy is the branch of materials science and engineering that focuses on the study of the chemical and physical behavior of metals, their compounds with other metals and alloys. At the same time it is also a technology focused on metals, more precisely the science in the production of metals and their compounds and alloys for manufacturers and general consumers. The word metallurgy derives from the ancient Greek word metalourgos, which means 'worker in metal'. The history of metallurgy is an old one. The remains from ancient civilizations have yielded the fact that it has existed from the moment man discovered metal and fire and what could be done using both. One can say that mankind advanced through the ages on the foundation of metal resources. Metallurgy is itself divided into two branches - non-ferrous metallurgy and ferrous metallurgy which is based on iron and its compounds. The metallurgical process is a long one, starting from extracting from the ore to the end product needed, and needs specialized workers. Metallurgy is an essential part of human industry and economy and it cannot be denied that the skill needed in this field deserves a commendation.

This book is an effort to collate data pertinent to the field of metallurgy with research and statistics that underline the importance of it in modern industry. I would like to thank all those who have put an effort behind the research in this book. I would also like to thank my family and friends for their support in this endeavor.

Editor

Influence of the Slag Density on the Splashing Process in a Steelmaking Converter

Miguel A. Barron,[1] Dulce Y. Medina,[1] Isaias Hilerio,[1] and Gabriel Plascencia[2]

[1] *Departamento de Materiales, Universidad Autónoma Metropolitana Azcapotzalco, Avenida San Pablo 180, Colonia Reynosa-Tamaulipas, 02200 Mexico, DF, Mexico*
[2] *Centro de Investigacion e Innovacion Tecnologica-IPN, Cerrada Cecati S/N, Colonia Santa Catarina, 02250 Mexico, DF, Mexico*

Correspondence should be addressed to Miguel A. Barron; bmma@correo.azc.uam.mx

Academic Editors: S. Eckert and T. Yue

The way in which slag density influences the slag splashing phenomenon in an oxygen steelmaking converter is numerically analyzed in this work. Several values of the density of the slag are considered, and their effect on the global mass balance and slag average volume fraction on the sidewalls of the converter is studied using isothermal, two-dimensional transient computational fluid dynamics simulations. Diameter of the slag drops is determined from the slag density and the impact velocity of the nitrogen jet. Besides, the effect of the nitrogen jet Mach number on the slag splashing is simulated and discussed. A qualitative comparison between the computer simulations and results from the literature is made.

1. Introduction

It is known that the wear of refractory lining in an oxygen steelmaking converter for raw steel manufacturing is a factor which greatly influences the production costs. In the last two decades, slag splashing has emerged as a new technology to extend the lifetime of the converter refractory lining given that this technology reduces the wear associated to thermal and chemical attack by slag and mechanical impact [1, 2]. After the draining of steel, molten slag remaining at the converter bottom is splashed towards the converter sidewalls using a supersonic jet of gaseous nitrogen. Molten slag freezes at the converter walls and forms a protective coating that prevents the wear of the refractory lining. Nitrogen is injected into the converter through a water-cooled vertical lance which has several inclined convergent-divergent nozzles.

During the slag splashing process three main stages have been identified in the formation of the slag protective coating: transport of molten slag to the converter walls, adherence of the molten slag to the sidewalls, and freezing and hardening of the slag layer [3]. When the molten slag is transported to the converter sidewalls, two transport mechanisms are present: wash coating and ejection coating [4]. The first one occurs due to the bulk movement of the molten slag to rise above the initial level and the second one due to the ejection of slag droplets which adhere to the vessel sidewalls [5].

In recent years, several experimental studies on the slag splashing phenomenon have been reported. In these studies, physical scale models of the converter are employed, and cold water and air replace molten slag and nitrogen, respectively [4, 6, 7]. In [4] it is reported that large nozzle inclination and lance heights increase the splashing and the main mechanism of splashing changes from ejection to washing as the viscosity of the liquid is increased. In [6] it is reported that the amount of slag splashed is increased as lance is raised, but beyond a critical value of the lance height splashing decreases. In [7] it is shown that when the jet flow rate is increased or the liquid viscosity is decreased, the lower regions of the walls are splashed at a much greater rate. It is also shown that by changing the lance height, more liquid is deposited in the upper regions of the vessel at the expense of lower regions.

On the other hand, some numerical studies of gaseous jets impinging on a liquid surface are reported using computational fluid dynamics simulations [8–10]. Agreement is reported between numerical and water model results; unfortunately, these studies are mainly focused on the jet-surface interaction and the prediction of the surface topography, rather than on liquid splashing. Recently, two studies [11, 12] by some of the authors of the present work were reported on the simulation of the slag splashing phenomenon using computational fluid dynamics, considering the dimensions of actual industrial converters and the influence of properties of molten slag. Particularly in [12] the influence of the slag viscosity on the slag splashing process was studied, and its conclusions are as follows: (i) ejection mechanism becomes dominant as viscosity is decreased; (ii) washing mechanism becomes dominant as viscosity is increased; (iii) efficiency of the slag splashing process is increased as slag viscosity decreases.

In this work the influence of the slag density on the slag splashing process is analyzed using computational fluid dynamics simulations. Values of the slag density ranging from 2000 to 3000 kg m^{-3} are considered. This range is typical of CaO–SiO_2 molten slags found in steelmaking [13]. Physical dimensions of the converter correspond to an actual industrial steelmaking converter of 150 metric tons. The volume fraction of slag in the converter sidewalls is employed for quantitative evaluation of the slag splashing efficiency. In order to detect spitting of slag from the converter mouth, the global mass balance is determined. Besides, the effect of the nitrogen jet Mach number on the slag splashing is simulated and discussed. Finally, a qualitative comparison between the computational fluid dynamics simulations and results from the literature is made.

2. Mathematical Modeling

During the slag splashing process, molten slag located in the converter bottom is splashed to the converter sidewalls by means of a supersonic jet of gaseous nitrogen. The momentum of the nitrogen jet is transferred to the molten slag, which causes the slag to be stirred and ejected by the action of a standing wave and high shear forces, respectively [4]. Predominance of one of the coating mechanisms, that is, washing or ejection, depends on factors such as the jet characteristics (velocity, exit angle), the operating conditions (lance height, molten slag depth) and the slag properties (viscosity, density, and temperature). Inertial, gravitational, viscous, and interfacial forces are applied on both phases, that is, gaseous nitrogen and molten slag, causing a nonisothermal multiphase flow. To model the complex system here considered, equations which govern the fluid flow, the mass balance, the turbulence, and the multiphase flow are required.

The flow of an incompressible newtonian fluid is governed by the Navier-Stokes equations, which in vector form are expressed as [14]:

$$\rho\left(\frac{\partial \vec{u}}{\partial t} + \vec{u}\cdot\nabla\vec{u}\right) = -\nabla p + \mu_{\text{eff}}\nabla^2\vec{u} + \rho\vec{g} + F, \quad (1)$$

where ρ is the fluid density, $\vec{u}$ is the velocity vector, t is time, p is pressure, μ_{eff} is the effective fluid viscosity, $\vec{g}$ is the gravity vector, and F represents additional forces applied or acting on the fluid. To maintain the mass balance in the system, the continuity equation [14] must be solved:

$$\frac{\partial u_j}{\partial x_j} = 0, \quad (2)$$

where u_j and x_j are the jth components of the velocity vector and the coordinate system, respectively.

Turbulence was simulated by means of the classical two equations K-ε model [15]:

$$\rho v_j \frac{\partial K}{\partial x_j} = \frac{\partial}{\partial x_j}\left(\frac{\mu_t}{\sigma_K}\frac{\partial K}{\partial x_j}\right) + \mu_t \frac{\partial v_j}{\partial x_i}\left(\frac{\partial v_i}{\partial x_j} + \frac{\partial v_j}{\partial x_i}\right) - \rho\varepsilon,$$
$$\rho v_j \frac{\partial \varepsilon}{\partial x_j} = \frac{\partial}{\partial x_j}\left(\frac{\mu_t}{\sigma_\varepsilon}\frac{\partial \varepsilon}{\partial x_j}\right) + C_1\mu_t \frac{\varepsilon}{K}\frac{\partial v_j}{\partial x_i}\left(\frac{\partial v_i}{\partial x_j} + \frac{\partial v_j}{\partial x_i}\right) - C_2\frac{\varepsilon}{K}\rho\varepsilon, \quad (3)$$

where K is the turbulent kinetic energy, ε is the dissipation rate, and μ_t is the turbulent viscosity. Besides, $\sigma_K = 1.0$, $\sigma_\varepsilon = 1.3$, $C_1 = 1.44$, and $C_2 = 1.92$. The effective viscosity employed in the Navier-Stokes equations is determined from

$$\mu_{\text{eff}} = \mu_0 + \mu_t, \quad (4)$$

where μ_0 is the laminar viscosity. The turbulent viscosity is calculated from

$$\mu_t = \frac{\rho C_\mu K^2}{\varepsilon}, \quad (5)$$

where $C_\mu = 0.09$ and K and ε are determined from (3).

Boundary conditions at the inlet nozzles are determined as follows: for the velocity components, $u_x = U_{\text{in}}\sin(\alpha)$ and $u_y = U_{\text{in}}\cos(\alpha)$, where U_{in} is the nominal velocity and α is the inlet angle of the nitrogen jet which corresponds to the exit angle of the nozzles. The inlet values of K and ε are determined from [15, 16]

$$K_{\text{in}} = 0.01U_{\text{in}}^2,$$
$$\varepsilon_{\text{in}} = \frac{2K_{\text{in}}^{1.5}}{D}, \quad (6)$$

where D is the diameter of the inlet nozzle. At the converter walls the classical nonslip condition is applied.

When droplet breakaway occurs, there is a balance between the inertial (F_i), interfacial (F_γ), and gravitational forces (F_g) [5, 17]:

$$F_i = F_\gamma + F_g. \quad (7)$$

The drop diameter (d) derived from the balance of forces of (7) is given by [5, 17]:

$$d = \frac{3U^2}{8g}\left(1 - \left(1 - \frac{128\gamma g}{3\rho U^4}\right)\right), \quad (8)$$

where U is the magnitude of the jet velocity at the impact point and γ is the surface tension.

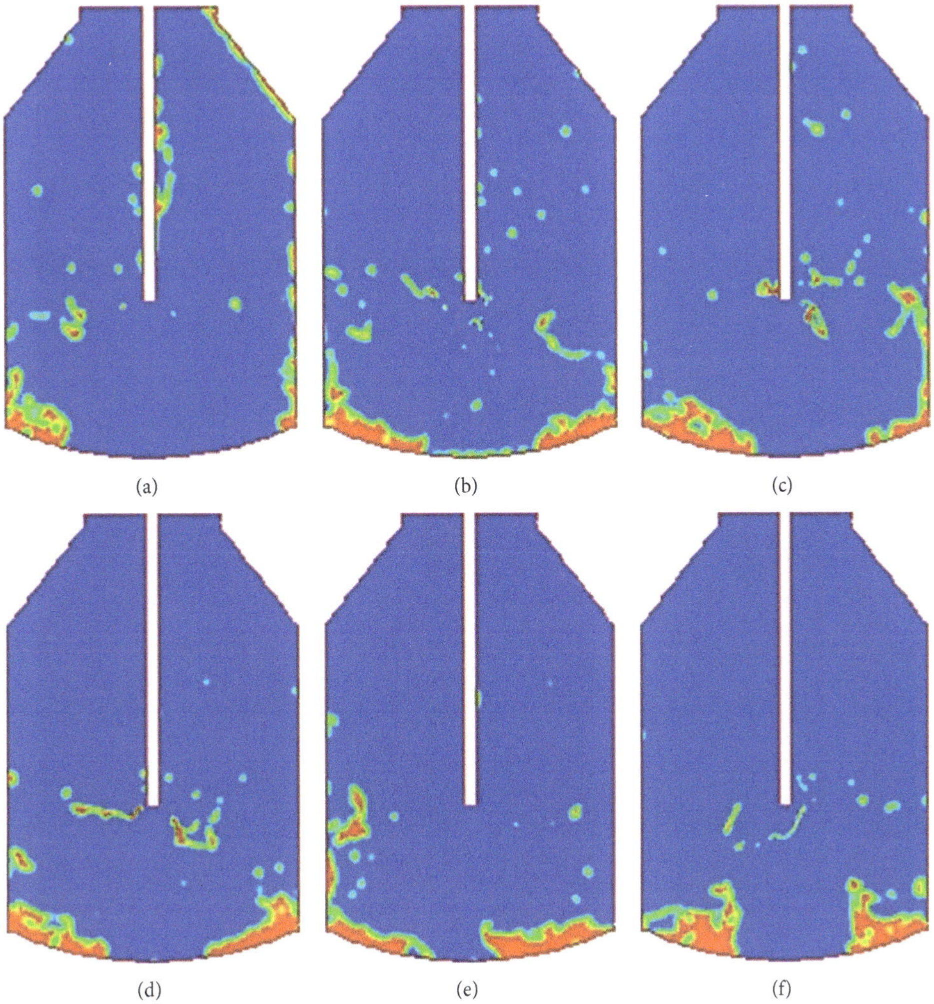

FIGURE 1: Distribution of phases for several values of the slag density (red is molten slag; blue is nitrogen). (a) 2000, (b) 2200, (c) 2400, (d) 2600, (e) 2800, and (f) 3000 kg m^{-3}.

The Volume of Fluid model is employed here to tackle the multiphase flow. This model is based on the assumption that two or more phases are not interpenetrating. Each ith phase has associated a volume fraction V_i, and in every control volume the sum of the volume fractions of all phases is equal to 1. In the Volume of Fluid Model the tracking of the interface between the phases is accomplished by solving the continuity equation for each phase [18]:

$$\frac{\partial V_i}{\partial t} + \vec{u} \cdot \nabla V_i = 0. \qquad (9)$$

To better define the slag-nitrogen interface, the computational fluid dynamics software (FLUENT) uses an interpolation scheme called Geometric Reconstruction Scheme.

3. Computer Simulations

Physical dimensions of the considered converter correspond to an actual industrial converter of 150 metric tons of capacity and were assumed as follows: height, 7.2 m; diameter, 4.8 m; lance diameter, 0.2 m; number of nozzles, 2; diameter of nozzles, 0.043 m; exit angle of nozzles, 10 degrees. The operating parameters were set as follows: lance height above the initial molten slag surface, 2 m; velocity of the supersonic nitrogen jet, Mach 1.5; initial molten slag depth, 0.5 m. Viscosity and surface tension of slag were assumed to be 0.5 kg m^{-1} s^{-1} and 0.4 N m^{-1}, respectively.

To reduce the computational effort, an isothermal two-dimensional converter was considered. The coupled Navier-Stokes equations, the continuity equation, the turbulence model, and the Volume of Fluid Model were numerically solved using computational fluid dynamics software (FLUENT). The Pressure Implicit Splitting Operation algorithm was employed in the transient simulations for the pressure-velocity coupling given that it maintains stability of the numerical scheme despite the larger time steps.

Transient isothermal two-dimensional computer simulations were carried out using a time step of 0.0001 s and a mesh consisting of 14075 trilateral cells. Due to the fast dynamics of the fluid flow, the integration time was 2 s; in

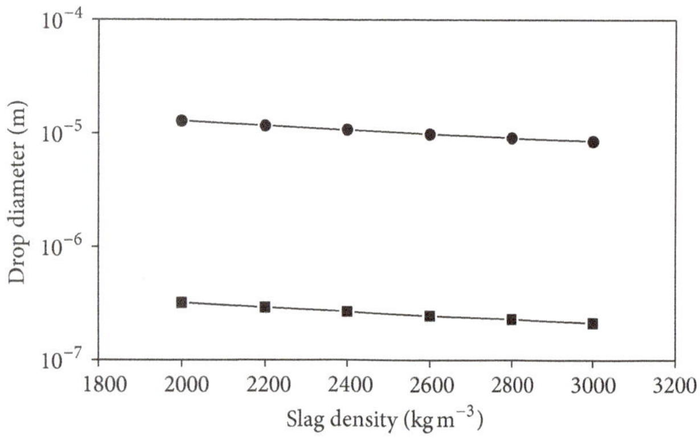

FIGURE 2: Diameter of the slag drops as a function of the slag density and the impact velocity: 50 m s^{-1} (circle) and 100 m s^{-1} (square).

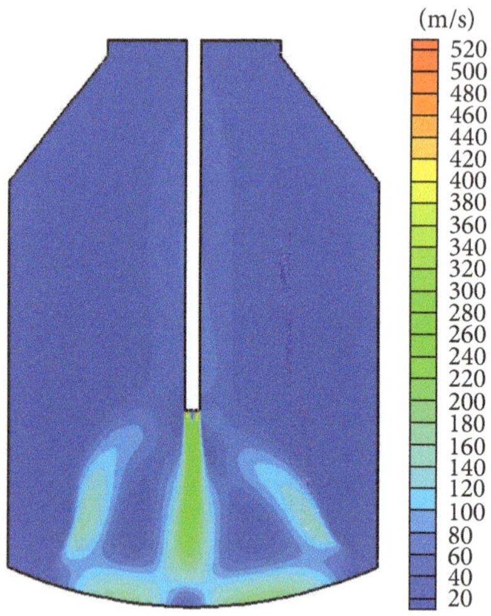

FIGURE 3: Contours of velocity magnitude corresponding to Figure 1(a).

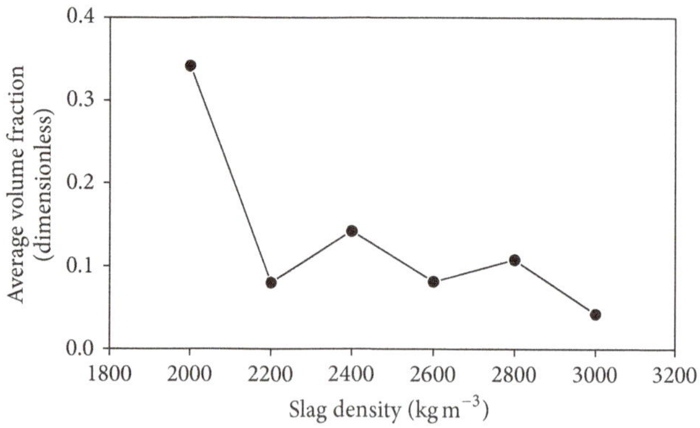

FIGURE 4: Average volume fraction on the converter side walls as a function of the slag density.

this time the multiphase flow is fully developed. Viscosity and surface tension of the molten slag was assumed to be 0.5 kg m^{-1} s^{-1} and 0.4 N m^{-1}, respectively. For the analysis, the slag density was varied from 2000 to 3000 kg m^{-3} using intervals of 200 kg m^{-3}.

4. Analysis of Results

Figure 1 depicts the distribution of phases (red phase is molten slag; blue phase is nitrogen) for several values of the slag density using the operating conditions and integration time described in Section 3. It is observed that the ejection height of the slag drops decreases as the slag density is increased. This is explained by the small forces required by the less dense slag drops to be broken and ejected compared with the relatively big forces required by the denser slag. The ejection forces are supplied by the momentum of the nitrogen jet. Figure 1(a) shows that for a density of 2000 kg m^{-3} the slag reaches the top of the converter, and as result of this, spitting of slag from the mouth converter occurs. This can be corroborated from a global mass balance. A negative value of the global mass balance indicates that molten slag is being expelled from the converter. The spitting phenomenon does not occur for values of slag density equal or greater than 2200 kg m^{-3} under the conditions described above.

In Figure 2 the diameter of the slag drops as function of the slag density and the velocity of the jet at the impact site is shown. The drop diameters were obtained using (8), which is strongly nonlinear. The velocity of the jet decays as it descends towards the converter bottom and the slag bath is impacted at a subsonic velocity. The impact velocity U of (8) was approximated from the contours of the velocity magnitude shown in Figure 3 corresponding to Figure 1(a). During the blowing the depth of the slag bath is reduced and the nitrogen jet covers a longer distance to impact the molten slag. Due to this the jet velocity suffers a significant decay below 200 m s^{-1}. Then, two impact velocities were considered: 50 and 100 m s^{-1}. It is observed that the drop diameter decreases in size as density is increased. A similar behavior is observed with respect to the impact velocity: the drop diameter diminishes as the impact velocity grows. Results of Figure 2 explain the finding that the height reached by the ejected slag drops decreases as the slag density is increased.

Efficiency of the splashing process can be evaluated in terms of the magnitude of the converter sidewalls coated by slag. Full wall adhesion of slag droplets was considered in the numerical simulations. In order to evaluate in a quantitative way the splashing efficiency as function of the slag density, the average volume fraction of slag (V_s) [12] in the sidewalls of the converter was determined. $V_s = 1$ indicates that the sidewalls became fully coated by slag, whereas $V_s = 0$ means that the sidewalls are not coated at all. Results are observed in Figure 4, where it is appreciated that the V_s is enhanced with low slag density whereas it is worsened by large slag density. This is consistent with the results of Figure 1, which shows that the upper sections of the sidewalls are more coated as the slag density is decreased. Generally speaking, from Figure 1 it can be said that low slag density promotes the coating of the upper converter walls, whereas high density provides the protective coating of the lower walls. In spite

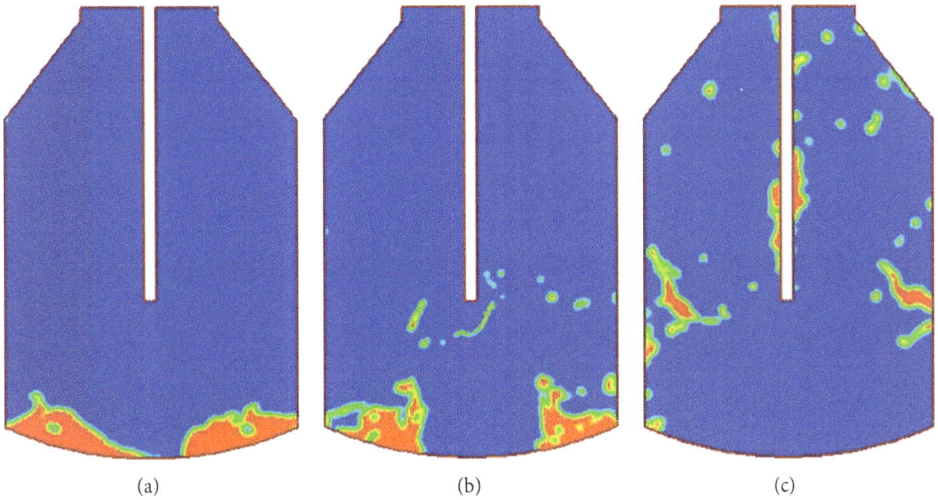

FIGURE 5: Distribution of phases for several values of the nitrogen jet velocity for a slag density of 3000 kg m^{-3} (red is molten slag; blue is nitrogen) (a) Mach 1, (b) Mach 1.5, and (c) Mach 2.

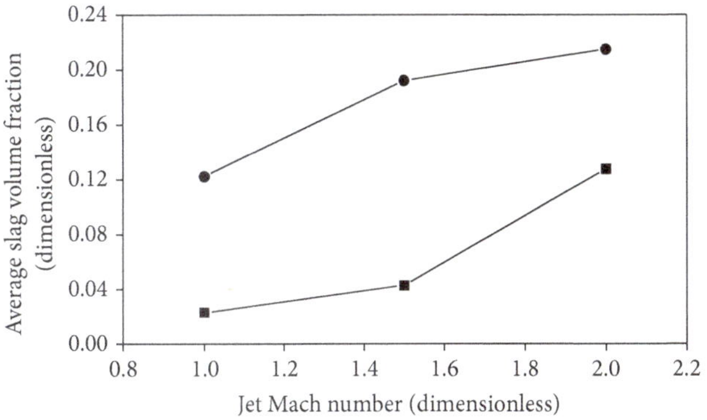

FIGURE 6: Average volume fraction on the converter side walls as function of the jet Mach number. Slag density: 2000 kg m^{-3} (circle), 3000 kg s^{-1} (square).

of the splashing efficiency is enhanced with low slag density, a serious operational drawback arises given that low density promotes the undesirable spitting phenomenon.

The negative effect of the high slag density on the splashing efficiency can be counteracted by increasing the nitrogen jet velocity given that high jet velocity provides additional momentum forces to enhance the formation small drops of the denser slag. Small drops are able to reach the top of the converter, promoting the coating of the upper sidewalls. This is visually appreciated in Figure 5, which illustrate that the ejection height of the slag drops increase as the nitrogen jet velocity is increased. Figure 6 shows that the volume fraction of slag at the converter sidewall is increased as the jet velocity is increased. However, this effect is magnified if the slag density is low. From an operational viewpoint this means that coating the extent of the upper sidewalls can be increased by (i) increasing the jet velocity or (ii) decreasing the slag density.

In accordance with [5], experimental studies report that the volume of slag splashed decreases with increasing slag density. This is explained by the fact that the Momentum number and the Weber number, which determine slag droplet generation, are dependent upon ρ^{-1} and $\rho^{-0.5}$, respectively. The experimental results reported in [5] are in agreement with the numerical results of this work, particularly with those results shown in Figure 1.

5. Conclusions

The slag splashing process in steelmaking converters was modeled and simulated by means of computational fluid dynamics. Particularly, in this work the effect of the slag density on the splashing efficiency is studied and discussed. The drop diameter of the resulting drops depends inversely on both the slag density and the jet impact velocity. Low density yields high efficiency of the considered process, unfortunately low density promotes the emergence of undesirable spitting phenomenon too. High density decreases the efficiency of the splashing process; however this negative effect can be counteracted by increasing the jet velocity given that additional inertial forces are provided.

Conflict of Interests

The authors declare that there is no conflict of interests regarding the publication of this paper.

References

[1] C. J. Messina and J. R. Paules, "The worldwide status of BOF slag splashing practices and performance," in *Proceedings of the 79th Steelmaking Conference*, pp. 153–155, Pittsburgh, Pa, USA, March 1996.

[2] K. M. Goodson, N. Donaghy, and R. O. Russell, "Furnace refractory maintenance and slag splashing," *Iron and Steelmaker*, vol. 22, no. 6, pp. 31–34, 1995.

[3] T. R. Galiullin, E. V. Protopopov, V. V. Sokolov, and A. G. Chernyatevich, "Gas-jet conditions in the slag coating of oxygen-converter linings," *Steel in Translation*, vol. 38, no. 2, pp. 97–100, 2008.

[4] K. D. Peaslee, "Physical modeling of slag splashing in the BOF," *Iron and Steel Engineer*, vol. 73, no. 11, pp. 33–37, 1996.

[5] K. C. Mills, Y. Su, A. B. Fox, Z. Li, R. P. Thackray, and H. T. Tsai, "A review of slag splashing," *ISIJ International*, vol. 45, no. 5, pp. 619–633, 2005.

[6] M. J. Luomala, T. M. J. Fabritius, E. O. Virtanen et al., "Physical model study of selective slag splashing in the BOF," *ISIJ International*, vol. 42, no. 11, pp. 1219–1224, 2002.

[7] K. D. Peaslee and W. Chen, "Important factors for effective slag splashing," in *Proceedings of the CIM Conference*, Edmonton, Canada, 2004.

[8] L. Salinas and R. Fuentes, "Interaction between a gaseous vertical jet and a liquid surface—a theoretical and experimental study," in *Proceedings of the Copper 99-Cobre 99 International Conference*, Phoenix, Ariz, USA, October 1999.

[9] A. Nguyen and G. Evans, "Computational fluid dynamics modelling of gas jets impinging onto liquid pools," *Applied Mathematical Modelling*, vol. 30, no. 11, pp. 1472–1484, 2006.

[10] M. Ersson, A. Tilliander, L. Jonsson, and P. Jönsson, "A mathematical model of an impinging air jet on a water surface," *ISIJ International*, vol. 48, no. 4, pp. 377–384, 2008.

[11] M. A. Barron and I. Hilerio, "Computer simulation of slag splashing in a steelmaking converter," in *Proceedings of the World Congress in Computer Science, Computer Engineering and Applied Computing*, Las Vegas, Nev, USA, 2011.

[12] M. A. Barron, D. Y. Medina, and I. Hilerio, "CFD analysis of influence of slag viscosity on the splashing process in an oxygen steelmaking converter," *Modeling and Numerical Simulation of Material Science*, pp. 90–93, 2013.

[13] R. J. Fruehan, *The Making, Shaping and Treating of Steel, Steelmaking and Refining Volume*, AISE Steel Foundation, Pittsburgh, Pa, USA, 11th edition, 1998.

[14] R. B. Bird, W. E. Stewart, and E. N. Lightfoot, *Transport Phenomena*, John Wiley & Sons, New York, NY, USA, 2nd edition, 2002.

[15] B. G. Thomas, Q. Yuan, S. Sivaramakrishnan, T. Shi, S. P. Vanka, and M. B. Assar, "Comparison of four methods to evaluate fluid velocities in a continuous slab casting mold," *ISIJ International*, vol. 41, no. 10, pp. 1262–1271, 2001.

[16] G. Solorio-Díaz, R. D. Morales, J. Palafax-Ramos, L. García-Demedices, and A. Ramos-Banderas, "Analysis of fluid flow turbulence in tundishes fed by a swirling ladle shroud," *ISIJ International*, vol. 44, no. 6, pp. 1024–1032, 2004.

[17] P. C. Mahapatra, S. Majumdar, R. Shrinivas et al., "Understanding slag splashing technology in BOF," *SEAISI Quarterly*, vol. 29, no. 4, pp. 54–60, 2000.

[18] C. W. Hirt and B. D. Nichols, "Volume of fluid (VOF) method for the dynamics of free boundaries," *Journal of Computational Physics*, vol. 39, no. 1, pp. 201–225, 1981.

2

Life Cycle Impact of Rare Earth Elements

P. Koltun[1] and A. Tharumarajah[2]

[1] *CSIRO Process Science and Engineering, Gate 5, Normanby Road, Clayton, VIC 3168, Australia*
[2] *EnviSolutions, 20 Galahad Crescent, Glen Waverley, Melbourne, VIC 3150, Australia*

Correspondence should be addressed to P. Koltun; paul.koltun@csiro.au

Academic Editors: M. Carboneras and P. Lukac

The diverse properties of rare earth elements have seen broad and growing applications in clean energy technologies, hybrid vehicles, pollution control, optics, refrigeration, and so on. This study presents a "cradle-to-gate" life cycle assessment of the energy use, resource depletion, and global warming potential resulting from the production of rare earth elements (REEs) using the Bayan Obo rare earth operation in Inner Mongolia, China, as a representative system. The study aggregates data from the literature, LCI databases, and reasonable estimations. A novel economic value-based allocation method for the multiple coproducts of the process is proposed. It is found that four of the high priced REEs scandium, europium, terbium, and dysprosium have very high GWPs from production relative to the rest. A mass-based allocation is also provided for comparison. Impacts on immediate local environment from waste streams that can be toxic are not included in this study.

1. Introduction

Rare earth elements (REEs) or RE metals are technically defined as the 15 elements in the lanthanide (La) series, yttrium (Y), and scandium (Sc). Y and Sc are considered REEs since they mostly occur in the same ore deposits and have similar chemical properties. A precise classification often used in extraction is as follows [1].

(i) Light rare earth elements (LREEs): lanthanum (La), cerium (Ce), praseodymium (Pr), neodymium (Nd), and promethium (Pm).

(ii) Medium rare earth elements (MREEs): samarium (Sm), europium (Eu), and gadolinium (Gd).

(iii) Heavy rare earth elements (HREEs): terbium (Tb), dysprosium (Dy), holmium (Ho), erbium (Er), thulium (Tm), ytterbium (Yb), lutetium (Lu), scandium (Sc), and yttrium (Y).

There are about 200 known rare-earth containing mineral deposits, mostly as carbonatites spread around the world. Contrary to a lay persons' understanding, REs are not rare in natural occurrence (cerium is more abundant than tin and yttrium is more abundant than lead), though REEs have much less tendency to become concentrated in exploitable ore deposits [2], in particular HREEs. This being so, only a few mineral species, such as bastnasite, monazite, and RE-bearing clay, have been recovered for commercial production. Bastnasite deposits in China and the United States constitute the largest percentage of the world's rare-earth economic resources [3].

REEs possess diverse nuclear, metallurgical, chemical, catalytic, electrical, magnetic, and optical properties. This has led to ever-increasing applications, some mundane (lighter flints, glass polishing) to others that are highly specific. Examples of the latter include catalysts in petroleum refining industry, as alloying agents (used to enhance the oxidation resistance of alloys) in metallurgical processes, in glass, phosphors, optics, permanent magnets, and electronics. Emerging and potential applications include using rare earths to absorb ultraviolet light in automotive glass, corrosion protection, and metal coatings in corrosive and salty environments. Futuristic applications of REE are in high-temperature superconductivity, safe storage, and transport of hydrogen for a posthydrocarbon economy [2]. For applications of individual REEs, see [4].

Only small quantities of REEs are used in most applications, for instance, 2 to 3 grams in a 20-watt phosphor lamp.

However, clean energy applications such as high capacity nickel-metal hydride batteries in hybrid cars (in a third-generation Toyota Prius) contain on average ~0.6 kg/battery of La [5] and RE magnets for electric motors as light weight alternative to iron magnets require around 1 kg of Nd [6, 7]. A utility scale wind turbine, another green power source, uses more than a ton of heavy-duty and lightweight magnets, 700 pounds of which is neodymium [8]. As the worldwide implementation of clean energy and other applications increase, many studies have identified the supply of particular REEs (dysprosium, neodymium, terbium, yttrium, and europium) to be at risk in the short and medium term [9, 10].

China currently accounts for 95% of global production of rare earths (about 90% of REs used in the US) and domestic industry consumes about 60% [18]. Bayan Obo mine (used as the reference site in this study) situated in Inner Mongolia is said to supply half of the total production [19]. However, this situation may change due to China's introduction of progressively reducing export quotas since 2004/2005 [5] for reasons of environmental concerns and conservation of resources [20]. This potential risk to supply may be somewhat absorbed by recycling of in-use stocks of REEs in products (an estimated 440,000 metric tons in 2007, [21]). In fact, this source can also help to reduce the impact on the environment from the production of virgin REEs.

Many environmental issues surround the production and use of REEs. REEs having similar chemical structures are difficult to separate [22] compounded by extremely low yield (net recovery from the ore at Bayan Obo is around 0.6% [23, 24]). Thus, apart from the electricity, acids, water, and resources expended in production, there can be huge amounts of waste that can be toxic with potential damage to the ecosystem.

One of the most important environmental issues for producers of rare earths is the problem posed by the radioactivity of thorium-containing monazite and xenotime ores. Australian monazite exports declined precipitously between 1989 and 1992 in part because of the radioactivity concern and since 1992, Rhodia, at that time the world's largest rare earth company, has used Chinese instead of Australian rare earth concentrates for production of individual rare earth oxides [25].

Other concerns of RE pollution relate contamination of the local environment. These include potential bioaccumulation in the food chain from waste or dust contamination of water ways and soil from mines when used in downstream industrial processes and in agricultural fertilisers [27, 28].

This study, with the exception of impact on the local environment, investigates the global warming impact and intensity of use of energy and resources in the production of RE oxides (REOs) and REEs. Each of the distinct stages from mining, separation of REOs, and reduction to REEs is modelled and the input of resources (energy, water, chemicals) and wastes is examined. GHG impacts and resource depletion potential are computed from the derived data.

It is the objective of this study to disseminate knowledge of the GHG and other impacts of producing REEs, in particular at this time when their application is growing and such knowledge is not readily available. Further, it is believed that such data will be valuable and provide a basis for constructing and comparing GHG impact profiles of rare earth based products for comparison purposes.

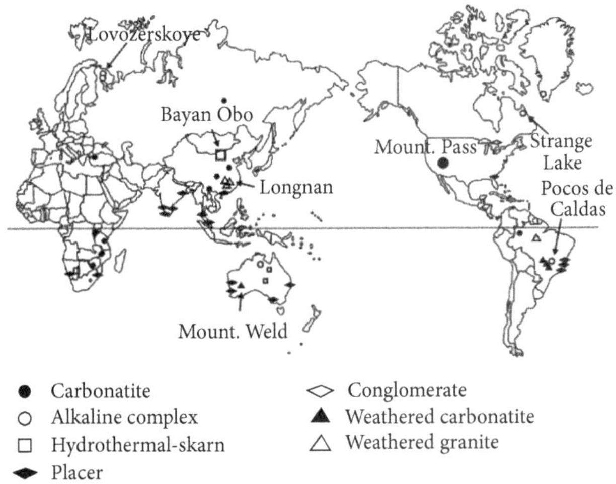

Figure 1: World distribution of rare earths [26].

2. Extraction and Production of Rare Earths

2.1. Sources and Distribution of Rare Earth Minerals. Only a few mineral species, such as bastnasite (a rare earth fluorocarbonate (Ce, La)(CO_3)F), monazite (a rare earth phosphate (Ce, La, Y Th) PO_4)), xenotime (YPO_4), and RE-bearing clay have been recovered for commercial production. Of these, bastnasites constitute the largest percentage (in US and China). Monazite deposits in Australia, Brazil, China, India, Malaysia, South Africa, Sri Lanka, Thailand, and the United States constitute the second largest segment. World distribution is shown in Figure 1. Undiscovered resources are thought to be very large relative to expected demand [3].

Around 95% of the world's rare earth metals are produced in China, and as mentioned before Bayan Obo mine supplies about half of the total production. This mine is estimated to contain at least 1.5 billion metric tons of iron (average grade 33% of iron oxide), 89 million tons of RE oxides (REOs) (average grade 6%), and 1 million tons of niobium (average grade 0.13%) [23]. The principal RE minerals in Bayan Obo are bastnasite and monazite accompanied by RE-Nb minerals; further details of the geological distribution are given in [23]. This study uses Bayan Obo operations as a representative system for rare earth extraction and production. Rare earth production at Mount Pass in USA ceased operation in 2002 due to environmental concerns (production has recommenced since 2008). The projected production in 2012 is expected to be between 8,000 and 10,000 metric tons [29]. Mining at the Mount Weld deposit in Western Australia began in 2007, and an estimated 98,000 cubic meters of ore has been stockpiled awaiting the completion of a concentration plant at the mine site. The concentrates will be exported to an advanced materials plant being built in Malaysia [30].

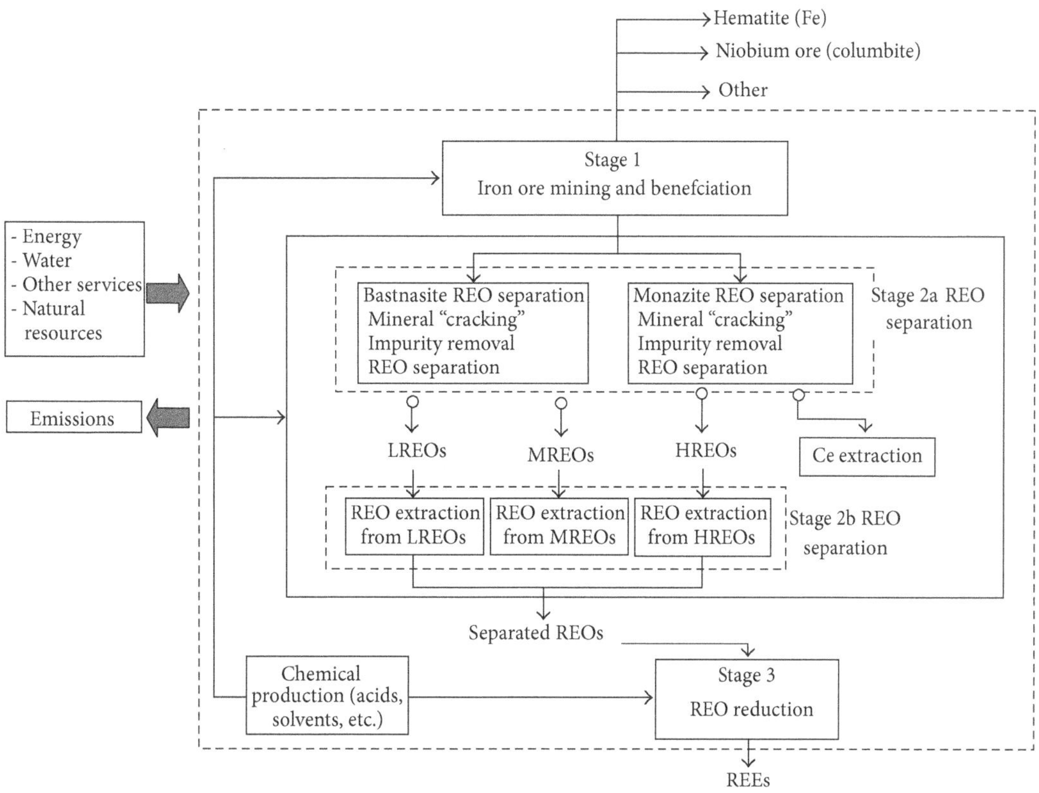

FIGURE 2: System boundary: mining and production of RE metals in China.

REEs usually have very small differences in solubility and complex formation and hence their separation can be difficult [22]. Thus, separation processes are often chemically intensive using ion exchange methods, solvent extraction, or fractional crystallization. The average recovery rate from RE containing ores is extremely low at 10% [24]. This is compounded by the low grade of REO in mined minerals at Bayan Obo (around 6% [23]). Thus, the net recovery is about 0.6% and results in high resource inputs (electricity, acids, water and resources expended in production) as well as huge amounts of waste that can be toxic with potential damage to the ecosystem.

A representative system of extraction and production in Bayan Obo is shown in Figure 2. This representation is used as the system boundary for this LCA study and is explained briefly.

2.2. Stage 1: Mining and Beneficiation. Bayan Obo mineral Fe-REO-Nb deposit in China is the source of both iron-ore and RE containing minerals bastnasite and monazite. In separating the concentrates of bastnasite and monazite, a combination of separation processes (after grinding) are used including wet-magnetic separation and froth floatation. Since mining and beneficiation are established technologies, data is mostly based on our previous studies in mining and beneficiation [31, 32].

2.3. Stage 2: Purification and Separation of REO. The concentrates of bastnasite and monazite (approximately 85% concentration by weight within the slurry [24]) from mining and beneficiation separately undergo further processing (shown as Stage 2a in Figure 2). This stage purifies and separates the light, medium, and heavy oxide groups using mineral cracking, acid leaching, impurity removal, and precipitation. While Stage 2a processes are similar for both concentrates, due to the substantial amount of radioactive elements in monazite (together thorium and uranium account about 6.5% of monazite) and phosphorus (phosphorus oxide accounts about 28% of monazite), the monazite processing is more energy and chemical intensive than bastnasite.

The extraction of RE oxide groups from both bastnasite and monazite produces cerium and separate streams of light, medium, and heavy RE oxides. The separated streams are further processed to produce the individual REOs. To extract the individual REOs, the oxide groups are subject to multistage extraction process using extractants in acidic medium. LREO separation involves 22 stages: 8 for extraction, 8 for scrubbing, and 6 for stripping [33]. MREOs have 29 extraction stages and 44 scrubbing stages [34]. Currently heavy RE oxides are separated by using ion exchange process with reagent impregnated resins [35].

2.4. Stage 3: Reduction of REEs. The industrial reduction to REEs from light REOs (La, Nd) including mischmetal

TABLE 1: Bayan Obo iron ore composition and relative economic value of coproducts.

Commodity	Fraction in Bayan Obo deposit, %	P_c—Price per tone*, US$	C_c—Normalised fraction, %	X_c—Share in economic value, %
Iron ore (hematite)	33	120.35	84.33	2.87
RE oxides (RE_xO_y)	6	215400.00	15.33	93.31
Niobium ore (columbite)	0.13	41000.00	0.33	3.82
Fluorite (CaF_2)	2	—	—	0
Thorium oxide (ThO_2)	0.16	—	—	0
Phosporus oxide (PO_4)	0.86	—	—	0
Rest	57.85	—	—	0

*Prices are shown for the year 2009-2010. Prices estimated for iron ore [11], REO [12], and niobium ore [13].

(which are a mixture of light lanthanide metals) uses fused-salt electrolysis (fused salt process is used when the reactivity of the metal does not allow electrowinning from aqueous solutions).

For MREEs and HREEs, metallothermic process using electrolysis from an aqueous solution or simple carbon based pyrometallurgy is employed due to the high electropositive nature of RE elements in these groups [36].

Processes for disposing toxic and other waste from the mining and processing are not considered in this study.

3. LCA Study Methodology and Assumptions

The principal goal of this study is to investigate the cradle-to-gate energy use (assessed as electricity use, heating fuel use, and total nonrenewable energy use), water use, and global warming impact of REEs produced in Bayan Obo, China. It is envisaged that the impact data of REEs created in this study could provide a basis for constructing and comparing impact profiles of rare earth based products.

On defining the scope of this study, the system boundary definition for ascertaining the impact is shown in Figure 2 and is explained in Section 2.2. Other topics of scope such as allocation, data sources, and uncertainty treatment are discussed below. Results of impact assessment are given in Section 4.

One aspect of scope that is given careful attention is the allocation of impacts. Normally, extraction of REEs is a multifunctional product system, meaning a number of outputs as saleable products are produced at the different stages of production. For instance, in Bayan Obo mining operation, hematite, niobium, and RE containing ores are produced as saleable coproducts of mining. The separated RE containing ore is then processed into saleable REOs or can be further reduced to produce individual REEs. Thus, properly allocating from "cradle to gate" (i.e., from mining to when REOs or reduced REEs are sold) is important.

Allocation among the coproducts uses both mass fractions (i.e., physical relationship) as well as an economic model. In fact, while mass-based allocation is sufficient as recommended by ISO 14044 [37], an additional economic model that combines mass and price basis (market value) is proposed here. This model reflects the underlying causality between economic reasons for producing coproducts (i.e., REEs from iron ore extraction) and their environmental impact as noted by [38]. It has also been suggested that economic allocations may be suited for coproducts that vary in prices [39] as is the case for REOs.

At Stage 1 iron ore mining (see Figure 2), three economically significant commodities (coproducts) are mined: hematite (FeO), columbite (niobium ore), and REO bearing ore (i.e., bastnasite and monazite minerals). Mass-based allocation uses the mass fractions of each product (see Table 1).

Economic allocation among these ores is performed on the basis of share of contribution of each product according to (1), where C_c, P_c, and X_c are, respectively, the mass fraction, price, and computed environmental share of commodity c. Mass allocation uses C_c as the basis. Share of allocation for each of the mined coproducts is shown in Table 1. The mass fractions (C_c) are adjusted (normalized) across the economically valuable ores (iron ore, REO, and niobium):

$$X_c = \frac{P_c * C_c}{\sum_c (P_c * C_c)}. \quad (1)$$

In Stages 2a and 2b, the plant produces a mixture of REOs as coproducts. This would require calculation of impact that can be assigned to each of the individual REOs. A combined mass and economic allocation model similar to Stage 1 is developed and applied for this as described by the following equations:

$$C_i = c_b b_i + c_m m_i, \quad (2)$$

$$X_i = \frac{P_i * C_i}{\sum_i (P_i * C_i)}. \quad (3)$$

In (2), the average mass composition (C_i, used for mass-based allocation) of REO_i is determined using itsmass composition in bastnasite (b_i) and monazite (m_i) weighted according to the ratio of bastnasite (c_b) and monazite (c_m) processed. The proportion of bastnasite and monazite in Bayan Obo mineral deposit is taken as 3 : 1, respectively, with each containing 60% of RE oxides [40].

Equation (3) computes the share of the environmental burden (X_i) using price per kg (P_i) of REO_i as the weighting factor normalised over all REOs. This way, both mass and economic proportions are combined to derive the final allocation given to each REO.

TABLE 2: Share of environmental impact allocated for REOs.

REE	REO	i	b_i, %	m_i, %	C_i, %	P_i, US$/kg	X_i, %
Lanthanum	La_2O_3	La	23.00	23.00	23.00	13.00	7.75
Cerium	CeO_2	Ce	50.00	46.00	49.00	12.00	15.24
Praseodymium (+3)	Pr_6O_{11}	Pr	6.20	5.00	5.90	95.00	14.53
Neodymium	Nd_2O_3	Nd	18.50	19.00	18.63	77.00	37.18
Promethium	Pm_2O_3	Pm	—	—	0.00	0.00	0.00
Samarium	Sm_2O_3	Sm	0.80	3.00	1.35	23.50	0.82
Europium	Eu_2O_3	Eu	0.20	0.10	0.18	2150.00	9.75
Gadolinium	Gd_2O_3	Gd	0.70	1.70	0.95	75.00	1.85
Terbium	Tb_4O_7	Tb	0.10	0.16	0.12	1750.00	5.22
Dysprosium	Dy_2O_3	Dy	0.10	0.50	0.20	975.00	5.05
Holmium	Ho_2O_3	Ho	0.00	0.00	0.00	—	0.00
Erbium	Er_2O_3	Er	0.00	0.13	0.03	77.00	0.06
Thulium	Tm_2O_3	Tm	0.00	0.00	0.00	—	0.00
Ytterbium	Yb_2O_3	Yb	0.00	0.06	0.02	69.12	0.03
Lutetium	Lu_2O_3	Lu	—	—	0.00	—	0.00
Scandium	Sc_2O_3	Sc	—	0.04	0.01	7200.00	1.87
Yttrium	Y_2O_3	Y	0.00	2.00	0.50	50.00	0.65

References used: b_i—[14]; m_i—[1], Sc_2O_3—[15]; P_i—[16], Sm_2O_3, Yb_2O_3—[17].

The values of pertinent parameters in (2) and (3) used in calculating the share of the environmental burden assigned to the REOs are given in Table 2. References used appear as notes in the table. The following REOs are not considered.

Tm and Ho are least abundant and are found with other rare earths such as gadolinite and other minerals containing rare earths, and Lu is the rarest of the rare earths [41]. Due to data availability, these are not considered here.

Pm is the only radioactive rare-earth metal of transition Group IIIb of the periodic table and not detected in nature [42]. It has a low period of half decay and its main route of production is artificial synthesis.

In Stage 3 REO-REE reduction, the mass fraction of individual REE present in the corresponding RE oxide is used for allocation.

The manner in which the above methods are used to calculate the impact allocated to REO or REE products, covering mining to production life cycle stages, is described below.

First, for the production of REO products, a reference flow for producing 1 kg mixture of REO products is considered at Stage 2. To obtain this, 166.7 kg of iron ore has to be mined (using average REE grade of 6% in the ore and REO recovery rate of 10%). Impact of mining this (Stage 1) is allocated among Fe, Nb, and RE containing concentrates (i.e., bastnasite and monazite) as described.

For Stage 2a (separation of REO groups), share of allocation derived for each REO is used (Table 2). In the case of Stage 2b (separation of REOs from REO groups), the same model as used for Stage 2a is used, though it is normalized for the REOs for the oxides occurring within each group. Impact from mining to REO production is the sum of the Stages 1-2 for the fraction of REOs in the reference flow. Finally, the impact per kg of individual REO is calculated.

In deriving the allocation for each REE, the impact of REO-REE reduction stage is first assigned on a mass basis, that is, the mass fraction of individual REE present in the corresponding RE oxide. "Cradle to gate" impact per kilogram of individual REE is determined by adding the impact of corresponding REO and this stage.

Inventory data for the processes is obtained from publicly available information sources (cited appropriately under each section) and by estimation. These include review of technological processes for RE production and analogous processes, environmental data pertaining to materials and chemicals mainly from EcoInvent life cycle inventory (LCI) database [43], and other LCI databases combined with modeling and estimation. Table 3 provides the basic data for energy and water use at each stage from mining to REO extraction. The diverse sources of data require uncertainty analysis, such as Pedigree Matrix [44] combined with Monte Carlo simulation available in SimaPro software [45]. However, such analysis will be performed as a future extension of this study.

The major assumptions of this study include the following.

(a) Distance between mining pit and beneficiation plant and beneficiation and REO separation plant are assumed as 15 km and 30 km, respectively. REO reduction and subsequent reduction to REE occur at the same place, that is, Bayan Obo Township.

(b) In separating REOs from monazite concentrates radioactive wastes can occur. Treatment of such waste can pose additional environmental effects including toxicity and soil and ground water contamination. The treatment processes of such waste are considered out of scope.

TABLE 3: Life cycle inventory and impact of production stages for the production of 1 kg of RE oxide.

Stage	Process	Energy consumption, MJ		Water consumption, kL	Environmental impact	
		Electricity	Heat energy		GHG emissions, kg CO_2 eq.	Resource depletion, MJ surplus
Stage 1	Mining and beneficiation	4.64	10.11	3.24	3.95	5.46
Stage 2a	REO extraction from bastnasite	5.60	90.00	19.09	10.5	13.9
	REO extraction from monazite	55.60	11.90	18.15	18.3	19.6
Stage 2b	Light REO separation	12.60	0	0.58	3.3	2.80
	Medium REO separation	5.20	0	0.16	1.3	1.29
	Heavy REO separation	15.50	0	8.52	5.5	5.57

(c) Average grid-mix of electricity in China is 75% from coal power stations and the rest is from hydro [31].

(d) Purity range of REOs produced (to derive the life cycle inventory data) is 98.0%–99.9% associated with current production technology in China.

(e) A steady-state process with a constant rate of materials and energy flows is assumed.

(f) Chemical and other materials used in the processes are assumed to be imported from Europe and hence, the energy consumption for their production is mostly based on data taken from the European LCI databases ([43, 45]).

4. Environmental Impact

The life cycle inventory (LCI) of energy (both electricity and heat) and water consumption and corresponding global warming impact for mining and separating REOs (i.e. stages 1 and 2 in Figure 2) to produce 1 kg of mixed REOs are shown in Table 3. Assignment of impacts among products uses the allocation methods described already. The widely used Eco-indicator 99 impact assessment methodology [46] is used in selecting and computing the appropriate environmental damage indicators. SimaPro software [45] is used in deriving the indicator scores.

Two indicators selected for this study are GHG for global warming potential and resource depletion. Resource depletion is defined in the Eco-indicator 99 methodology report [46]. It is measured in terms of surplus energy (MJ) required to obtain the same quality of resource as they become depleted. The calculation of this indicator uses geostatistical models to analyse the concentration of a mining resource at a reference year followed by assumptions on future depletion rates. The somewhat arbitrary and limiting nature of assumptions and calculation makes this indicator useful only as a comparative measure (the methodology report in [46] provides more information). Nevertheless, in the absence of more objective indicators resource depletion is used here.

Examining Table 3, one can see that energy consumption and GHG emissions for the REO extraction (Stage 2a) are much more for monazite than bastnasite, due to the required additional separation of uranium, thorium, and phosphorous for the former. In Stage 2b, the impact is the highest for separation of HREOs due to ion exchange process used. While solvent extraction process is used for both LREOs and MREOs with the latter being more difficult, impact from LREO extraction is higher due to the higher mass contribution (1 kg mixed REOs contains about 96% LREOs). Out of all the stages, Stage 2a (REO extraction, see Table 3) has the most resource deletion of around 33.5 MJ surplus

Impact for each of the separated REOs from Stages 1 and 2 is computed using both mass- and price-based allocation methods described in the previous section. Price-based allocation of impact per kilogram of each REO is shown in Table 4.

Comparison of environmental impacts among the REOs shows that LREOs (La-Nd) have lower impact than the rest of the REOs with La and Ce being minimal within the LREO group. The impact for other REOs varies substantially and cannot be attributed to their classification as MREOs and HREOs.

The impacts assigned to REOs are influenced by the concentration and price; the lower the percentage concentration and the higher the price, the higher the share of impact per kg. Among the REOs, scandium oxide (Sc_2O_3) has the highest GHG impact at ~6,332 kg of CO_2 eq./kg due to its very high price (USD 7,200 per kg) and very low occurrence (0.01%). The next highest impact is for europium oxide (Eu_2O_3) at ~1,884 kg of CO_2 eq./kg. Its concentration is comparatively higher and price is lower than Sc_2O_3. Other notable high-impact REOs are terbium oxide (Tb_4O_7, ~ 1,539 kg of CO_2 eq./kg) and dysprosium oxide (Dy_2O_3, ~ 857 kg of CO_2 eq/kg).

When mass-based allocation is used, a separated REO carries the same impact assigned to an RE group. This is due

TABLE 4: Life cycle inventory and impact for the production of 1 kg RE oxides using price-based allocation.

RE oxide	Classification	Energy consumption		Water consumption, kL	Environmental impact	
		Electricity, MJ	Heat, MJ		GHG emissions, kg CO_2 eq.	Resources depletion, MJ surplus
La_2O_3	Light	26.13	34.39	12.82	11.16	12.52
CeO_2	Light	24.12	31.74	11.83	10.30	11.56
Pr_6O_{11}	Light	190.97	251.28	93.69	81.53	91.52
Nd_2O_3	Light	154.78	203.67	75.94	66.09	74.18
Pm_2O_3	Medium	0.00	0.00	0.00	0.00	0.00
Sm_2O_3	Medium	37.95	62.16	22.74	17.73	20.59
Eu_2O_3	Medium	3472.18	5686.98	2080.32	1622.04	1883.96
Gd_2O_3	Heavy	121.12	198.38	72.57	56.58	65.72
Tb_4O_7	Heavy	2826.83	4628.93	1718.02	1325.42	1538.92
Dy_2O_3	Heavy	1574.95	2578.98	957.18	738.45	857.40
Ho_2O_3	Heavy	0.00	0.00	0.00	0.00	0.00
Er_2O_3	Heavy	124.38	203.67	75.59	58.32	67.71
Tm_2O_3	Heavy	0.00	0.00	0.00	0.00	0.00
Yb_2O_3	Heavy	111.65	182.83	67.86	52.35	60.78
Lu_2O_3	Heavy	0.00	0.00	0.00	0.00	0.00
Sc_2O_3	Heavy	11630.4	19044.8	7068.42	5453.15	6331.55
Y_2O_3	Heavy	80.77	132.26	49.09	37.87	43.97

TABLE 5: Life cycle inventory and impact for the production of 1 kg RE oxides using mass-based allocation.

Classification	Energy consumption		Water consumption, kL	Environmental impact	
	Electricity, MJ	Heat, MJ		GHG emissions, kg CO_2 eq.	Resources depletion, MJ surplus
Light REOs	74.08	102.51	38.06	32.29	36.50
Medium REOs	66.68	102.51	37.64	30.29	34.99
Heavy REOs	76.98	102.51	37.64	34.49	39.27

to the allocation being normalized by group and derived on a per kg basis. Table 5 shows the mass-based impact by RE group. Here, the values in the table are affected by Stage 2b, which is particular to the groups. Hence, MREOs, being lower than others, have lower inventory and impact values.

A comparative GHG profile of REOs with allocations based on price and mass is shown in Figure 3. As noted, GHG allocation using mass-based allocation is the same for REOs in the same group.

Separated REOs are further processed in a reduction stage to extract the individual REEs. This study assumes that the reduction plants are located in Bayan Obo and no or minimal transport takes place. An average impact is calculated for electrolytic process (for LREEs) and metallothermic process (HREEs and MREEs) based on [47] and [48], respectively. Thus, the impact calculated for this stage alone is affected by the process yield of 80% for LREEs and 95% for other REEs; the impacts for reduction are, respectively, 5.64 and 5.04 kg of CO_2 eq./kg of REE.

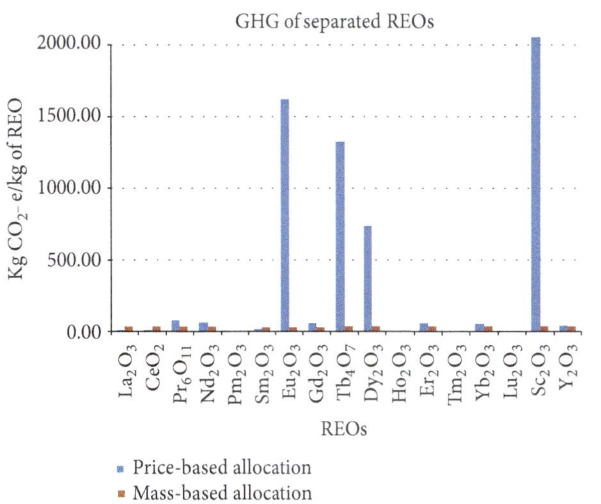

FIGURE 3: Comparative GHG of rare earth oxides for mass- and price-based allocations.

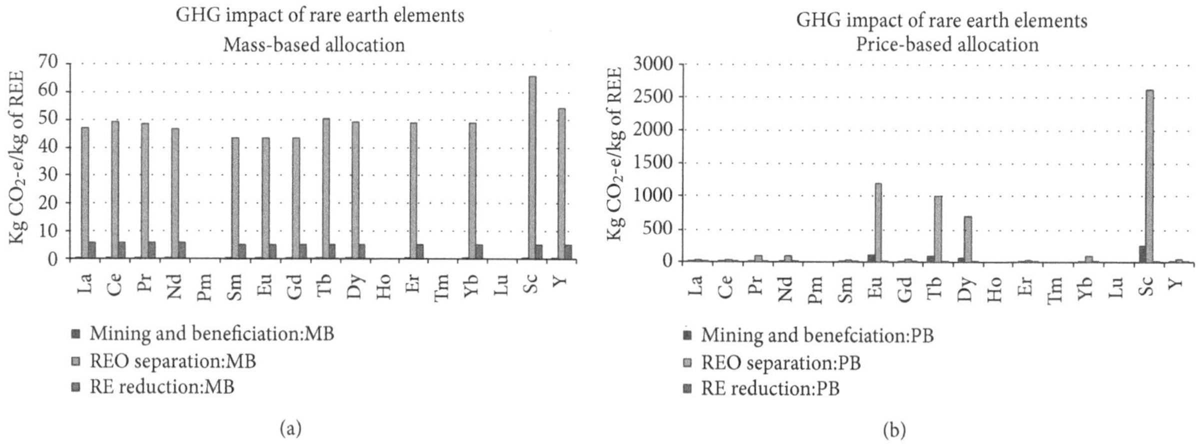

FIGURE 4: Cradle-to-gate GHG emissions of rare earth elements.

5. Discussion

GHG emissions from "cradle-to-gate" life cycle stages of production (i.e., mining, REO separation, and REE reduction) based on mass- and price-based allocation for extracting 1kg of individual REE are displayed in Figure 4. Mass-based approach allocates more emissions to more abundant elements, though when computed per kilogram of REE, the mass proportion becomes insensitive. Thus, the factor that influences the comparative allocation is the energy and materials expended (differs between RE groups) and the extractable mass of REE from its oxide. This is apparent in Figure 4(a) where Sc has the highest impact followed by Y. The extractable mass of REE from the 25 oxide for these two is, respectively, 65% and 78% compared to others that are in the range of 81%–85%.

Unlike mass-based allocation, price-based allocation model proposed here uses both mass concentrations in mineral and price. The comparative impacts allocated are hence influenced by: (1) energy and materials consumed at each stage; (2) extractable mass of REE in the minerals, that is concentrations in bastnasite and monazite and mass proportion in the respective oxides; and (3) price used in calculating the share of environmental burden to be allocated.

In this model, scandium, europium, terbium, and dysprosium have relatively very high impact (Figure 4(b)). The corresponding REOs of these are high too for reasons cited in the last section. In reducing them to REEs, a further increase in the case of scandium occurs due to its lower extractable mass from its oxide of ~65% compared to ~86% for europium, ~85% for terbium, and ~87% for dysprosium. Thus, to produce 1 kg of Sc would require 30% more Sc_2O_3 than other oxides. Additionally, Sc_2O_3, Tb_4O_7, and Dy_2O_3 are HREOs, the separation of which requires more energy. Economic allocation is influenced by price fluctuations. While the prices of REOs have been volatile dropping as much as 40%–60% in 2011-12 [16], nevertheless, the comparative prices of the above four high impact REOs have been consistently higher than other REOs. Thus, they remain the high impact rare earths relative to others. In any case, the impact of individual REEs can vary with price fluctuations.

Comparison of the two allocation methods shows that mass-based allocation is not sensitive to the scarcity of the rare earth in the mineral (per kilogram basis). The suggested allocation based on combination mass, concentration, and price shows such sensitivity; for example, allocated GHG emissions for low concentration REEs and those that are highly priced tend to be high.

Further, Bayan Obo is unique as an iron ore mine that also produces rare earths, and though similar, Mountain Pass does not produce iron ore or niobium. This means that in this case only part of the total energy and environmental impacts associated with mining stage is assigned to rare earths. While the impact of this stage may be smaller than other stages, this peculiarity of the model has to be recognised.

6. Conclusion

This study has in some detail investigated the cradle-to-gate GHG impact of a rather complex route of producing REEs starting from extraction of rare earth minerals, separation of oxides, and final reduction of REEs.

Both mass- and price-based allocation models have been employed in estimating the impact. The former is only sensitive to the extractable mass concentrations, whereas price based allocation model proposed in this study is additionally sensitive to price. Thus, where the price of coproducts of product systems varies widely as in the case of rare earths, it tends to amplify the impact of highly priced rare earths that have lower concentrations. This information can be useful in focussing efforts to improve process efficiency and recycling to increase supply.

Recycling is an attractive pathway given the increasing prices of rare earths (while 2011-12 prices were down, they are generally higher than 2010 prices) combined with the recent clamp down on exports by China. The route to recycling can be closed-loop, meaning the recovery of the original

RE alloys with minimum loss of property for similar applications. Such direct recycling, however, has its challenges in collecting, sorting, separating components, and finding suitable processes. An open-loop recycling, where REEs can be recovered from alloys for use in other applications, is also viable. Investigation of GHG impacts of the various routes to recycling is planned as the extension of this study.

An additional concern is the impact on the environment from processing waste. In particular, the large amount of tailings is produced in beneficiation and extraction of bastnasite and monazite concentrates. Tailings of both of these, monazite in particular, contain naturally occurring radionuclides and the release of this to the environment by air, wastewater, and rain leaching can have longer-term health effects to humans and ecosystems of the local environment. An assessment of these and other impacts from waste processing and disposal is seen as supplementing this study.

Conflict of Interests

The authors declare that there is no conflict of interests regarding the publication of this paper.

References

[1] Australian Industry Commission, *New and Advanced Materials*, Australian Government Publishing Service, Melbourne, Australia, 1995.

[2] U.S. Geological Survey, "Rare Earth Elements-Critical Resources for High Technology," Fact Sheet 087-02, 2002, http://pubs.usgs.gov/fs/2002/fs087-02/.

[3] U.S. Geological Survey, "Mineral Commodity Summaries—Rare Earths," 2008, http://minerals.usgs.gov/minerals/pubs/commodity/rare_earths/mcs-2008-raree.pdf.

[4] HEFA Rare Earth Canada Co. Ltd, Rare Earth products by element, http://www.baotou-rareearth.com/.

[5] U.S. Department of Energy, "Critical Materials Strategy," December 2011, http://energy.gov/pi/office-policy-and-international-affairs/downloads/2011-critical-materials-strategy.

[6] S. Gorman, "As hybrid cars gobble rare metals, shortage looms," Reuters, 30th August 2009, http://www.reuters.com/article/GCA-GreenBusiness/idUSTRE57U02L20090831.

[7] S. Gorman, "California mine digs in for, "green" gold rush," Reuters, 30th August 2009, http://www.reuters.com/article/idUSTRE57U02I20090831?sp=true.

[8] A. Pasternack, "China Tightens Grasp on Rare Earth Metals Vital for Green Technologies," Treehugger (Business & Politics), August 2009, http://www.earth-stream.com/outpage.php?s=18&id=201861.

[9] D. Bauer, D. Diamond, J. Li, D. Sandalow, P. Telleen, and B. Wanneret, "Critical Materials Strategy, US Department of Energy," 2010, http://energy.gov/sites/prod/files/edg/news/documents/criticalmaterialsstrategy.pdf.

[10] T. E. Graedel, "On the future availability of the energy metals," *Annual Review of Materials Research*, vol. 41, pp. 323–335, 2011.

[11] "Iron Ore Monthly Price," November 2012, http://www.indexmundi.com/commodities/?commodity=iron-ore&months=12.

[12] D. Kingsnorth, "Rare earths: is supply critical in 2013?" in *Proceedings of the Critical Minerals Conference Perth (AusIMM '13)*, Curtin Graduate School of Business; Curtin University & Industrial Minerals Company of Australia Pty Ltd, Western, Australia, June 2013.

[13] U.S. Geological Survey, "Niobium (Columbium) and Tantalum Statistics and Information," Mineral Commodity Summaries—Niobium (Columbium) 2012, http://minerals.usgs.gov/pubs/commodity/niobium/mcs-2012-niobi.pdf.

[14] U.S. Geological Survey, Rare Earths Statistics and Information 2010, Minerals Yearbook 2010—Rare Earth Contents of Major and Potential Source Minerals, 2010, http://minerals.usgs.gov/minerals/pubs/commodity/rare_earths/myb1-2010-raree.xls.

[15] C. Yan, J. Jia, C. Liao, S. Wu, and G. Xu, "Rare earth separation in China," *Tsinghua Science and Technology*, vol. 11, no. 2, pp. 241–247, 2006.

[16] MineralPrices, December 2012, http://www.mineralprices.com/.

[17] AsianMetal, "Rare Earth Prices," December 2012, http://www.asianmetal.com/price/initPriceListEn.am?priceFlag=8&productInfoIMG=PrNd+Mischmetal%E3%80%8299%EF%BC%85min+Nd+75%EF%BC%85+China%E3%80%82RMB%2FMT#.

[18] The China Post, "China will limit its exports of rare earth metals, official says," 3rd August 2009, http://www.chinapost.com.tw/business/asia/b-china/2009/09/03/223211/China-will.htm.

[19] W. M. Morrison and R. Tang, "China's Rare Earth Industry and Export Regime: Economic and Trade Implications for the United States," Congressional Research Service 7-5700 R42510, April 2012, http://www.fas.org/sgp/crs/row/R42510.pdf.

[20] ChinaDaily, "China announces rare earth export quotas," July 2011, http://www.chinadaily.com.cn/china/2011-07/14/content_12906607.htm.

[21] X. Du and T. E. Graedel, "Global in-use stocks of the rare earth elements: a first estimate," *Environmental Science and Technology*, vol. 45, no. 9, pp. 4096–4101, 2011.

[22] A. M. Helmenstine, "Rare Earth Properties," 2009, http://chemistry.about.com/od/elementgroups/a/rareearths.htm.

[23] L. J. Drew, M. Qingrun, and S. Weijun, "The Bayan Obo iron-rare-earth-niobium deposits, Inner Mongolia, China," *LITHOS*, vol. 26, no. 1-2, pp. 43–65, 1990.

[24] C. W. Sinton, "Study of the Rare Earth Resources and Markets for the Mt," Weld Complex for Lynas Corporation. BCC Research, Washington, DC, USA, 2008, http://www.lynas-corp.com/content/upload/files/press_releases/BCC_FINAL_REPORT.pdf.

[25] R. Will, A. Eric, and N. Takei, "Chemical Economics Handbook—Rare Earth Minerals and Products," SRI Consulting, Menlo Park, Calif, USA, 2010, http://www.sriconsulting.com/CEH/Public/Reports/765.5000/.

[26] Y. Kanazawa and M. Kamitani, "Rare earth minerals and resources in the world," *Journal of Alloys and Compounds*, vol. 408-412, pp. 1339–1343, 2006.

[27] G. X. Xing and H. M. Chen, "Environmental impacts of metal and other inorganics on soil and groundwater in China," in *Soils and Groundwater Pollution and Remediation: Asia, Africa, and Oceania*, P. M. Huang and I. K. Iskandar, Eds., pp. 167–200, CRC Press, 1999.

[28] S. Zhang and X.-Q. Shan, "Speciation of rare earth elements in soil and accumulation by wheat with rare earth fertilizer application," *Environmental Pollution*, vol. 112, no. 3, pp. 395–405, 2001.

[29] MolyCorp, "Mountain Pass Production," 2011, http://www.molycorp.jp/about-us/current-future-production/.

[30] Geoscience Australia, "Rare Earth Production and Exports," December 2012, http://www.ga.gov.au/minerals/mineral-resources/rare-earth-elements.html.

[31] S. Ramakrishnan and P. Koltun, "Global warming impact of the magnesium produced in China using the Pidgeon process," *Resources, Conservation and Recycling*, vol. 42, no. 1, pp. 49–64, 2004.

[32] A. Tharumarajah and P. Koltun, "Is there an environmental advantage of using magnesium components for light-weighting cars?" *Journal of Cleaner Production*, vol. 15, no. 11-12, pp. 1007–1013, 2007.

[33] C. A. Morais and V. S. T. Ciminelli, "Process development for the recovery of high-grade lanthanum by solvent extraction," *Hydrometallurgy*, vol. 73, no. 3-4, pp. 237–244, 2004.

[34] L. Wenli, G. D. ' Ascenzo, R. Curini et al., "Simulation of the development automatization control system for rare earth extraction process: combination of ESRECE simulation software and EDXRF analysis technique," *Analytica Chimica Acta*, vol. 417, no. 1, pp. 111–118, 2000.

[35] C. Liao, C. Yan, and J. Jia, "A Novel Solvent Extraction System for Tm, Yb and Lu," China Patent CN, 98100226.9, 2008, http://www.sciencedirect.com/science/article/pii/S1007021406701833.

[36] P. F. Duby, *Kirk-Othmer Encyclopedia of Chemical Technology*, John Wiley & Sons, 2005.

[37] International Standards Organisation, "Environmental management—Life cycle assessment—Requirements and guidelines," ISO 14044, 2006.

[38] K. Kodera, *Analysis of allocation methods of bioethanol LCA, Internship at CML [M.S. thesis]*, Faculty of Earth and Life Science, Leiden University, Leiden, The Netherlands, 2007.

[39] F. Ardente and M. Cellura, "Economic allocation in life cycle assessment: the state of the art and discussion of examples," *Journal of Industrial Ecology*, vol. 16, no. 3, pp. 387–398, 2012.

[40] J. Ren, S. Song, A. Lopez-Valdivieso, and S. Lu, "Selective flotation of bastnaesite from monazite in rare earth concentrates using potassium alum as depressant," *International Journal of Mineral Processing*, vol. 59, no. 3, pp. 237–245, 2000.

[41] Development Corp., Rare Earth Elements, http://www.candldevelopment.com/rare_earth_detail.htm.

[42] C. R. Hammond, *The Elements, Handbook of Chemistry and Physics*, CRC press, 81st edition, 2000.

[43] EcoInvent Centre, http://www.ecoinvent.org/database/.

[44] B. P. Weidema and M. S. Wesnæs, "Data quality management for life cycle inventories—an example of using data quality indicators," *Journal of Cleaner Production*, vol. 4, no. 3-4, pp. 167–174, 1996.

[45] Pre Consultants, 2011, http://www.pre.nl/.

[46] Pre Consultants, "A Damage oriented method for Life Cycle Assessment, Methodology Report," 2nd edition, 2000, http://teclim.ufba.br/jsf/indicadores/holan%20ecoindicator%2099.pdf.

[47] E. S. Shedd, J. D. Marchant, and T. A. Henrie, "Electrowinning and tapping of lanthanum metal," Bureau of Mines Report of Investigations 6882, U.S. Department of the Interior, Washington, DC, USA, 1966.

[48] R. A. Sharma, "Mettalothermic reduction of rare earth chlorides," US Patent 4680055, General Motors Corporation, http://www.freepatentsonline.com/4680055.pdf.

Aluminising of Mild Steel Plates

Udaya Bhat Kuruveri, Prashanth Huilgol, and Jithin Joseph

Department of Metallurgical and Materials Engineering, NITK, Surathkal 575025, India

Correspondence should be addressed to Udaya Bhat Kuruveri; udayabhatk@gmail.com

Academic Editors: Y. Yamabe-Mitarai and T. Yue

Hot dip aluminising of low carbon steel was done at temperatures 690°C and 750°C for dipping time ranging from 300 to 2400 seconds. During aluminising a mixture of $ZnCl_2$ and NH_4Cl was used as flux. During aluminising components of the flux decomposed and zinc formed interacted with the Fe and Al. The aluminised samples were characterised for iron-aluminium intermetallic layer formation, morphology, and local composition. It was observed that intermetallic layer was predominantly Fe_2Al_5 and $FeAl_3$ at 690°C and at 750°C coating consisted of $FeAl_3$ layer and a layer with Al/Fe ratio greater than 3.26. For both temperatures, coating thickness increased with increase in time. For a given dipping time, deposition was less at higher temperature and this is attributed to changes in the kinetics of growth of individual layers due to dissolved zinc in the aluminium, at 750°C. Also, spalling of intermetallic layers was observed at elevated temperatures and longer dipping times.

1. Introduction

Steel remains the material of choice for numerous applications, largely because of its low cost and wide variety of properties available through alloy design, heat treatment, and amenability for forming operations [1]. The oxide layer forming on the steel substrate is a noncompact one and due to this conventional mild steel exhibits poor oxidation and corrosion resistance properties. To improve corrosion properties, the steel can be alloyed with various elements such as chromium and nickel, but this increases material cost and limits width of the processing window [1]. Alternate route is to modify the surface of the steel appropriately, so that it can replace expensive alloy steel material [2]. During surface modification, the surface of the steel material is modified over a relatively shallow depth, using an appropriate method. When a suitable technique is selected and correctly done, the following properties of the steel can be increased: resistance to wear, chemical resistance, mechanical resistance, and oxidation resistance [3, 4]. Among various surface modification methods available for steel, the hot dipping process has drawn much attention because it is an effective and inexpensive process. In this context, use of aluminium as a coating material is highly relevant because, aluminium, when applied to steel provides corrosion resistance layer [5]. Also, aluminium bearing steels can be easily nitrided and nitrided steel has better wear and fatigue properties. Continuous alumina layer is highly resistant to oxidation, sulfidation, and degradation in chloride containing aqueous solutions [6]. Hence, hot dip aluminising is widely used for coating of steel with aluminium to increase corrosion, oxidation, and hardness of the steel components [5, 7].

Fe–Al is a complex system and its phase diagram is shown in Figure 1. It shows five types of intermetallic compounds, namely, Fe_3Al, FeAl, $FeAl_2$, Fe_2Al_5, and $FeAl_3$. Beneficial application of aluminising depends on minimising detrimental brittle phases and promoting growth of relatively ductile phases. There are investigations on the formation and growth kinetics of the intermetallic phases, under both isothermal and nonisothermal conditions. Shatynski et al. [9] modelled the formation of intermetallic phases from purely diffusional perspective. They found that growth of intermetallics follows parabolic relation. In the case of Cr–Ni stainless steel, Dybkov [10] experimentally proved that the intermetallics follows paralinear (initially linear followed by parabolic) growth kinetics. Eggeler et al. [11] have studied interaction between pure solid steel with liquid aluminium and observed that near aluminium side $FeAl_3$ forms and

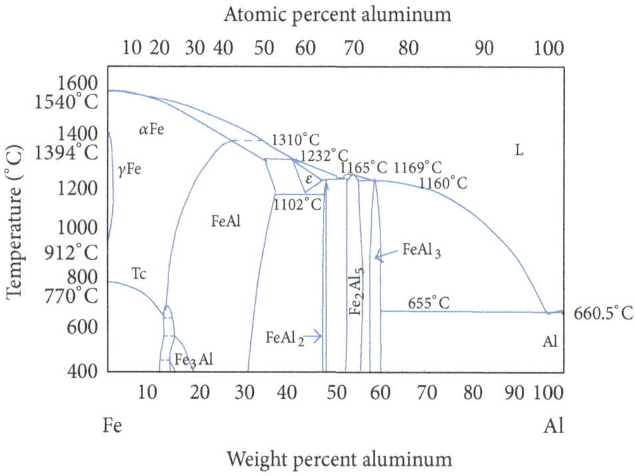

FIGURE 1: Phase diagram of Fe–Al [8].

FIGURE 2: Microstructure of steel consisting of proeutectoid ferrite and pearlite.

near iron side Fe_2Al_5 intermetallic forms. Heumann and Dittrich [12] tried to explain the tongue like morphology of F_2Al_5 layer. Bouayad et al. [13] studied the kinetics of various intermetallic layer formation in pure solid iron and pure liquid aluminium and found that $FeAl_3$ follows linear growth kinetics and Fe_2Al_5 follows parabolic kinetics. They did experiments at temperatures 700, 800, and 900°C. Naoi and Kajihara [14] did Al/Fe interdiffusion experiments over temperature range 550–640°C and found that only Fe_2Al_5 forms at the interface. They did experiments for fairly long time (120 hours) before arriving at this conclusion. Springer et al. [15] concluded that Fe_2Al_5 grows faster at lower temperature. They covered a temperature range of 600–675°C to cover solid-solid and solid-liquid diffusion couples. Maki et al. [16] have observed alloying reaction in aluminised steel when it was exposed for long durations at 600°C. Kobayashi and Yakou [17] have done experiments with a view to control intermetallic layers at interface between medium carbon steel and aluminium during diffusion experiments over the temperature range from 700 to 900°C for 300 s. They observed that interface is exclusively of Fe_2Al_5. It is stated that the growth of intermetallic layers depends on the diffusion of Fe atoms into the Al rich intermetallic layers [17]. It is also stated that intermetallic compounds with high Al content

TABLE 1: Composition of the steel strip (wt.%).

C	Mn	Si	S	P	Fe
0.186	0.47	0.23	0.027	0.018	Bal.

(namely, $FeAl_2$, Fe_2Al_5, and $FeAl_3$) are not desirable due to their brittleness. Hwang et al. [18] investigated the role of carbon in the dissolution of the substrate and formation of intermetallics. They concluded that carbon effectively lowers the rate of intermetallic layer formation.

Aluminum is a highly reactive metal and gets oxidised very fast, especially, when it is molten. Shielding molten aluminium is very essential. Also, the steel surface has to be cleaned for effective aluminisation of the surface. Mixture of zinc chloride and ammonium chloride is a promising one. It is a violent flux used in galvanisation [19], but its use in aluminising is less explored. Here an effort was made to understand the effectiveness of the zinc chloride and ammonium chloride as the flux and its role on the formation of Fe–Al based intermetallic layers.

2. Materials and Experimental Methods

The substrate material used in this investigation was low carbon steel, purchased in the form of a rolled strip. The composition of the strip is given in Table 1. The strip was cut into coupons of 3 cm length and 1 cm wide. Coupons were annealed at 900°C for two hours to remove the effect of prior work. Coupon surfaces were ground and polished. The coupons were precleaned using 10% HCl, washed in distilled water, and dried just before they were dipped in liquid aluminium.

Aluminium melt for hot dipping was prepared by heating aluminium ingot in a clay crucible heated by resistance heating. The aluminium was melted and the set temperature was attained. Fluxing was done using a eutectic mixture of ammonium chloride and zinc chloride. Fluxing treatment was done in order to enhance the wetting and dissolution of the oxide coating on the steel, in the molten aluminium bath. Fluxing was done by maintaining a layer on the top of the Al bath. Steel coupons were dipped in aluminium melt maintained at 690°C and 750°C for the time intervals of 300, 600, 900, 1500, and 2400 seconds.

The specimens were cut across the cross-section, metallographically polished, and etched using 3% nital solution. Microstructures were observed in scanning electron microscope (SEM) and coating morphology was studied. Intermetallic layer thickness was measured. Local composition was measured using EDS (energy dispersive spectroscopy) attached to SEM.

3. Results and Discussion

Figure 2 shows the microstructure of the low carbon steel used in the experiment. Microstructure consists of proeutectoid ferrite and eutectoid pearlite. The area fractions of these features confirm that the substrate used is a low carbon steel. Figures 3 and 4 show the microstructure of the aluminised

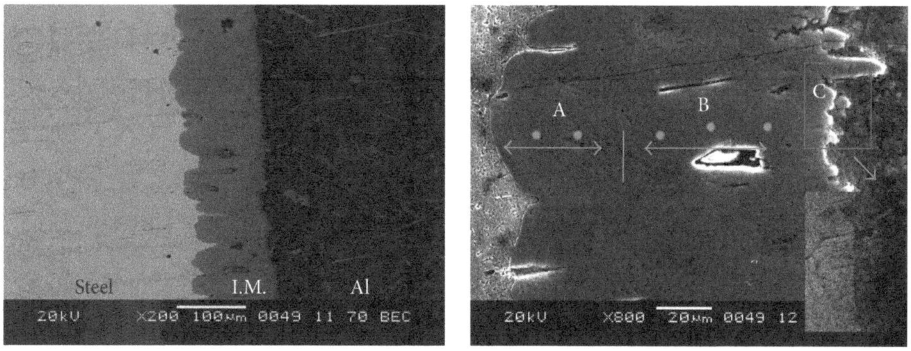

(a) (b)

FIGURE 3: (a) Intermetallic (I.M) formation at the interface (690°C and 300 s). (b) Magnified from (a) EDS analysis at circled points indicates 3 distinct zones: (A) Fe_2Al_5, (B) $FeAl_3$, and (C) Fe : Al~1 : 4.6.

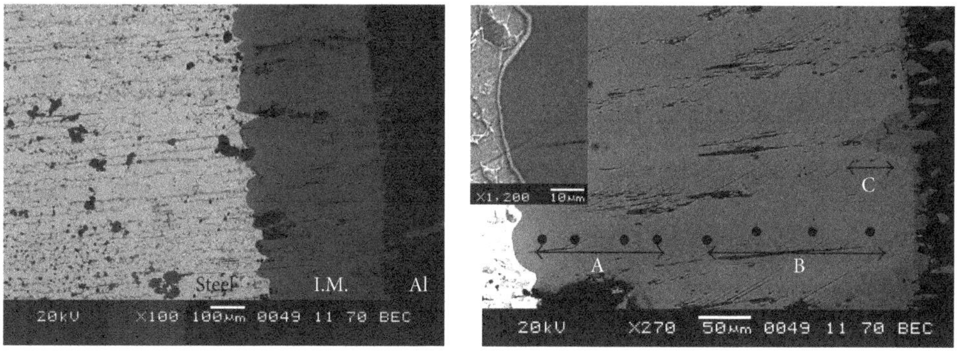

(a) (b)

FIGURE 4: (a) Intermetallic (I.M.) formation at the interface (690°C and 2400 s). (b) Magnified from (a) EDS analysis at circled regions indicates 3 distinct zones: (A) Fe_2Al_5, (B) $FeAl_3$, and (C) Al/Fe ratio 4 : 1.

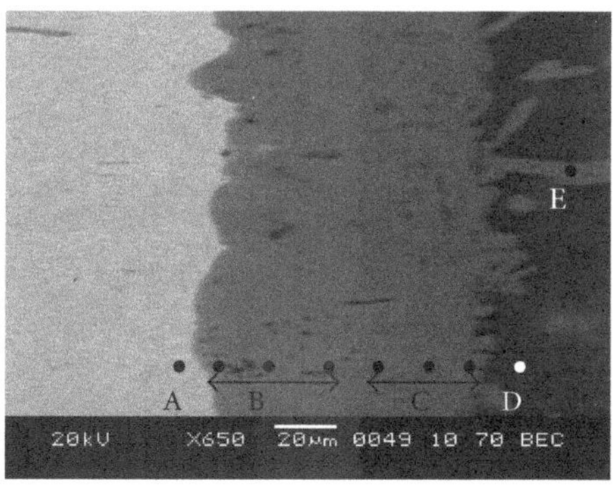

FIGURE 5: Intermetallic layer formation at 750°C for 300 s. At (%) at various locations are (A) 99.3%Fe, 0.7Al, (B) $FeAl_3$ ratio with Zn 3 to 3.4%, (C) Al : Fe ratio in the range 3.6 : 1 to 4.1 : 1, Zn 0.4 to 1.3, (D) 0.4Fe, 93.3Al, 5.4Zn, (E) 80.4Al, 18.8Fe, 0.9Zn.

steel, aluminised at 690°C for 300 s and 2400 s, respectively. We see a layer of intermetallic phases (I.M.) between steel and aluminium. Intermetallic layer has a characteristic morphology, which is referred to in the literature as "tongue" like morphology. Intermetallic layer is continuous but its thickness is not constant. The interface between the Al and intermetallic layer is not smooth but the unevenness is on a smaller length scale. On the contrast, the interface between steel and intermetallic layer is more uneven and appears like wave or tongue like. Wave or tongue like morphology for intermetallic layers during aluminising is reported in the literature [15, 17]. Thickness was measured over a length of 1 mm. For 300 s exposure time, the interface thickness was in the range of 95 to 120 μm with an average value of 112 micrometers. Intermetallic layer was showing predominantly single contrast, but periodic EDS measurements on this layer showed two distinct regions A and B, whose chemistry was matching with Fe_2Al_5 and $FeAl_3$, respectively.

There exists another layer C (Figures 3(b) and 4(b)) where the Al/Fe ratio was much higher than 3.26 which is the upper limit of the atomic ratio for the phase, $FeAl_3$ [9]. When the dipping time was large, a thin layer D at Fe/Fe_2Al_5 interface is clearly visible (inset in Figure 4(b)). EDS analysis on it at selected places indicates Fe/Al ratio as 1 : 2 indicating it most probably as $FeAl_2$ compound. It may be mentioned that Fe_2Al_5 and $FeAl_3$ are not line phases (refer to phase diagram shown in Figure 1), indicating the possibility of different Fe/Al ratios at different locations. This is also observed during EDS analysis.

Figure 5 shows microstructural features under the conditions of 750°C and 300 seconds. We see two distinct regions

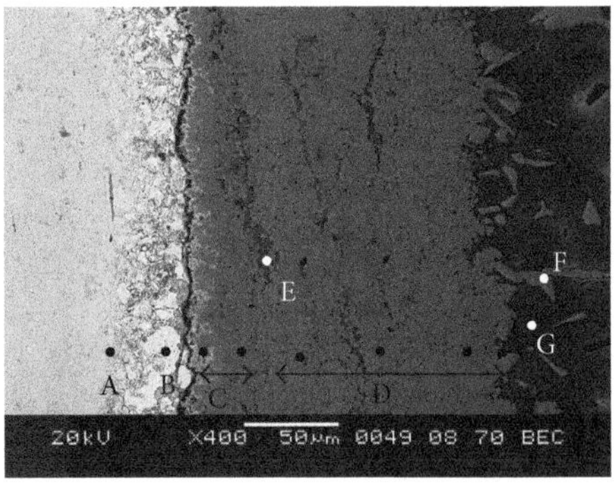

(a)

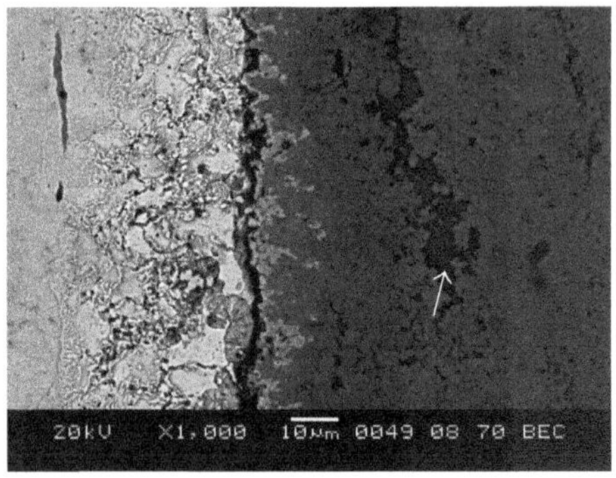

(b)

FIGURE 6: (a) Intermetallic layer formation at 750°C for 2400 s. At (%) at various locations are (A) 100Fe, (B) 93Fe7Zn, (C) $FeAl_3$ ratio, Zn 1.4 to 3, (D) Al : Fe ratio 3.6 : 1 to 4.2 : 1, Zn up to 1.2, (G) Fe : Al : Zn = 1 : 94 : 5, (F) Fe : Al : Zn = 80.4 : 18.8 : 0.8.3b: Enlarged at point (E). EDS at arrow point in (b) indicates a ternary system with Fe : Al : Zn = 85.5 : 6.6 : 7.9.

corresponding to $FeAl_3$ and another layer with Al : Fe ratio greater than 3.26 : 1. Both show dissolved zinc of various scales. There is no layer corresponding to Fe_2Al_5 which is distinctly clear in 690°C. Figures 6(a) and 6(b) show microstructures in the case of 750°C bath temperature and holding time of 2400 seconds. Here also no layer corresponding to Fe_2Al_5 is visible. A layer corresponding to solid solution of Al in iron is clearly visible. All layers show dissolved zinc in them.

We see many polyhedric crystals peeled off from the intermetallic layers in the aluminium matrix. This is due to decrease in the iron solubility in aluminium during cooling. The occurrence of these crystals is more favoured when the cooling rate is increased and the interaction time is long (Figures 4(a) and 5(a)). These crystals have a composition close to that existing in the intermetallic layer close to Al side. Such peeling off is reported by Bouayad et al. [13].

Formation of various phases in the intermetallic layer under different conditions is explained as follows. In the initial stages of dipping, the intermetallic layer forms are $FeAl_3$ and Fe_2Al_5 [17]. The $FeAl_3$ layer is initially formed at the interface between the aluminium and steel substrate by the diffusion of iron atoms into aluminium layer and the Fe_2Al_5 layer is then formed at the interface between the $FeAl_3$ layer and steel substrate. Formation of this layer depends on the relative values of diffusivities (iron into aluminium and aluminium into iron). Under normal conditions, diffusion of iron towards aluminium is larger [20]. It favours Fe_2Al_5 formation at Fe-$FeAl_3$ interface. This is happening at lower temperature. At higher temperature, the zinc increased the activity of Al in the bath (for details refer to the next paragraph). Increased activity leads to increased diffusion of Al towards iron-$FeAl_3$ interface. This reduced the chances of Fe_2Al_5 and increased the chances of $FeAl_3$ layer as well as a layer with Al/Fe ratio greater than 3.2 : 1.

Higher dipping temperature means higher amount of $ZnCl_2$ from the flux mixture reacts with metal oxides or oxygen in the environment. It gives ZnO which gets reduced by Al into Zn. This explains higher levels of Zn in the intermetallic layer formed at elevated temperatures. Higher levels of dissolved zinc in Al favour formation of $FeAl_3$ intermetallic than Fe_2Al_5. The two-phase equilibrium of Fe_2Al_5 and $FeAl_3$ cannot appear in this study as it involves diffusion of species under chemical potential gradient. Presence of Zn in the bath increases activity of Al, leading to a layer with more Al concentration. Such observations are also reported in the literature [21]. This explains the observation of $FeAl_3$ and higher Al concentration layers. The eutectic reaction at 655°C in Fe–Al system [8] is bypassed during experimental investigation. Another aspect is liquid aluminium-iron layer on the surface is rapidly cooled giving metastable phases. Skinner et al. [22] have reported possibility of a number of metastable phases, namely, Al_6Fe, Al_5Fe, and Al_9Fe_2 under rapid cooling conditions. All these metastable phases have Al concentration more than that in $FeAl_3$. In our experiments also, the condition at the surface is rapid cooling leading to metastable phases.

Figure 7 shows variation of thickness as a function of dipping time. It shows an increase in the deposit thickness as a function of time. To explore the parabolic relationship between intermetallic layer thickness and dipping time, a plot of thickness versus square root of dipping time is drawn (Figure 8). A straight line relationship is observed indicating that total intermetallic layer follows parabolic growth behaviour. At 690°C, increase in thickness is mainly from thickness of Fe_2Al_5 and $FeAl_3$. Though $FeAl_2$ layer is visible during longer dipping times, its thickness is negligible compared to overall deposition thickness. Amongst Fe_2Al_5 and $FeAl_3$, Fe_2Al_5 undergoes parabolic growth behaviour and $FeAl_3$ follows quasilinear behaviour [13]. Gibbs free energy for formation of $FeAl_3$ is lower than Fe_2Al_5 and due to this $FeAl_3$ forms instantaneously during dipping. But from kinetics point of view Fe_2Al_5 is more favoured [23]. Diffusion of Al and Fe takes place across the $FeAl_3$ layer, promoting formation of Fe_2Al_5 [24]. Due to this, for lower dipping times, $FeAl_3$ and for longer dipping times, Fe_2Al_5 are major constituents. Because of this, thickness seems to be nonzero even at the start of the dipping. Such observations are also reported by Cheng and Wang [23] and Li et al. [25]. It is also to be mentioned the smallest experimental duration during

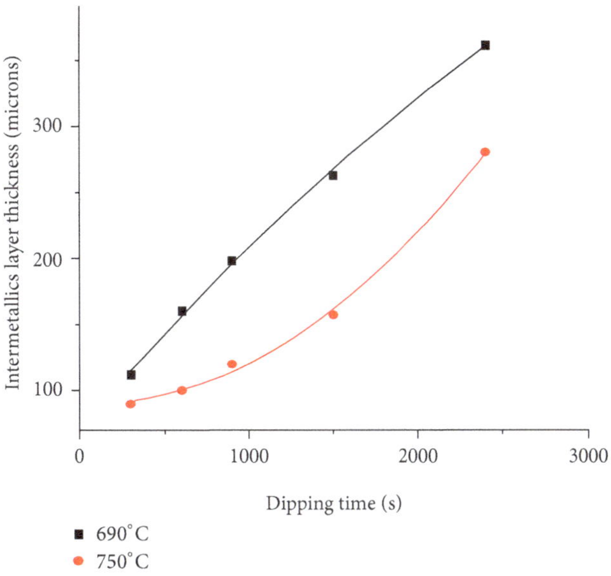

FIGURE 7: Plot shows variation of intermetallic layer thickness as a function of dipping time for temperatures 690°C and 750°C.

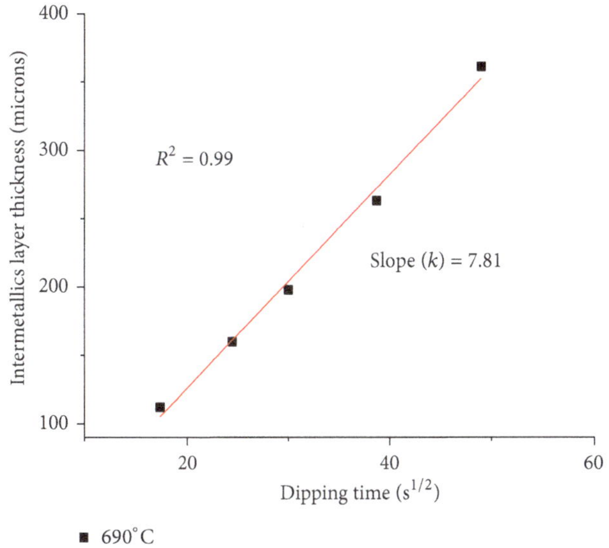

FIGURE 8: Plot showing total intermetallic layer thickness versus square root of time. A good (linear) fit indicates that the total intermetallic layer growth follows parabolic growth behaviour.

our experiments is 300 seconds. It is reported that by this time all the transient nature associated with this reactive diffusion is crossed [25].

At 750°C, the deposit consists of $FeAl_3$ and another layer $FeAl_{3.26+x}$ where x is a positive decimal number. These individual layers have different growth kinetics and hence at 750°C the overall thickness is less compared to that at 690°C. At long interaction times and high temperatures, the spalling of the intermetallic layer takes place and it affects the measurement of the layer thickness [11].

4. Conclusions

The experimental investigations lead to the following conclusions. Use of a mixture of $ZnCl_2$ and NH_4Cl as a flux for aluminizing leads to some amount of dissolved zinc in the Al bath, especially when bath temperature is high.

At 690°C, the intermetallic layer consisted of predominantly, Fe_2Al_5, and $FeAl_3$ and at 750°C, the intermetallic layer consists of $FeAl_3$, and a layer with higher levels of Al concentration (Al/Fe ratio more than 3.26) forming due to rapid cooling conditions. This change is attributed to increased Al activity due to dissolved zinc.

The growth of Fe_2Al_5 layer and total intermetallic layer follows parabolic growth behaviour with respect to dipping time.

References

[1] G. Krauss, *Properties and Selection: Iron, Steel and High Performance Slloys*, vol. 1 of *ASM Metals Hand Book*, Materials Park, Ohio, USA, 1990.

[2] J. R. Davis, *Surface Engineering*, vol. 5 of *ASM Metals Hand Book*, Materials Park, Ohio, USA, 1990.

[3] Z. Ahmad, *Principles of Corrosion Engineering and Corrosion Control*, Elsevier, London, UK, 2006.

[4] T. Burakowski and T. Weirzchok, *Surface engineering of Metals-Principles, Equipments, Technologies*, CRC Press, London, UK, 2000.

[5] D. Wang and Z. Shi, "Aluminizing and oxidation treatment of 1Cr18Ni9 stainless steel," *Applied Surface Science*, vol. 227, no. 1–4, pp. 255–260, 2004.

[6] K. Murakami, N. Nishida, K. Osamura, and Y. Tomota, "Aluminization of high purity iron by powder liquid coating," *Acta Materialia*, vol. 52, no. 5, pp. 1271–1281, 2004.

[7] A. Bahadur and O. N. Mohanty, "Structural studies of hot dip aluminized coatings on mild steel," *Materials Transactions, JIM*, vol. 32, no. 11, pp. 1053–1061, 1991.

[8] N.Y. Tang, "Determination of liquid phase boundaries in Zn-Fe-Mx systems," *Journal of Phase Equilibria*, vol. 21, no. 1, pp. 167–176, 2000.

[9] S. R. Shatynski, J. P. Hirth, and R. A. Rapp, "A theory of multiphase binary diffusion," *Acta Metallurgica*, vol. 24, no. 12, pp. 1071–1078, 1976.

[10] V. I. Dybkov, "Reaction diffusion in heterogeneous binary systems—part 1 Growth of the chemical compound layers at the interface between two elementary substances: one compound layer," *Journal of Materials Science*, vol. 21, no. 9, pp. 3078–3084, 1986.

[11] G. Eggeler, W. Auer, and H. Kaesche, "On the influence of silicon on the growth of the alloy layer during hot dip aluminizing," *Journal of Materials Science*, vol. 21, no. 9, pp. 3348–3350, 1986.

[12] T. Heumann and S. Dittrich, "Uber die Kinetic der Reaction von Festemund Flussigem Aluminum mit Eisen," *Zeit fi Afetallkunde*, vol. 10, pp. 617–624, 1959.

[13] A. Bouayad, C. Gerometta, A. Belkebir, and A. Ambari, "Kinetic interactions between solid iron and molten aluminium," *Materials Science and Engineering A*, vol. 363, no. 1-2, pp. 53–61, 2003.

[14] D. Naoi and M. Kajihara, "Growth behavior of Fe_2Al_5 during reactive diffusion between Fe and Al at solid-state temperatures," *Materials Science and Engineering A*, vol. 459, no. 1-2, pp. 375–382, 2007.

[15] H. Springer, A. Kostka, E. J. Payton, D. Raabe, A. Kaysser-Pyzalla, and G. Eggeler, "On the formation and growth of intermetallic phases during interdiffusion between low-carbon steel and aluminum alloys," *Acta Materialia*, vol. 59, no. 4, pp. 1586–1600, 2011.

[16] J. Maki, M. Suehiro, and Y. Ikematsu, "Alloying reaction of aluminized steel sheet," *ISIJ International*, vol. 50, no. 8, pp. 1205–1210, 2010.

[17] S. Kobayashi and T. Yakou, "Control of intermetallic compound layers at interface between steel and aluminium by diffusion treatment," *Materials Science and Engineering A*, vol. 338, pp. 44–53, 2002.

[18] S. H. Hwang, J. H. Song, and Y. S. Kim, "Effects of carbon content of carbon steel on its dissolution in to molten aluminium alloy," *Materials Science and Engineering A*, vol. 390, pp. 437–443, 2005.

[19] M. Gilles, D. Leychkis, Y. Grot, and R. Pankert, "Preparation of steel surfaces for single dip aluminium rich zinc galvanising," US Patent no. 7160581, January 2007.

[20] G. Neumann and A. D. Le Claire, *Diffusion in Solid Metals and Alloys, Numerical Data and Functional Relationships in Science and Technology*, vol. 26, Springer, 1990.

[21] M. V. Akdeniz and A. O. Mekhrabov, "The effect of substitutional impurities on the evolution of Fe-Al diffusion layer," *Acta Materialia*, vol. 46, no. 4, pp. 1185–1192, 1998.

[22] D. J. Skinner, K. Okazaki, and C. M. Adam, "Physical metallurgy and mechanical properties of Al alloys containing 8 to 12%Fe rapidly solidified powder alumium alloys," in *Rapidly Solidified Powder Aluminum Alloys*, M. E. Fine and E. A. Starke, Eds., ASTM STP 890, pp. 211–236, American Society for Testing and Materials, Philedelphia, Pa, USA, 1986.

[23] W.-J. Cheng and C.-J. Wang, "Characterisation of intermetallic layer formation in aluminde/nickel duplex coating on mild steel," *Materials Characterisation*, vol. 69, pp. 63–70, 2012.

[24] K. Bouché, F. Barbier, and A. Coulet, "Intermetallic compound layer growth between solid iron and molten aluminium," *Materials Science and Engineering A*, vol. 249, no. 1-2, pp. 167–175, 1998.

[25] Y. J. Li, J. Wang, and X. Holly, "X-ray diffraction and TEM analysis of Fe-Al alloy layer in coating of new hot dip aluminised steel," *Materials Science and Technology*, vol. 19, no. 5, pp. 657–660, 2003.

4

Electrochemical Fabrication of Niobium Silicon Alloys from Oxide Powder Mixtures

Fanke Meng and Huimin Lu

School of Materials Science and Engineering, Beihang University, No.37 Xueyuan Road, Haidian District, Beijing 100191, China

Correspondence should be addressed to Fanke Meng; fankemeng0208@gmail.com

Academic Editors: Y. S. Chang, P. Lukac, A. Pinkerton, and Y.-d. Wang

NbSi alloys were prepared by direct electrochemically reducing four mixed Nb_2O_5 and SiO_2 powders (Nb-10Si, Nb-20Si, Nb-30Si, and Nb-37.5Si) in molten $CaCl_2$ electrolyte at 900°C. The samples were characterized with scanning electron microscope (SEM), X-ray diffraction (XRD), and energy-dispersive X-ray spectroscopy (EDX). No oxidized phases were remained by XRD tests. Under SEM, Nb phase was scattered in Nb_5Si_3 phase for the samples of Nb-10Si, Nb-20Si, and Nb-30Si. For the sample of Nb-37.5Si, only Nb_5Si_3 was obtained after electrolysis.

1. Introduction

Nickel alloys are the mostly used superalloys in aviation industry. Due to the demand of exploring new generation aircraft turbine engines, scientists and engineers are devoted to investigate less expensive superalloys with higher melting points to substitute the Ni-based alloys [1, 2]. Following this idea, NbSi alloys (0.5–37.5 at % Si) are attracting more and more attention because of the high melting points of the two composed phases (melting points: Nb, 2469°C; Nb_5Si_3, 2520°C) in the alloy system [1]. Moreover, NbSi alloys still have three other excellent properties: close coefficient of heat expansion of the two compositions (Nb, $7.3 \times 10\,k^{-1}$; Nb_5Si_3, $6.1 \times 10\,k^{-1}$), relative lower density (7.16 g cm^{-3}), and reasonable strength at high temperature. Thus, NbSi alloys are considered to be a potential candidate of the next generation superalloys. Until now, the mainly used techniques for fabricating NbSi alloys are orientation solidification [3], smelting with argon arc [4], physical vapor deposition (PVD), as well as some other methods of powder metallurgy. Generally, all of these techniques have the following procedures: first, extracting pure Nb and Si from minerals; second, melting and mixing molten Nb and Si by using arc or electron beam melting method; third, casting the mixed molten alloys to ingots. However, the notorious weakness of the process is resulted from high energy consumption and complex production procedures. As a typical example, the only commercial process for the production of metallic Nb is reducing K_2NbF_7 or Nb_2O_5 with sodium reduction or carbothermal reduction, respectively [5]. For production of silicon, the conventional technique is reducing SiO_2 with carbothermal reduction technique. The typical reduction temperature is as high as 1700°C, which therefore needs intensive energy and releases carbon oxides as wasted gases in the process [6]. Also, another critical disadvantage should be paid more attention: it is difficult to homogenize Nb and Si in any melting process due to the very high melting point of metallic Nb.

During the recent decade, in order to produce high-melting-point metals and superalloys, many new skills were studied by scientists and engineers. Among them, direct electrochemical reduction of metal oxides or mixed oxides in molten electrolytes was an effective method for production of metals and alloys. The typical examples were FFC Cambridge Process and the derived methods. In this process, metal oxides were pressed to solid pellets as a cathode and the graphite were acted as an anode, which were soaked into the molten $CaCl_2$ at about high temperatures (700–900°C) for electrolysis simultaneously. Consequently, a potential was loaded between the two electrodes. Generally, this technique was used in production of many metals and nonmetals for example, Cr, Si, Nb, Ni, and Zr [7–12]. The whole reaction process can be described with the following equations:

Cathode

$$AO_X + 2xe^- = A + xO^{2-} \quad (1)$$

Anode

$$2O^{2-} - 4e^- + C = CO_2 \quad \text{or} \quad O^{2-} - 2e^- + C = CO \quad (2)$$

In these equations, the A is a symbol of metal or nonmetal elements. If the cathode was mixed oxides, the reduction production might be oxygen-free alloys. Therefore, this electrochemical method can be effectively used for refractory metals, high-melting-point nonmetals, and superalloys.

Herein, we used this electrochemical reduction method for the production of NbSi alloys in the molten $CaCl_2$ at high temperature (900°C). The processors were mixed Nb_2O_5 and SiO_2 powders of four various atomic ratios of Nb and Si. After electrolysis, the NbSi alloys were successfully obtained. After scanning electron microscope (SEM), X-ray diffraction (XRD), and energy-dispersive X-ray spectroscopy (EDX) analysis, the NbSi alloys were composed of different phases and almost free of oxygen. Therefore, this electrochemical reduction method had paved a way for the NbSi alloys production, which is also able to be applied to fabricate other refractory alloys.

2. Experimental Details

2.1. Materials Preparation. In this research work, the electrolyte was $CaCl_2$. In this work, $CaCl_2$ (≥99.0%), Nb_2O_5 (>99.9%), and SiO_2 (>99.8%) powders were purchased from Chengde Optical and Electrical Materials Ltd., Hebei Province, China. The pellets preparation was following the below steps. First, $CaCl_2$ powders were dried at 80°C overnight and then pressed to small cylindrical blocks with a pressure of 30 MPa. Second, the cylindrical blocks were dried again at 80°C for several hours and then pulverized to small $CaCl_2$ pellets with an average sized less than 1 mm. For the preparation of the Nb_2O_5/SiO_2 pellets, the process was identical as that used for $CaCl_2$ pellets preparation. The Nb_2O_5 and SiO_2 powders were mixed with the atomic ratios of Nb-10Si (Nb, 90 at %; Si, 10 at %), Nb-20Si (Nb, 80 at %; Si, 20 at %), Nb-30Si (Nb, 70 at %; Si, 30 at %), and Nb-37.5Si (Nb, 62.5 at %; Si, 37.5 at %) between Nb and Si. The total mixed powder was 200 g. The graphite crucibles (inner diameter: ~9.0 cm, height: ~18.0 cm, wall thickness: ~1.0 cm) were cleaned with ethanol (>95%, Beijing Keshi Co. Ltd. Beijing, China) for at least 5 times until the inner walls of the graphite crucible were almost free of graphite powders. At last, the mixed Nb_2O_5/SiO_2 pellets were placed in the cleaned crucible and then the $CaCl_2$ pellets (~2000 g) were placed on the top of the Nb_2O_5/SiO_2 pellets.

2.2. Potentiostatic Electrolysis. The process of electrolysis was conducted in a programmable temperature-controlled furnace produced by China Iron & Steel Research Institute Group. The design of the instrument was shown in the schematic of Figure 1. The graphite crucible filled with Nb_2O_5/SiO_2 and $CaCl_2$ was placed in a graphite stage in the furnace. As an anode, a graphite rod (diameter: 6.0 cm) was just hanged above the crucible vertically by a mechanic instrument, which can push the graphite rod move up and down to ensure that the rod can contact with the molten electrolyte during the whole electrolysis process. The direct current for the electrolysis was supplied by a KGHF-1000A/6V rectifier, made by Beijing Chunshu Electric rectifier Ltd., Beijing, China. The temperature-controlled furnace was vacuumed and then filled with argon (≥99.9%, $O_2 \leq 10$ ppm, $H_2O \leq 15$ ppm, $N_2 \leq 5$ ppm) as soon as possible. It should be noted that the argon flow was maintained with a rate of 1.0–1.5 L min^{-1} and the outlet was adjusted with a valve to ensure that the gas pressure inside the furnace was about 0.01 MPa larger than the ambient pressure, which was helpful for avoiding air to ingress into the furnace in the electrolysis process. The heating process was of three steps: first, the temperature was heated to 780°C with a ramping rate of 10°C min^{-1}; second, kept at 780°C for 30 min, the $CaCl_2$ electrolyte was completely melted in the crucible; third, the temperature continued to rise up to 900°C with a heating rate of 10°C min^{-1}. Consequently, the potential was loaded and then stabilized at 3.0 V for electrolysis between the two electrodes. The electrolysis would be lasted for about 4-5 hours, and the graphite crucible with the products was removed out the furnace as the temperature in the furnace was decreased to room temperature after electrolysis. The product was obtained by smashing the crucible.

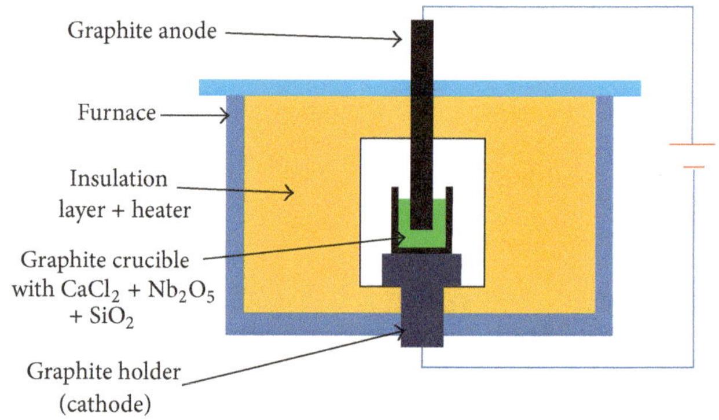

FIGURE 1: Schematic of eletrolysis system.

2.3. Characterizations of Morphology and Composition. Micrograph and composition analyses were conducted with a VEGA\\LMU SEM system equipped with a CCD camera, from TESCAN. EDX equipment was purchased from OXFORD and analysis software was INCA. The phase composition of the NbSi alloys was identified by a Rigaku D/Max 2200 PC X-ray diffractometer.

3. Results and Discussion

After electrolysis at 900°C in molten $CaCl_2$, the samples were obtained. Figure 2 showed the pictures of the obtained

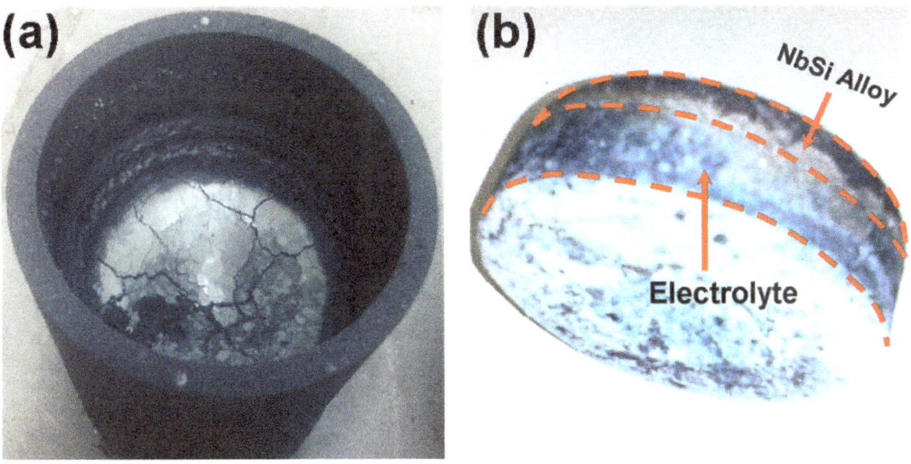

(a) (b)

FIGURE 2: (a) top view of the crucible after electrolysis; (b) the sample with electrolyte after electrolysis.

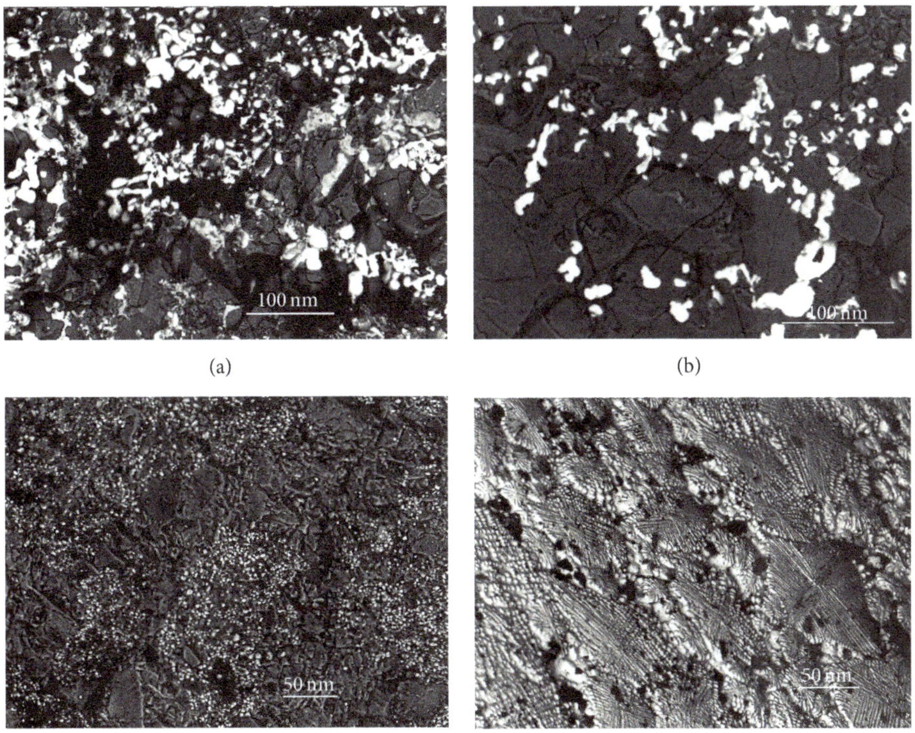

(a) (b) (c) (d)

FIGURE 3: SEM images of four samples with different Nb/Si ratio: (a) Nb-10Si; (b) Nb-20Si; (c) Nb-30Si; (d) Nb-37.5Si.

product after electrolysis. As shown in Figure 2(a), the $CaCl_2$ electrolyte and the sample were molten together after electrolysis. By smashing the crucible, a block of ingot was obtained in Figure 2(b). Clearly, the ingot had two layers of different materials. A metal-like layer with a color of grayish was rich of alloys and other layer was mainly composed of electrolyte.

It should be noted that Figure 2 only showed typical pictures after electrolysis of the processor powder for the alloy of Nb-37.5Si. The other three samples were also of a similar picture, which were not shown here. The morphologies of the alloys were shown in the SEM images in Figure 3. The white small particles with an average size about 10–20 nm had Nb concentration of 98.2 at % in Table 1. On the contrary, the concentrations of the Si and the O were only 1.6 at % and 0.2 at %, respectively. Therefore, it was clearly to show that the white small particles were mainly composed of Nb. In the grey areas, the concentrations of NB and Si were 89.4 and 10.1 at % respectively, which was almost equated to the ratio of Nb-10Si

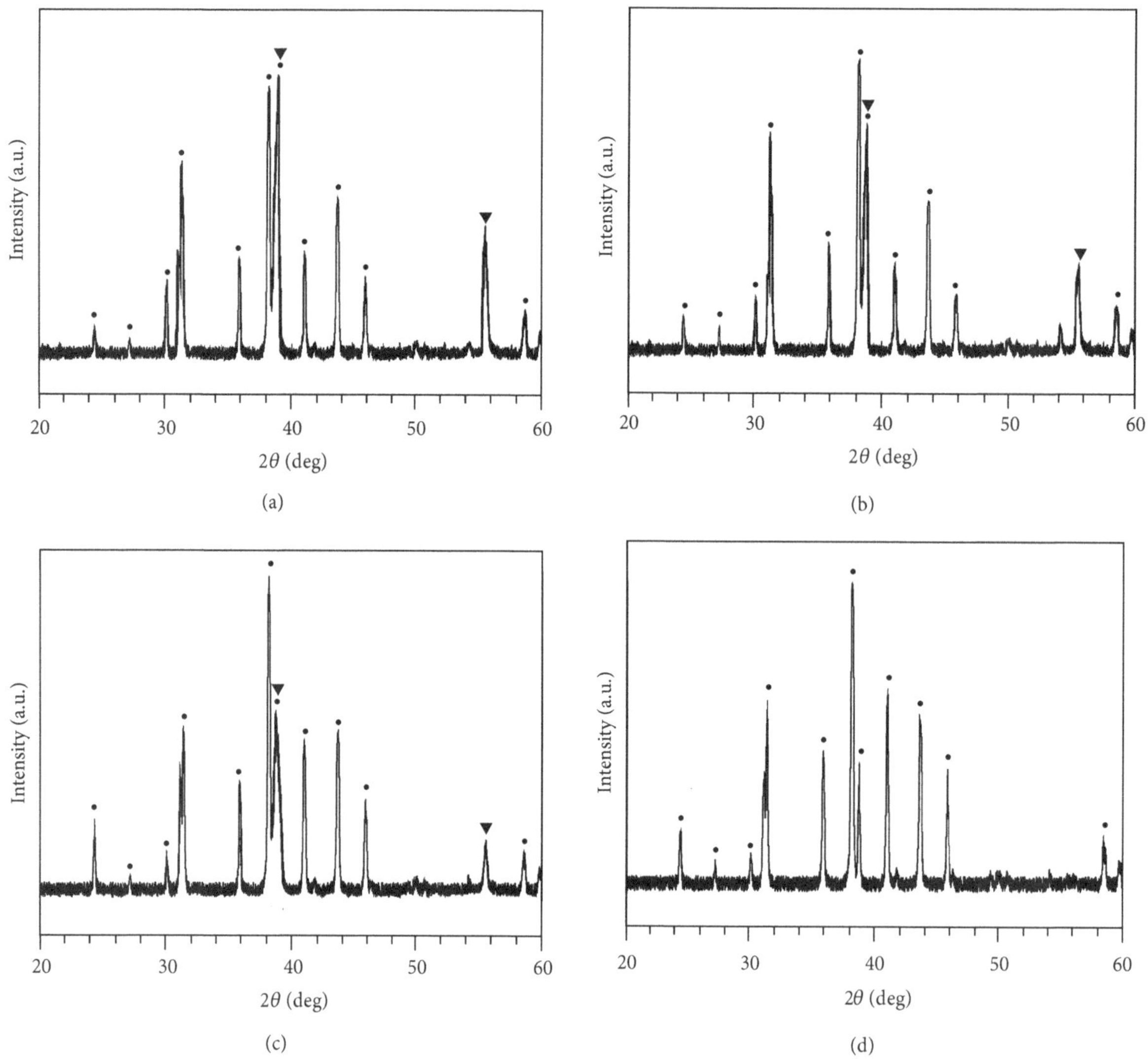

FIGURE 4: XRD patterns of four samples with different Nb/Si ratio: (a) Nb-10Si; (b) Nb-20Si; (c) Nb-30Si; (d) Nb-37.5Si.

alloy. It should be noted that only a very small part of O was still remained in both the white small particles and the grey areas. Although the O was not completely electrochemical reduced, it can be removed by other metallurgical methods in the consequent process. Typically, the mixed Nb_2O_5/SiO_2 powder was almost completely electrodeoxidized in molten $CaCl_2$ at 900°C. For the processor powder mixture of Nb-20Si alloy, the product after electrolysis the SEM image was shown in Figure 3(b), which was similar to the image in Figure 3(a). The EDX results in the Table 1 also showed that the white particles were Nb and the grey areas were NbSi alloy (Nb-20Si). The similar SEM image and EDX results were shown for the sample Nb-30Si alloy. Compare the three alloys (Nb-10Si, Nb-20Si, and Nb-30Si), two conclusions were obtained. First, the grey areas in all the three images had Nb and Si concentrations close to 62.5 and 37.5 at %, respectively, which was the atomic ratio in the Nb_5Si_3 alloy. Second, the white small particles were almost mainly composed of metallic Nb. Third, although O was remained in a little part in all the three samples, most of the O were still removed electrochemically. For the powder mixture of Nb-37.5Si alloy, the product after electrolysis, the SEM image, and EDX result were shown in Figure 3(d) and Table 1, respectively. Clearly, the product in the SEM image showed a homogenous phase with a Nb/Si ratio of 62.2/37.0, which was also close to the atomic ratio of Nb5Si3.

To address the phase composition of the products in the four figures, the XRD patterns were obtained in Figure 4. All of the samples have a phase of Nb_5Si_3 (JCPDS: 30-0874) as indicated with solid circle. Only the alloys of the Nb-10Si, Nb-20Si, and Nb-30Si had a phase of Nb (JCPDS: 35-0789) shown in Figures 4(a), 4(b), and 4(c). Based on the data of the XRD, SEM, and EDX, three critical conclusions can be drawn. First, all the samples were composed of two phases: Nb and Nb_5Si_3, which were the two phases in NbSi alloys. Second, in

TABLE 1: EDX results of the four samples in Figure 3.

Positions	Nb (at %)	Si (at %)	O (at %)
a-white	98.2	1.6	0.2
a-grey	62.1	37.6	0.3
b-white	99.1	0.8	0.1
b-grey	62.6	37.4	—
c-white	98.7	1.1	0.2
c-grey	62.7	37.0	0.3
d	62.2	37.5	0.4

a-white, b-white, and c-white showed the concentrations of Nb, Si and O in the positions of white small particles in Figures 3(a), 3(b), and 3(c), respectively. Similarly, a-grey, b-grey, and c-grey showed the concentrations of Nb, Si and O in the positions of grey areas in Figures 3(a), 3(b), and 3(c), respectively. The concentrations of Nb, Si, and O in the sample Nb-37.5Si alloy in Figure 3(d) were shown in position d.

Figures 3(a), 3(b) and 3(c), the small white particles were Nb phase, and grey phase was Nb_5Si_3. Third, the Nb-37.5Si sample was completely composed of Nb_5Si_3 phase after electrolysis.

4. Conclusion

We used four mixed powders of different Nb-Si compositions (Nb-10Si, Nb-20Si, Nb-30Si, and Nb-37.5Si) for NbSi alloys preparation. NbSi alloys can be prepared by direct electrochemical reduction method in molten $CaCl_2$ at 900°C, which is an easy and one-plot procedure. After electrolysis, Nb phase showed white scattered particles in the grey Nb_5Si_3 phase under SEM. XRD, and EDX results have proved that the NbSi alloys were successfully obtained, which had paved a creative way for production of alloys with direct electrochemical reduction method.

Acknowledgments

The authors are grateful for the Science and Technology Ministry of China for the financial support to the national high technology project.

References

[1] M. Singh and D. R. Behrendt, "Reactive melt infiltration of silicon-niobium alloys in microporous carbons," *Journal of Materials Research*, vol. 9, pp. 1701–1708, 1994.

[2] C. L. Ma, A. Kasama, R. Tanaka, and S. Hanada, "Development of Nb-Nb-silicide in-situ composites," *Transactions of Metal Heat Treatment*, vol. 21, p. 2, 2000.

[3] M. G. Mendiratta and D. M. Dimiduk, "Microstructures and mechanical behaviors of 2-phase niobium silicide-niobium alloys," in *Proceedings of the Material Research Society Symposia Proceeding*, vol. 133, MRS, Pittsburgh, Pa, USA, 1989.

[4] W. Y. Kim, H. Tanaka, A. Kasama, and S. Hanada, "Microstructure and room temperature fracture toughness of Nbss/Nb_5Si_3 in situ composites," *Intermetallics*, vol. 9, no. 9, pp. 827–834, 2001.

[5] X. Y. Yan and D. J. Fray, "Production of niobium powder by direct electrochemical reduction of solid Nb_2O_5 in a eutectic $CaCl_2$-NaCl melt," *Metallurgical and Materials Transactions B*, vol. 33, pp. 685–693, 2002.

[6] T. Nohira, K. Yasuda, and Y. Ito, "Pinpoint and bulk electrochemical reduction of insulating silicon dioxide to silicon," *Nature Materials*, vol. 2, no. 6, pp. 397–401, 2003.

[7] G. Z. Chen, D. J. Fray, and T. W. Farthing, "Direct electrochemical reduction of titanium dioxide to titanium in molten calcium chloride," *Nature*, vol. 407, no. 6802, pp. 361–364, 2000.

[8] F. K. Meng and H. M. Lu, "Study on a new electrolysis technology of preparing high Nb bearing TiAl alloy from metal oxides," in *EPD Congress Proceedings*, pp. 711–717, 2009.

[9] G. Z. Chen, E. Gordo, and D. J. Fray, "Direct electrolytic preparation of chromium powder," *Metallurgical and Materials Transactions B*, vol. 35, pp. 223–233, 2004.

[10] F. Meng and H. Lu, "Direct electrochemical preparation of NbSi alloys from mixed oxide preform precursors," *Advanced Engineering Materials*, vol. 11, no. 3, pp. 198–201, 2009.

[11] A. J. M. Wood, R. C. Copcutt, G. Z. Chen, and D. J. Fray, "Electrochemical fabrication of nickel manganese gallium alloy powder," *Advanced Engineering Materials*, vol. 5, no. 9, pp. 650–653, 2003.

[12] K. Dring, R. Bhagat, M. Jackson, R. Dashwood, and D. Inman, "Direct electrochemical production of Ti-10W alloys from mixed oxide preform precursors," *Journal of Alloys and Compounds*, vol. 419, no. 1-2, pp. 103–109, 2006.

5

Optimization of GMAW Process Parameters Using Particle Swarm Optimization

P. Sreeraj,[1] T. Kannan,[2] and Subhashis Maji[3]

[1] *Department of Mechanical Engineering, Valia Koonambaikulathamma College of Engineering and Technology, Kerala 692574, India*
[2] *Department of Mechanical Engineering, SVS College of Engineering, Coimbatore, Tamilnadu 642109, India*
[3] *Department of Mechanical Engineering, IGNOU, Delhi 110068, India*

Correspondence should be addressed to T. Kannan; kannan_kct@yahoo.com

Academic Editors: M. Carboneras, C. Panagopoulos, and S. C. Wang

To improve the corrosion-resistant properties of carbon steel cladding process is usually used. It is a process of depositing a thick layer of corrosion resistant material-over carbon steel plate. Most of the engineering applications require high strength and corrosion resistant materials for long-term reliability and performance. By cladding, these properties can be achieved with minimum cost. The main problem faced in cladding is the selection of optimum combinations of process parameters for achieving quality clad and hence good clad bead geometry. This paper highlights an experimental study to optimize various input process parameters (welding current, welding speed, gun angle, contact tip to work distance, and pinch) to get optimum dilution in stainless steel cladding of low-carbon structural steel plates using gas metal arc welding (GMAW). Experiments were conducted based on central composite rotatable design with full-replication technique and mathematical models were developed using multiple regression method. The developed models have been checked for adequacy and significance. Using particle swarm optimization (PSO) the parameters were optimized to get minimal dilution.

1. Introduction

Prevention of corrosion is a major problem in industries. Even though it cannot be eliminated completely, it can be reduced to some extent. A corrosion resistant protective layer is made over the less corrosion resistant substrate by a process called cladding. This technique is used to improve life of engineering components but also reduce their cost. This process is mainly used in industries such as chemical, textiles, nuclear, steam power plants, food processing, and petro-chemical industries [1].

Most accepted method of employed in weld cladding is GMAW. It has got the following advantages.

(i) High reliability;
(ii) all position capability;
(iii) ease to use;
(iv) low cost;
(v) high productivity;
(vi) suitable for both ferrous and non ferrous metals;
(vii) high deposition rate;
(viii) absences of fluxes;
(ix) cleanliness and ease of mechanization.

The mechanical strength of clad metal is highly influenced by the composition of metal but also by clad bead shape. This is an indication of bead geometry. Figure 1 shows the clad bead geometry. It mainly depends on wire feed rate, welding speed, arc voltage, and so forth. Therefore it is necessary to study the relationship between in process parameters and bead parameters to study clad bead geometry. Using mathematical models it can be achieved.

This paper highlights the study carried out to develop mathematical and PSO models to optimize clad bead geometry, in stainless steel cladding deposited by GAMAW. The experiments were conducted based on four factor five level central composite rotatable designs with full replication technique [2]. The developed models have been checked for

TABLE 1: Chemical composition of base metal and filler wire.

Materials	Elements wt%								
	C	SI	Mn	P	S	Al	Cr	Mo	Ni
IS 2062	0.150	0.160	0.870	0.015	0.016	0.031	—	—	—
ER 308L	0.03	0.57	1.76	0.021	0.008	—	19.52	0.75	10.02

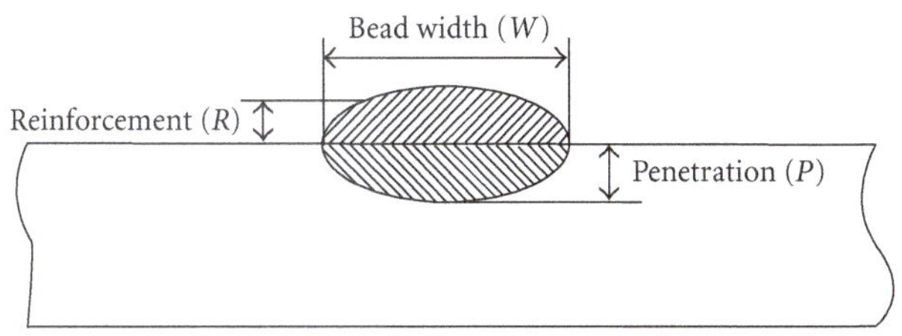

FIGURE 1: Clad bead geometry. Percentage dilution $(D) = [B/(A+B)] \times 100$.

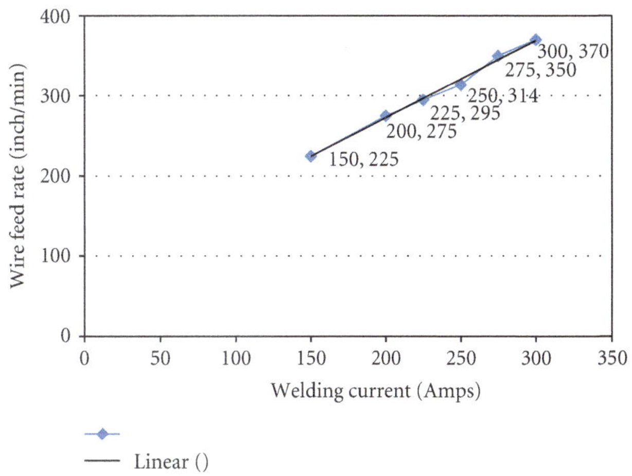

FIGURE 2: Relationship between current and wire feed rate.

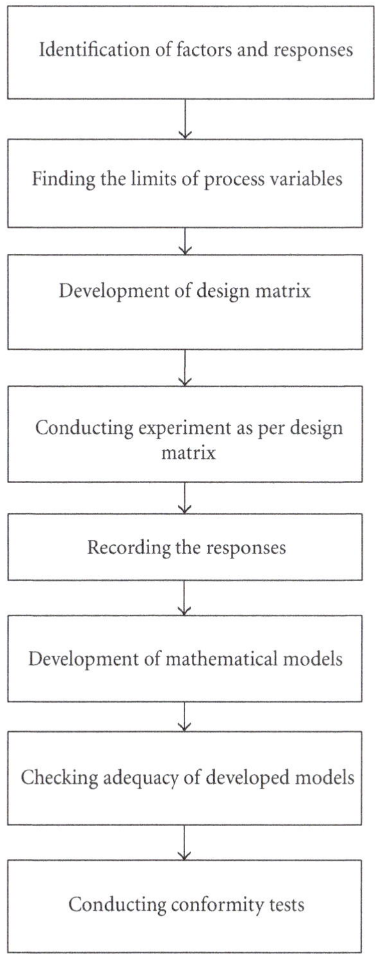

FIGURE 3: Experimental design procedures.

their adequacy and significance. Again using PSO, the bead parameters were optimized.

2. Experimental Procedure

The following machines and consumables were used for the purpose of conducting experiments.

(1) A constant current gas metal arc welding machine (Invrtee V 350-PRO advanced processer with 5–425 amps output range);

(2) welding manipulator;

(3) wire feeder (LF-74 Model);

(4) filler material Stainless Steel wire of 1.2 mm diameter (ER-308 L).;

(5) gas cylinder containing a mixture of 98% argon and 2% of oxygen;

(6) mild steel plate (grade IS-2062).

Test plates of size 300 × 200 × 20 mm were cut from mild steel plate of grade IS-2062 and one of the surfaces is cleaned to remove oxide and dirt before cladding. ER-308 L stainless steel wire of 1.2 mm diameter was used for depositing the clad beads through the feeder. Argon gas at a constant flow rate of 16 litres per minute was used for shielding. The properties of base metal and filler wire are shown in Table 1. The important and most difficult parameter found from trial run is wire feed rate. The wire feed rate is proportional to current [3].

Wire feed rate must be greater than critical wire feed rate to achieve pulsed metal transfer. The relationship found from trial run is shown in (1). The formula derived is shown in Figure 2:

$$\text{wire feed rate} = 0.96742857 * \text{current} + 79.1. \quad (1)$$

The selection of the welding electrode wire based on the matching the mechanical properties and physical characteristics of the base metal, weld size, and existing electrode

Table 2: Welding parameters and their levels.

Parameters	Unit	Notation	Factor levels				
			−2	−1	0	1	2
Welding current	A	*I*	200	225	250	275	300
Welding speed	mm/min	*S*	150	158	166	174	182
Contact tip to work distance	mm	*N*	10	14	18	22	26
Welding gun angle	Degree	*T*	70	80	90	100	110
Pinch	—	*Ac*	−10	−5	0	5	10

Table 3: Design matrix.

Trial number	Design matrix				
	I	*S*	*N*	*T*	*Ac*
1	−1	−1	−1	−1	1
2	1	−1	−1	−1	−1
3	−1	1	−1	−1	−1
4	1	1	−1	−1	1
5	−1	−1	1	−1	−1
6	1	−1	1	−1	1
7	−1	1	1	−1	1
8	1	1	1	−1	−1
9	−1	−1	−1	1	−1
10	1	−1	−1	1	1
11	−1	1	−1	1	1
12	1	1	−1	1	−1
13	−1	−1	1	1	1
14	1	−1	1	1	−1
15	−1	1	1	1	−1
16	1	1	1	1	1
17	−2	0	0	0	0
18	2	0	0	0	0
19	0	−2	0	0	0
20	0	2	0	0	0
21	0	0	−2	0	0
22	0	0	2	0	0
23	0	0	0	−2	0
24	0	0	0	2	0
25	0	0	0	0	−2
26	0	0	0	0	2
27	0	0	0	0	0
28	0	0	0	0	0
29	0	0	0	0	0
30	0	0	0	0	0
31	0	0	0	0	0
32	0	0	0	0	0

I: welding current; *S*: welding speed; *N*: contact tip to work distance; *T*: welding gun angle; *Ac*: pinch.

inventory. A candidate material for cladding which has excellent corrosion resistance and weldability is stainless steel. These have chloride stress corrosion cracking resistance and strength significantly greater than other materials. These have good surface appearance, good radiographic standard quality, and minimum electrode wastage. Experimental design procedure used for this study is shown in Figure 3 and important steps are briefly explained.

3. Plan of Investigation

The research work was planned to be carried out in the following steps.

(1) Identification of factors and responses.
(2) Finding limits of process variables.
(3) Development of design matrix.
(4) Conducting experiments as per design matrix.
(5) Recording the responses.
(6) Development of mathematical models.
(7) Checking the adequacy of developed models.
(8) Conducting conformity tests.

4. Prediction of Clad Bead Geometry Using Regression Equation

The following independently controllable process parameters were found to be affecting output parameters. These are wire feed rate (*W*), welding speed (*S*), welding gun angle (*T*), contact tip to work to distance (*N*) and pinch (*Ac*), the responses chosen were clad bead width (*W*), height of reinforcement (*R*), depth of Penetration (*P*), and percentage of dilution (*D*). The responses were chosen based on the impact of parameters on final composite model.

The basic difference between welding and cladding is the percentage of dilution. The properties of the cladding is the significantly influenced by dilution obtained. Hence control of dilution is important in cladding where a low dilution is highly desirable. When dilution is quite low, the final deposit composition will be closer to that of filler material and hence corrosion resistant properties of cladding will be greatly improved. The chosen factors have been selected on the basis to get minimal dilution and optimal clad bead geometry [4].

Few significant research works have been conducted in these areas using these process parameters and so these parameters were used for experimental study. Working ranges of all selected factors are fixed by conducting trial runs. This was carried out by varying one of the factors while keeping the rest of them as constant values. Working range of each process parameters was decided upon by inspecting the

TABLE 4: Design matrix and observed values of clad bead geometry.

Trial no.	Design matrix					Bead parameters			
	I	S	N	T	Ac	W (mm)	P (mm)	R (mm)	D (%)
1	−1	−1	−1	−1	1	6.9743	1.67345	6.0262	10.72091
2	1	−1	−1	−1	−1	7.6549	1.9715	5.88735	12.16746
3	−1	1	−1	−1	−1	6.3456	1.6986	5.4519	12.74552
4	1	1	−1	−1	1	7.7635	1.739615	6.0684	10.61078
5	−1	−1	1	−1	−1	7.2683	2.443	5.72055	16.67303
6	1	−1	1	−1	1	9.4383	2.4905	5.9169	15.96692
7	−1	1	1	−1	−1	6.0823	2.4672	5.49205	16.5894
8	1	1	1	−1	−1	8.4666	2.07365	5.9467	14.98494
9	−1	−1	−1	1	−1	6.3029	1.5809	5.9059	10.2749
10	1	−1	−1	1	1	7.0136	1.5662	5.9833	9.707297
11	−1	1	−1	1	1	6.2956	1.58605	5.5105	11.11693
12	1	1	−1	1	−1	7.741	1.8466	5.8752	11.4273
13	−1	−1	1	1	1	7.3231	2.16475	5.72095	15.29097
14	1	−1	1	1	−1	9.6171	2.69495	6.37445	18.54077
15	−1	1	1	1	−1	6.6335	2.3089	5.554	17.23138
16	1	1	1	1	1	10.514	2.7298	5.4645	20.8755
17	−2	0	0	0	0	6.5557	1.99045	5.80585	13.65762
18	2	0	0	0	0	7.4772	2.5737	6.65505	15.74121
19	0	−2	0	0	0	7.5886	2.50455	6.4069	15.77816
20	0	2	0	0	0	7.5014	2.1842	5.6782	16.82349
21	0	0	−2	0	0	6.1421	1.3752	6.0976	8.941799
22	0	0	2	0	0	8.5647	3.18536	5.63655	22.94721
23	0	0	0	−2	0	7.9575	2.2018	5.8281	15.74941
24	0	0	0	2	0	7.7085	1.85885	6.07515	13.27285
25	0	0	0	0	−2	7.8365	2.3577	5.74915	16.63287
26	0	0	0	0	2	8.2082	2.3658	5.99005	16.38043
27	0	0	0	0	0	7.9371	2.1362	6.0153	15.18374
28	0	0	0	0	0	8.4371	2.17145	5.69895	14.82758
29	0	0	0	0	0	9.323	3.1425	5.57595	22.8432
30	0	0	0	0	0	9.2205	3.2872	5.61485	23.6334
31	0	0	0	0	0	10.059	2.86605	5.62095	21.55264
32	0	0	0	0	0	8.9953	2.72068	5.7052	19.60811

W: width; P: penetration; R: reinforcement; D: dilution %.

bead for smooth appearance without any visible defects. The upper limit of given factor was coded as −2. The coded value of intermediate values were calculated using

$$X_i = \frac{2\left[2x - (x_{\max} + x_{\min})\right]}{\left[(x_{\max} - x_{\min})\right]}, \quad (2)$$

where X_i is the required coded value of parameter X is any value of parameter from $X_{\min} - X_{\max}$. $X_{\min}$ is the lower limit of parameters and $X_{\max}$ is the upper limit parameters. The chosen level of the parameters with their units and notation are given in Table 2.

Design matrix chosen to conduct the experiments was central composite rotatable design. The design matrix comprises of full replication of $2^5(= 32)$, Factorial designs. All welding parameters in the intermediate levels (o) constitute the central points and combination of each welding parameters at either is highest value (+2) or lowest value (−2) with other parameters of intermediate levels (0) constitute star points. 32 experimental trails were conducted that make the estimation of linear quadratic and two way interactive effects of process parameters on clad geometry [5].

The experiments were conducted at SVS College of Engineering, Coimbatore, India. In this work thirty-two experimental runs were allowed for the estimation of linear quadratic and two-way interactive effects of corresponding each treatment combination of parameters on bead geometry as shown Table 3 at random. At each run settings for all parameters were disturbed and reset for next deposit. This is very essential to introduce variability caused by errors in experimental set up.

In order to measure clad bead geometry of transverse section of each weld overlays were cut using band saw from mid length [6]. Position of the weld and end faces were machined and grinded. The specimen and faces were

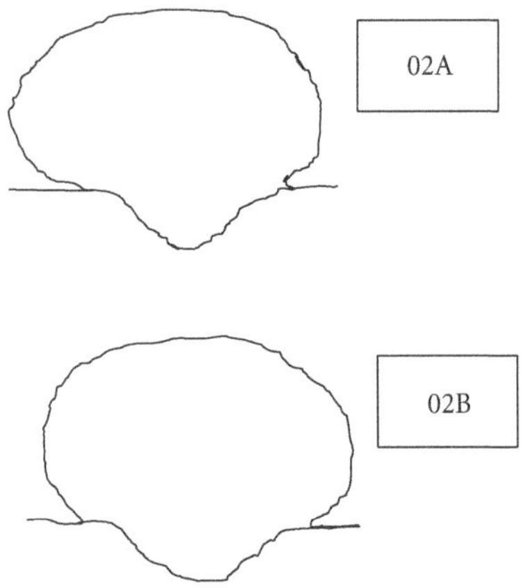

FIGURE 4: Traced Profiles (Specimen no. 2). 02A represents profile of the specimen (front side) and 02B represents profile of the specimen (rear side).

FIGURE 5: Clad deposited at optimal conditions.

polished and etched using a 5% nital solution to display bead dimensions. The clad bead profiles were traced using a reflective type optical profile projector at a magnification of ×10, in M/s Roots Industries Ltd. Coimbatore. Then the bead dimension such as depth of penetration height of reinforcement and clad bead width were measured. The traced bead profiles were scanned in order to find various clad parameters and the percentage of dilution with help of AUTO CAD software. This is shown in Figure 4 [7].

The measured clad bead dimension and percentage of dilution is shown in Table 4. Cladding deposited at optimal conditions is shown in Figure 5.

5. Development of Mathematical Models

The response function representing any of the clad bead geometry can be expressed as [8],

$$Y = f(A, B, C, D, E), \quad (3)$$

where Y = response variable, A = welding current (I) in amps, B = welding speed (S) in mm/min, C = contact tip to work distance (N) in mm, D = welding gun angle (T) in degrees, E = pinch (Ac).

The second order surface response model equals can be expressed as below [9].

$$\begin{aligned} Y = \beta_0 + {} & \beta_1 A + \beta_2 B + \beta_3 C + \beta_4 D + \beta_5 E \\ & + \beta_{11} A^2 + \beta_{22} B^2 + \beta_{33} C^2 \\ & + \beta_{44} D^2 + \beta_{55} E^2 + \beta_{12} AB \\ & + \beta_{13} AC + \beta_{14} AD + \beta_{15} AE \\ & + \beta_{23} BC + \beta_{24} BD + \beta_{25} BE \\ & + \beta_{34} CD + \beta_{35} CE + \beta_{45} DE, \end{aligned} \quad (4)$$

where β_0 is the free term of the regression equation, the coefficients β_1, β_2, β_3, β_4 and β_5 are linear terms, the coefficients β_{11}, β_{22}, β_{33}, β_{44} and β_{55} quadratic terms, and the coefficients β_{12}, β_{13}, β_{14}, β_{15}, and so forth are the interaction terms. The coefficients were calculated using Quality America six sigma software (DOE-PC IV). After determining the coefficients, the mathematical models were developed. The developed mathematical models are given as follows.

$$\begin{aligned} & \text{Clad Bead Width } (W), \text{mm:} \\ & = 8.923 + 0.701A + 0.388B \\ & \quad + 0.587C + 0.040D + 0.088E \\ & \quad - 0.423A^2 - 0.291B^2 - 0.338C^2 \\ & \quad - 0.219D^2 - 0.171E^2 + 0.205AB \\ & \quad + 0.405AC + 0.105AD + 0.070AE \\ & \quad - 0.134BC + 0.225BD + 0.098BE \\ & \quad + 0.26CD + 0.086CE + 0.012DE. \end{aligned} \quad (5)$$

$$\begin{aligned} & \text{Depth of Penetration } (P), \text{mm:} \\ & = 2.735 + 0.098A - 0.032B \\ & \quad + 0.389C - 0.032D - 0.008E \\ & \quad - 0.124A^2 - 0.109B^2 - 0.125C^2 \\ & \quad - 0.187D^2 - 0.104E^2 - 0.33AB \\ & \quad + 0.001AC + 0.075AD + 0.005AE \\ & \quad - 0.018BC + 0.066BD + 0.087BE \\ & \quad + 0.058CD + 0.054CE - 0.036DE. \end{aligned} \quad (6)$$

TABLE 5: Analysis of variance for testing adequacy of the model.

Parameter	1st order terms		2nd order terms		Lack of fit		Error terms		F-ratio	R-ratio	Whether model is adequate
	SS	DF	SS	DF	SS	DF	SS	DF			
W	36.889	20	6.233	11	3.513	6	2.721	5	1.076	3.390	Adequate
P	7.810	20	0.404	11	0.142	6	0.261	5	0.454	7.472	Adequate
R	1.921	20	0.572	11	0.444	6	0.128	5	2.885	3.747	Adequate
D	506.074	20	21.739	11	6.289	6	15.45	5	0.339	8.189	Adequate

SS: sum of squares; DF: degree of freedom; F ratio (6, 5, 0.5) = 3.40451; R ratio (20, 5, 0.05) = 3.20665.

$$\begin{aligned}
&\text{Height of Reinforcement } (R), \text{mm:} \\
&= 5.752 + 0.160A - 0.151B \\
&\quad - 0.060C + 0.016D - 0.002E \\
&\quad + 0.084A^2 + 0.037B^2 - 0.0006C^2 \\
&\quad + 0.015D^2 - 0.006E^2 \\
&\quad + 0.035AB + 0.018AC \\
&\quad - 0.008AD - 0.048AE \\
&\quad - 0.024BC - 0.062BD \\
&\quad - 0.003BE + 0.012CD - 0.092CE \\
&\quad - 0.095DE.
\end{aligned} \tag{7}$$

$$\begin{aligned}
&\text{Percentage Dilution } (D): \\
&= 19.705 + 0.325A + 0.347B \\
&\quad + 3.141C - 0.039D - 0.153E \\
&\quad - 1.324A^2 - 0.923B^2 - 1.012C^2 \\
&\quad - 1.371D^2 - 0.872E^2 - 0.200AB \\
&\quad + 0.346AC + 0.602AD + 0.203AE \\
&\quad + 0.011BC + 0.465BD + 0.548BE \\
&\quad + 0.715CD + 0.360CE \\
&\quad + 0.137DE.
\end{aligned} \tag{8}$$

6. Checking Adequacy of the Developed Models

The adequacy of the developed model was tested using the analysis of variance (ANOVA) technique. As per this technique, if the F-ratio values of the developed models do not exceed the standard tabulated values for a desired level of confidence (95%) and the calculated R-ratio values of the developed model exceed the standard values for a desired level of confidence (95%) then the models are said to be adequate within the confidence limit [10]. These conditions were satisfied for the developed models. The values are shown in Table 5.

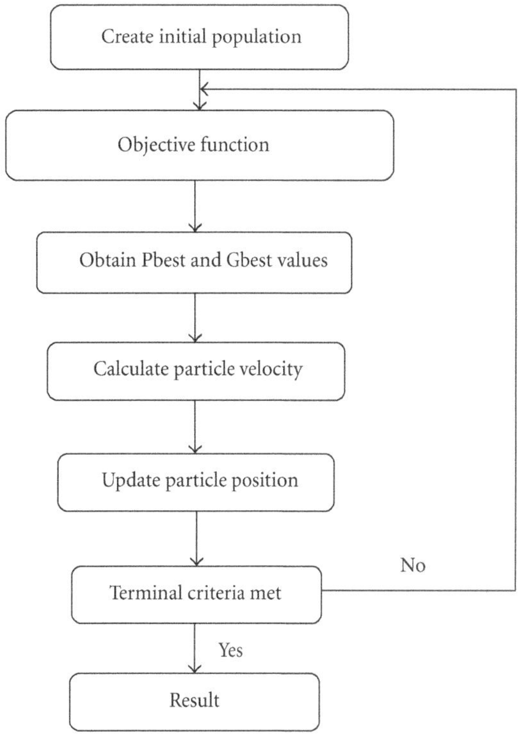

FIGURE 6: Procedure for proposed PSO to optimize GMAW process parameters.

7. Optimization of GMAW Process Parameters by PSO

Heuristic technique PSO is proposed to optimize clad bead geometry of stainless steel cladding deposited by GMAW. This is for achieving good cladding. The optimization procedure is shown in Figure 6. Initial populations is the possible number of solutions (particles) of the optimization problem and each possible solution is called an individual. In this study possible number of solutions is formed by the values of welding current, welding speed, contact tip to work distance, welding gun angle, and pinch. The objective of this study is to minimize percentage of dilution taken from (8).

Initial population is created using the cladding parameters (welding current, welding speed, contact tip to work distance, welding gun angle, and pinch) and for the current populations using objective function. The best fitness value is stored as Pbest from history. From all random solutions reobtained for percentage of dilution. Fitness function values

Table 6: Parameters for PSO optimization.

Population size	30
Dimension size	5
Inertia weight	0.4–0.9
Velocity factors C1, C2	1.4
Number of iteration allowed	100

for each individual (particles) are calculated particles choose the best fitness value called Gbest [11]. For each particle, calculate the velocity of the particle by

Velocity[] = wVelocity[] + C1rand1(Pbest[]− present[]) + C2xrand2(Gbest[]− present[]).

Particle position is updated by

Present[] = Present[] + velocity[].

rand1 and rand2 are two random functions in the range [0, 1] where C1 and C2 are two positive constants named learning factors taken as 2 and "w" is the inertial weight taken as 0.5.The parameters used for PSO optimization are shown in Table 6.

7.1. Method for Developing PSO Model

(i) Initiate each particle.

(ii) Calculate fitness value of each particle. If the fitness value is better than the best fitness value (Pbest) in history. Set the current value as new Pbest.

(iii) Calculate Gbest.

(iv) For each particle calculate the particle velocity.

7.2. Numerical Illustration for Developed PSO Model. The numerical illustration for the developed model to find optimal parameters for percentage of dilution as summarised in below.

ïĆû Welding current $I = I_{min} + (I_{max} - I_{min}) - \text{rand}()$

ïĆû Welding speed $S = S_{min} + (S_{max} - S_{min}) - \text{rand}()$

ïĆû Contact tip to work distance $N = N_{min} + (N_{max} - N_{min}) - \text{rand}()$

ïĆû Welding gun angle $T = T_{min} + (T_{max} - T_{min}) - \text{rand}()$

ïĆû Pinch $P = P_{min} + (P_{max} - P_{min}) - \text{rand}()$

These values are substituted in (8) and dilution is obtained Consider.

ïĆû $X(1) = A$ = Welding current (I) in Amps,

ïĆû $X(2) = B$ = Welding Speed (S) in mm/min,

ïĆû $X(3) = C$ = Contact to work piece distance (N) in mm,

ïĆû $X(4) = D$ = Welding gun angle (T) in degree,

ïĆû $X(5) = E$ = Pinch (Ac).

Table 7: pbest value.

I	S	T	N	AC
0.6241	1.1490	−1.9241	−0.3936	−0.5707
0.6213	1.1414	−1.9139	−0.4406	−0.5562
−0.7617	1.2651	−1.8204	−0.3466	−0.5956
1.1993	0.4661	−1.4140	0.4173	−0.1746
0.6213	1.1368	−1.9224	−0.4496	−0.5725
1.2525	1.5234	−0.1270	−0.9208	0.6041
0.6188	1.1379	−1.9242	−0.4524	−0.5750
0.6196	1.1364	−1.9284	−0.4547	−0.5780
0.6230	1.1385	−1.9187	0.4487	−0.5717
0.6223	1.1368	−1.9227	−0.4507	−0.5761
0.6182	1.1378	−1.9227	−0.4507	−0.5761
0.6182	1.1378	−1.9221	−0.4511	−0.5754
0.6256	1.1393	−1.9129	−0.4459	−0.5675
0.6230	1.1385	−1.9187	−0.4487	−0.5717

Objective function for percentage of dilution which must be minimized was derived from 5–8. The constants of welding parameters are given Table 2 [12].

Subjected to bounds:

$$\begin{aligned} 200 &\le X(1) \le 300, \\ 150 &\le X(2) \le 182, \\ 10 &\le X(3) \le 26, \\ 70 &\le X(4) \le 110, \\ -10 &\le X(5) \le 10. \end{aligned} \tag{9}$$

7.3. Objective Function

$$\begin{aligned} f(x) = {} & 19.75 + 0.325 * x(1) + 0.347 * x(2) \\ & + 3.141 * x(3) - 0.039 * x(4) - 0.153 * x(5) \\ & - 1.324 * x(1)^2 - 0.923 * x(2)^2 \\ & - 1.012 * x(3)^2 - 1.371 * x(4)^2 \\ & - 0.872 * x(5)^2 - 0.200 * x(1) * x(2) \\ & + 0.346 * x(1) * x(3) + 0.602 * x(1) * x(4) \\ & + 0.203 * x(1) * x(5) \\ & + 0.011 * x(2) * x(3) + 0.465 * x(2) * x(4) \\ & + 0.548 * x(2) * x(5) + 0.715 * x(3) * x(4) \\ & + 0.360 * x(3) * x(5) + 0.137 * x(4) \\ & * x(5). \end{aligned} \tag{10}$$

This is the percentage of dilution.

Table 8: Velocity of particles.

I	S	T	N	Ac
−0.0766	−0.0000	0.0000	0.0000	0.0000
−0.0737	−0.0000	0.0000	0.0000	0.0000
−0.0278	−0.0953	−0.0960	−0.3871	−0.3927
0.0369	−0.0000	−0.6304	−0.5643	−0.6966
−0.611	−0.1051	−0.1121	−0.4074	−0.4324
−0.0732	−0.0000	0.0000	0.0000	0.0000
−0.0000	−0.0000	0.0000	0.0000	0.0000
−0.0195	−0.0000	0.0000	0.0000	0.0000
−0.0517	−0.0940	−0.1307	−0.3794	−0.4224
−0.0296	−0.0000	0.0000	0.0000	0.0000
−0.0344	−0.0000	0.0000	0.0000	0.0000
−0.0766	−0.1135	−0.0974	−0.4006	−0.4197
−0.0211	−0.0000	0.0000	0.0000	0.0000

Table 9: Optimal process parameters.

Parameters	Range (coded value)	Actual
Welding current (I)	0.6196	265 A
Welding speed (S)	−1.1364	174 mm/min
Contact tip to work distance (N)	−1.9284	10 mm
Welding gun angle (T)	−0.4547	88 degree
Pinch (Ac)	−0.5780	−6

Dilution obtained is 8.5% and optimal process parameters shown in Table 9

7.4. Constraint Equations

$$\begin{aligned} W = {} & (8.923 + 0.701 * x(1) + 0.388 * x(2) \\ & + 0.587 * x(3) + 0.040 * x(4) + 0.088 * x(5) \\ & - 0.423 * x(1)^2 - 0.291 * x(2)^2 \\ & - 0.338 * x(3)^2 - 0.219 * x(4)^2 - 0.171 \\ & * x(5)^2 + 0.205 * x(1) * x(2) + 0.405 \\ & * x(1) * x(3) + 0.105 * x(1) * x(4) + 0.070 \\ & * x(1) * x(5) - 0.134 * x(2) * x(3) + 0.2225 \\ & * x(2) * x(4) + 0.098 * x(2) * x(5) + 0.26 \\ & * x(3) * x(4) + 0.086 * x(3) * x(5) \\ & + 0.12 * x(4) * x(5)) - 3. \end{aligned} \quad (11)$$

(Clad bead width (W) mm lower limit)

$$\begin{aligned} P = {} & (2.735 + 0.098 * x(1) - 0.032 * x(2) \\ & + 0.389 * x(3) - 0.032 * x(4) - 0.008 * x(5) \\ & - 0.124 * x(1)^2 - 0.109 * x(2)^2 \\ & - 0.125 * x(3)^2 \\ & - 0.187 * x(4)^2 - 0.104 * x(5)^2 \\ & - 0.33 * x(1) * x(2) + 0.001 * x(1) * x(3) \\ & + 0.075 * x(1) * x(4) \\ & + 0.005 * x(1) * x(5) \\ & - 0.018 * x(2) * x(3) \\ & + 0.066 * x(2) * x(4) + 0.087 * x(2) * x(5) \\ & + 0.058 * x(3) * x(4) \\ & + 0.054 * x(3) * x(5) \\ & - 0.036 * x(4) * x(5)) - 3. \end{aligned} \quad (12)$$

(Depth of penetration (P) upper limit),

$$\begin{aligned} P = {} & (2.735 + 0.098 * x(1) - 0.032 * x(2) \\ & + 0.389 * x(3) - 0.032 * x(4) - 0.008 * x(5) \\ & - 0.124 * x(1)^2 - 0.109 * x(2)^2 \\ & - 0.125 * x(3)^2 - 0.187 * x(4)^2 \\ & - 0.104 * x(5)^2 - 0.33 * x(1) * x(2) \\ & + 0.001 * x(1) * x(3) + 0.075 * x(1) * x(4) \\ & + 0.005 * x(1) * x(5) - 0.018 * x(2) * x(3) \\ & + 0.066 * x(2) * x(4) \\ & + 0.087 * x(2) * x(5) + 0.058 * x(3) * x(4) \\ & + 0.054 * x(3) * x(5) \\ & - 0.036 * x(4) * x(5)) + 2. \end{aligned} \quad (13)$$

(Depth of penetration (P) lower limit),

$$\begin{aligned} W = {} & (8.923 + 0.701 * x(1) + 0.388 * x(2) + 0.587 * x(3) \\ & + 0.040 * x(4) + 0.088 * x(5) \\ & - 0.423 * x(1)^2 - 0.291 * x(2)^2 - 0.338 * x(3)^2 \\ & - 0.219 * x(4)^2 - 0.171 * x(5)^2 \\ & + 0.205 * x(1) * x(2) + 0.405 * x(1) * x(3) \\ & + 0.105 * x(1) * x(4) + 0.070 * x(1) * x(5) \\ & - 0.134 * x(2) * x(3) + 0.225 * x(2) * x(4) \\ & + 0.098 * x(2) * x(5) + 0.26 * x(3) * x(4) \\ & + 0.086 * x(3) * x(5) + 0.012 * x(4) * x(5)) - 10. \end{aligned} \quad (14)$$

(Clad bead width (W) upper limit),

$$\begin{aligned} R = {} & (5.752 + 0.160 * x(1) - 0.151 * x(2) - 0.060 * x(3) \\ & + 0.016 * x(4) - 0.002 * x(5) + 0.084 * x(1)^2 \\ & + 0.037 * x(2)^2 - 0.0006 * x(3)^2 + 0.015 * x(4)^2 \\ & - 0.006 * x(5)^2 + 0.035 * x(1) * x(2) \\ & + 0.018 * x(1) * x(3) - 0.008 * x(1) * x(4) \\ & - 0.048 * x(1) * x(5) - 0.024 * x(2) * x(3) \\ & - 0.062 * x(2) * x(4) - 0.003 * x(2) * x(5) \\ & + 0.012 * x(3) * x(4) - 0.092 * x(3) \\ & * x(5) - 0.095 * x(4) * x(5)) - 6. \end{aligned} \quad (15)$$

(Height of reinforcement (R) lower limit),

$$\begin{aligned} R = {} & (5.752 + 0.160 * x(1) - 0.151 * x(2) - 0.060 \\ & * x(3) + 0.016 * x(4) - 0.002 * x(5) + 0.084 \\ & * x(1)^2 + 0.037 * x(2)^2 - 0.0006 * x(3)^2 \\ & + 0.015 * x(4)^2 - 0.006 * x(5)^2 + 0.035 \\ & * x(1) * x(2) + 0.018 * x(1) * x(3) - 0.008 \\ & * x(1) * x(4) - 0.048 * x(1) * x(5) \\ & - 0.024 * x(2) * x(3) - 0.062 * x(2) * x(4) - 0.003 \\ & * x(2) * x(5) + 0.012 * x(3) * x(4) - 0.092 \\ & * x(3) * x(5) - 0.095 * x(4) * x(5)) + 6. \end{aligned} \quad (16)$$

(Heights of reinforcement (R) upper limit),

$$\begin{aligned} & f(x) - 23 \\ & -f(x) + 8. \end{aligned} \quad (17)$$

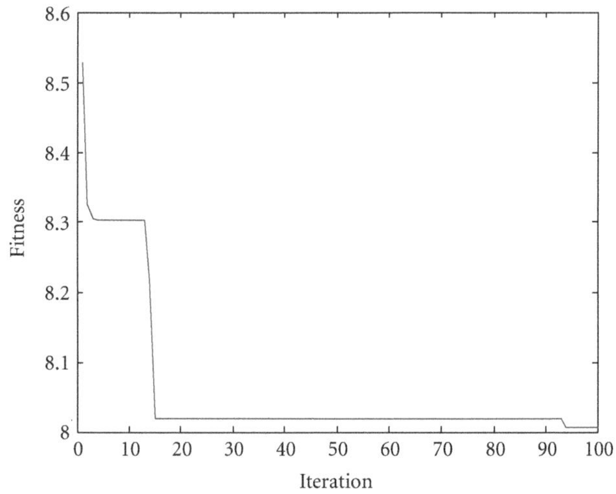

Figure 7: Convergence of PSO for optimal dilution.

(Dilution Upper and lower limit),

$$\begin{aligned} & x(1), x(2), x(3), x(4), x(5) \leq 2, \\ & x(1), x(2), x(3), x(4), x(5) \geq -2. \end{aligned} \quad (18)$$

7.5. Calculation of Pbest Value. The minimum percentage of dilution for each individual solution is considered as Pbest value. This is the best value for the particular solution only [13, 14].

7.6. Calculation of Gbest Value. The minimum dilution for initial solution or the whole iteration is considered as Gbest value. Table 7 shows Pbest values and Table 8 shows velocity of particle. A program on MATLAB 7 is created and run to get optimal dilution. Figure 7 shows convergence of PSO for optimal dilution.

8. Results and Discussions

Experiments were conducted using GMAW to produce cladding of austenitic stainless steel material. From the experimental results a mathematical model was developed using regression model. Further to enhance scope of work PSO model was developed to optimize clad bead geometry.

In this study, a PSO model to optimize clad bead geometry was developed. To ensure accuracy of model developed evolutionary computing technique PSO invoked to optimize the parameters of GMAW, for optimal weld quality. Convergence of developed model for optimal dilution is shown in Figure 7. From the figure it is evident that dilution is increasing up to 15th iteration then it is constant up to 92 and iterations and decreasing and then constant.

PSO maintains internal memory to store the Gbest and Pbest solutions. Each individual in the population will try to emulate the Gbest and Pbest solutions in the memory through updating PSO equations. Hence effectiveness of finding the global solutions is s very effective.

It can be see that PSO models can be effectively used to model cladding parameters. These optimized values can be directly used in automatic cladding in the forms of programs and for real time quality control and for the entire cladding process control application to improve bead geometry.

9. Conclusions

(i) A PSO model has been developed from the experimental data to achieve desired clad bead geometry. PSO models are capable of making optimization of clad bead geometry with reasonable accuracy.

(ii) The developed models are able to optimize process parameters required to achieve the desired clad bead geometry of stainless steel cladding deposited by GMAW with reasonable accuracy.

(iii) In this study, the following steps were applied for prediction of stainless steel clad bead geometry using GMAW: (a) data collection using experimental studies, (b) analysing and processing of data, (c) prediction using regression equation, (d) development of PSO model, and (e) optimizing using PSO algorithm.

(iv) The results showed that PSO models can be used as an alternative tool according to the present conventional calculation methods.

References

[1] P. K. Palani and N. Murugan, "Prediction of delta ferrite content and effect of welding process parameters in claddings by FCAW," *Materials and Manufacturing Processes*, vol. 21, no. 5, pp. 431–438, 2006.

[2] T. Kannan and N. Murugan, "Prediction of Ferrite Number of duplex stainless steel clad metals using RSM," *Welding Journal*, vol. 85, no. 91, p. 99, 2006.

[3] N. Murugan and V. Gunaraj, "Prediction and control of weld bead geometry and shape relationships in submerged arc welding of pipes," *Journal of Materials Processing Technology*, vol. 168, no. 3, pp. 478–487, 2005.

[4] I. S. Kim, K. J. Son, Y. S. Yang, and P. K. D. V. Yaragada, "Sensitivity analysis for process parameters in GMA welding processes using a factorial design method," *International Journal of Machine Tools and Manufacture*, vol. 43, no. 8, pp. 763–769, 2003.

[5] W. G. Cochran and G. M. Coxz, *Experimental Design*, John Wiley & Sons, New York, NY, USA, 1987.

[6] S. Karaoglu and A. Seçgin, "Sensitivity analysis of submerged arc welding process parameters," *Journal of Materials Processing Technology*, vol. 202, no. 1–3, pp. 500–507, 2008.

[7] P. K. Ghosh, P. C. Gupta, and V. K. Goyal, "Stainless steel cladding of structural steel plate using the pulsed current GMAW process," *Welding Journal*, vol. 77, no. 7, pp. 307s–314s, 1998.

[8] V. Gunaraj and N. Murugan, "Prediction and comparison of the area of the heat-affected zone for the bead-on-plate and bead-on-joint in submerged arc welding of pipes," *Journal of Materials Processing Technology*, vol. 95, no. 1–3, pp. 246–261, 1999.

[9] D. C. Montgamery, *Design and Analysis of Experiments*, John Wiley & Sons, 2003.

[10] T. Kannan and J. Yoganandh, "Effect of process parameters on clad bead geometry and its shape relationships of stainless steel claddings deposited by GMAW," *International Journal of Advanced Manufacturing Technology*, vol. 47, pp. 1083–1095, 2010.

[11] R. Poli, J. Kennedy, and T. BlackWell, "Particle swarm optimization. An Overview," *Swaran Intelligence*, vol. 1, no. 1, pp. 33–57, 2007.

[12] F. Madadi, F. Ashrafizadel, and M. Shamanian, "Optimization of pulsed TIG welding process of satellite alloy on carbon steel using RSM," *Journal of Alloys and Compounds*, vol. 510, pp. 71–77, 2012.

[13] R. Mudkerjee, S. Chakraborty, and S. Samanta, "Selection of wire electrical discharge machining process parameters using non traditional optimization algorithms," *Applied Soft Computing*, vol. 12, no. 8, pp. 2506–2516, 2012.

[14] N. Yusup, A. M. Zain, and S. Z. M. Hashim, "Evolutionary techniques in optimizing machining parameters: review of recent applications," *Expert Systems with Applications*, vol. 39, no. 10, pp. 9909–9927, 2012.

High-Temperature (1700–1800°C) Electrochemical Preparation of Metallic Ti from Rutile: A Pathway of Step-by-Step Electrodeoxidization

Fanke Meng and Huimin Lu

School of Materials Science and Engineering, Beihang University, No. 37 Xueyuan Road, Haidian District, Beijing 100191, China

Correspondence should be addressed to Huimin Lu; lhm0862002@gmail.com

Academic Editors: K. Ameyama, M. Carboneras, T. Czujko, P. Markovsky, and J.-Y. Uan

The mechanism of producing metallic titanium by electrochemically reducing rutile (TiO_2) at high temperatures was studied. First, the oxygen was successfully electroremoved from TiO_2 at temperatures 1700, 1750, and 1800°C in molten CaF_2 under a stable electrolytic potential of 2.5 V. Second, the electrodeoxidization process was studied with cyclic voltammetry (CV) tests at 1750°C. It was found that the electrochemical reduction for preparing metallic Ti from TiO_2 at the high temperatures can be divided into several steps. In other words, the oxygen in TiO_2 was electro-removed as a step-by-step pathway ($TiO_2 \rightarrow Ti_4O_7 \rightarrow Ti_3O_5/Ti_2O_3 \rightarrow TiO \rightarrow Ti$) at different electrolytic potentials. It unraveled the mechanism of electrochemical reduction of TiO_2 at the high temperatures, which is helpful for monitoring the reduction procedure.

1. Introduction

Titanium and its alloys play an essential role in industry because of their excellent properties of high ratio of strength/weight, fine plasticity, and superior corrosion-resistance [1]. Although Ti has an attractive future in industry, it cannot be widely used due to the high cost and complex production procedure. Until now, the mainly used commercial procedure for Ti production is Kroll process [2], but the Kroll process needs large energy and long time. In the recent decade, FFC Cambridge process was proposed and considered to be an effective alternative method of direct electrochemical reduction of rutile (TiO_2) to Ti [3, 4]. In this process, the cathode of TiO_2 pellets was immersed into molten $CaCl_2$ and reduced to Ti electrochemically at 900°C under the electrolytic potentials of 2.8–3.2 V. The electrolytic reactions given by this process can be summarized as the following equations:

$$MO_X + 2Xe^- = M + XO^{2-} \quad \text{at the cathode} \quad (1)$$

$$2O^{2-} - 4e^- + C = CO_2 \quad \text{or}$$
$$O^{2-} - 2e^- + C = CO \quad \text{at the anode} \quad (2)$$

This process has two main advances compared with the traditional Kroll method: first, FFC process is a one-plot technique, thereby less time and less energy are spent; second, it can be applied to produce other metals or alloys by this electrochemical reduction method. For example, until now, many metals and alloys, such as, Cr, Si, Nb, Ni, Zr and Ni-Mn-Ga, Ti-W, and Nb-Si, were produced from their oxides by the electrochemical reduction method [5–14]. However, the most critical challenge to sabotage broad industrial applications of FFC process is low current efficiency and slow reaction rate. For example, the current efficiency was between 30% and 50% in most cases; in only a few experiments it can reach 80% [15]. Moreover, only the TiO_2 contacted with the wrapped electrode wires could be completely deoxidized to metallic Ti after electrochemical reduction for several tens of hours [3]. Perovskite phase ($1.49 < O/Ti < 1.89$) still exists in the areas which are not directly contacted with the electrode wires [16]. The incomplete and time-consuming electrolysis is attributed to relatively lower electrochemical reduction activity at lower experimental temperatures (850–950°C). Moreover, the produced metallic Ti could be covered on the surface of the TiO_2, which would seal the free pathways for oxygen releasing, leading to the termination of the electroreduction process [17, 18].

Recently, elevating electrochemical reduction temperature is considered to be an effective method to increase the efficiency of electrochemical reduction of TiO_2 to produce metallic Ti [19]. Not only was the reaction rate increased by elevating reaction temperature, but also the current efficiency was remarkably rocketed up to 90% [19]. In addition, the electrochemical reduction method at high temperatures (>1500°C) was even used for alloys production, especially for some superalloys [14, 20]. Therefore, the electrochemical reduction method at the high temperatures might be widely used in the future for the production of Ti and other metals or alloys in industry. However, taking the Ti production by high-temperature electrolysis as an example, the mechanism is still unclear. It should be pointed out that the mechanism of electrochemical reduction at high temperature is critical for producing metals or for extracting alloys directly from their oxides.

In this paper, it is the first time to study the mechanism of the electrochemical reduction of TiO_2 to Ti at high temperatures (1700–1800°C). By the cyclic voltammetry (CV) measurement and thermodynamic analysis, we have proven that the electrochemical reduction of TiO_2 has a step-by-step pathway. At an applied experimental temperature, the electrochemical reduction process was dependant on the applied reduction potential.

2. Experimental

2.1. Materials Preparation. In our experiments, CaF_2 is used as electrolyte instead of $CaCl_2$ because the electrolytic temperatures are high [19]. As it is known, the electrolyte can be decomposed if the potential applied is high enough. According to the thermodynamic data in [29], we can calculate that the electrochemical decomposition potential of the $CaCl_2$ is lower than 2.2 V if the electrolysis temperature goes up to 1700°C. The electrochemical decomposition potential of CaF_2, however, is still higher than 4.5 V even in our experimental temperatures (1700–1800°C). In electrolysis, because of unavoidable IR drop, we should select an electrolyte (CaF_2) with a high electrochemical decomposition potential to ensure the fine stability of electrolyte. In this work, CaF_2 (≥99.0%, Chengde Optical & Electrical Materials Ltd., Hebei, China) was prepared as in the following steps: first, CaF_2 powders were die-pressed into small cylindrical pellets under the pressure of 30 MPa; second, the pellets were dried at 70°C in an oven overnight; third, the pellets were pulverized to small grains with an average size less than 5 mm. TiO_2 pellets were prepared from rutile TiO_2 powders (≥99.6%, Chengde Optical & Electrical Materials Ltd., Hebei, China) with an identical method as that used in preparation of the CaF_2 pellets. Then the TiO_2 pellets were pulverized to smaller grains of size less than 1 mm. Graphite crucibles (inner diameter: 9.0 cm, height: 18.0 cm, and wall thickness: 1.0 cm) were flushed with industrial ethanol (>95%, Beijing Keshi Co. Ltd., Beijing, China) for 3–5 times until the inner surfaces were almost free of graphite powders. At last, the TiO_2 grains were placed in the graphite crucible and then covered with CaF_2 grains. Each crucible had about 200 g of TiO_2 and 2000 g of CaF_2.

2.2. Potentiostatic Electrolysis. The CaF_2 was used as electrolyte in both TiO_2 electrolysis and subsequent cyclic voltammetry (CV) measurements. As a cathode, a graphite crucible with TiO_2 and CaF_2 was placed in a temperature-controlled furnace made by China Iron & Steel Research Institute Group, as shown in Scheme 1(a). A graphite anode rod (diameter: 6.0 cm) was hung by an electromotor, which can move up and down vertically to ensure that the graphite rod can contact with the molten electrolyte during the electrolysis. KGHF-1000 A/6 V rectifier, made by Beijing Chunshu Electric rectifier Ltd., Beijing, China, was equipped with the furnace to supply a direct current for the electrolysis. The furnace was vacuumed, and then argon (≥99.9%, O ≤ 10 ppm, H_2O ≤ 15 ppm, and N_2 ≤ 5 ppm) was pumped into the furnace as soon as possible. Purged with the argon for 20 min, all valves of the furnace were turned off. Then the argon flow was kept at a rate of 1.0–1.5 L min^{-1}, and the inlet and outlet valves were adjusted to maintain a gas pressure in the furnace around 0.01 MPa larger than the ambient pressure to further prevent air ingress. Heated the furnace to 500°C with a ramping rate of 20°C min^{-1} and then raised the temperature to 1450°C at an average rate of 10°C min^{-1}. Kept at 1450°C for 30 min to ensure the electrolyte CaF_2 was completely melted in the crucible. Then the temperature continued to rise up to the experimental temperatures of 1700–1800°C with a heating rate of 10°C min^{-1}. Consequently, the potential was set to 2.5 V for electrolysis. It should be pointed out that preelectrolysis in this experiment was not applied as it was in other experiments [3–5, 21, 22]. Residual water and redox in the molten CaF_2 were much less than those in the molten $CaCl_2$, which would be proven by the stable electrolytic current with only mild vibration during electrolysis in the first few minutes. After electrolysis and consequent cooling to room temperature, the graphite crucible was taken out of the furnace. The samples after electrochemical reduction at different high temperatures were obtained by smashing the graphite crucibles.

2.3. CV Measurement. In this work, three-electrode system was used in the CV tests and its design was shown in Scheme 1(b). A Mo wire (diameter: 1.0 mm) and a graphite tablet (4 mm × 10 mm × 50 mm) worked as a quasireference and a counter electrode respectively. The working electrode was TiO_2 slices (2 mm × 10 mm × 40 mm), which were wrapped with a Mo wire (diameter: 0.5 mm). The three electrodes were linked to a CHI 604C Electrochemical System (Shanghai Chenhua Inc., Shanghai, China) with the Mo wires (diameter: 0.5 mm). For fixing the Mo wires, three holes were punched along the axial direction of a graphite rod (diameter: 6.0 cm), which had the same size as the rod used in electrolysis process. Thus, the three electrodes can be hung at the bottom of the graphite rod and move up and down together with the rod. The heating process was the same as that used in the former electrolysis. After reaching the experimental temperature, three electrodes were inserted into molten CaF_2 simultaneously by pushing down the graphite rod for CV measurements.

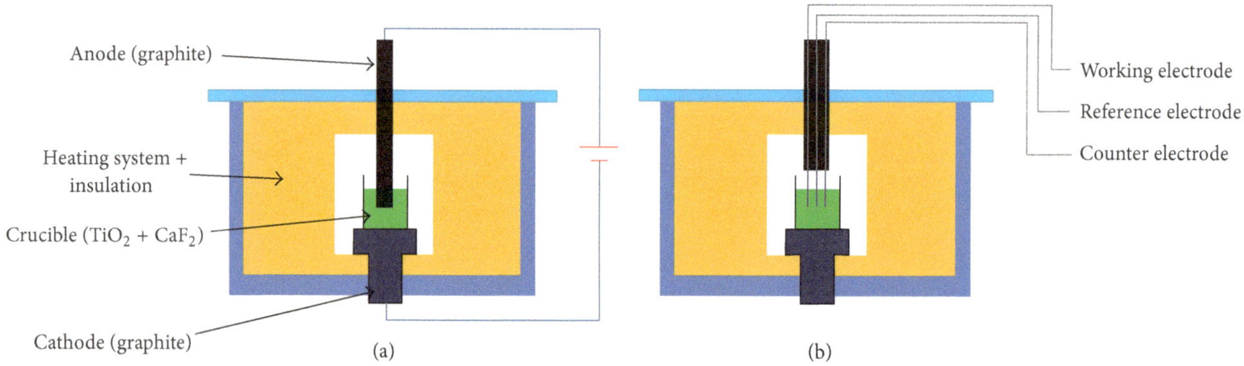

Scheme 1: (a) Schematic of electrolytic cell for TiO_2 electrolysis; (b) design of three-electrode system for cyclic voltammetry (CV) tests.

Figure 1: (a) SEM image of TiO_2 powders before electrolysis; SEM images of the samples after electrolysis at (b) 1700°C, (c) 1750°C, and (d) 1800°C, respectively. The inset in (c) shows two selected pieces of metallic Ti slices after electrolysis at 1750°C. Red crosses in (b) and (d) mark the areas for EDX measurement.

2.4. Characterizations of Morphology and Composition. Micrograph and composition analyses were performed with a VEGA\\LMU SEM system equipped with a CCD camera, from TESCAN. EDX equipment was purchased from OXFORD, and the analysis software was INCA.

3. Results and Discussion

Figure 1(a) shows the morphology of pure TiO_2, which were sintered and pulverized into small particles before electrolysis. Most of the particles had a size smaller than 100 μm. Absolutely, the sample before electrolysis was pure TiO_2 by EDX measurement in Table 1. As shown in Figure 1(b), after electrochemical reduction at 1700°C, a bright S-shaped stripe was formed at the bottom of graphite crucible. Left downside areas of the S-shaped stripe, which was close to the bottom of graphite crucible, had large dark areas of graphite. Many scattered small bright areas were located at the right upside of the S-shaped stripe. With EDX measurement of these bright areas in Table 1, the Ti content increased to 75.3 at % but the O content decreased to 9.4 at %, compared with the pure TiO_2. However, the S-shaped stripe still had some Ca and F. Clearly, they were from electrolyte CaF_2. Figure 1(c) shows the sample after electrochemical reduction at 1750°C. Dozens of metallic Ti were obtained by smashing the crucible after electrolysis, among which two selected pieces were shown in the inset of Figure 1(c). Each piece was about 1.5 g in weight. The EDX

TABLE 1: EDX results of samples before and after electrolysis at different temperatures.

Samples	Ti [at. %]	O [at. %]	Ca [at. %]	F [at. %]
Pure TiO_2 [a]	33.2	66.8	—	—
1700°C [b]	75.3	9.4	5.4	9.9
1750°C [c]	100.0	—	—	—
1800°C [b]	91.3	—	3.6	5.1

—Not obtained by EDX.
[a]The sample of pure TiO_2 was measured by EDX before electrolysis.
[b]EDX measurement was taken from the areas in Figures 1(b) and 1(d), respectively. Content of each element was the average value in the four randomly selected areas marked in Figures 1(b) and 1(d), respectively.
[c]EDX result was obtained by surface scanning over the area in Figure 1(c).

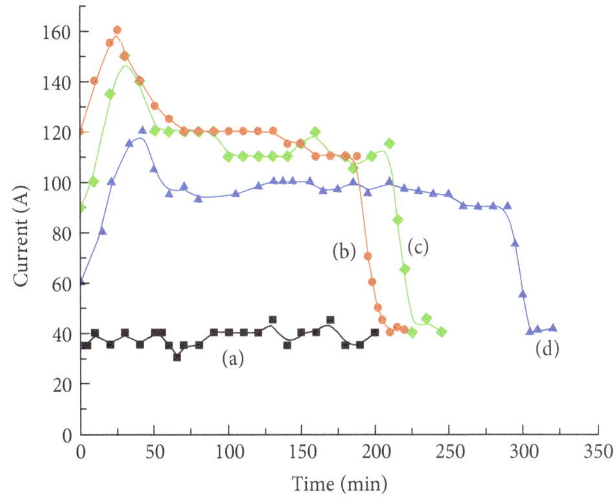

FIGURE 2: Chronoamperometric plots under constant electrolytic potential of 2.5 V at (b) 1800°C, (c) 1750°C, and (d) 1700°C. Plot (a) is a blank *I*-*t* curve, obtained under 2.5 V at 1750°C (only electrolyte CaF_2 in the electrolytic graphite crucible).

results in Table 1 show that the samples obtained after electrolysis at 1750°C were pure metallic Ti. Given that the electrolytic temperatures used in our experiments were just below the melting point of TiO_2 (1843°C) [23], the oxygen ions were easily moved out from TiO_2 crystal lattice. This sample had a porous structure as shown in the SEM image in Figure 1(c). Simultaneously, the porosity structure can supply more free pathways to enhance oxygen ions transport. It should be pointed out that even at lower electrolytic temperatures, the porous structure was also demonstrated [18, 19]. After electrolysis at 1800°C, shown in Figure 1(d), many bright white areas were scattered in grey areas. By checking with the EDX, the large grey areas were electrolyte CaF_2, while the white spots were rich in Ti (91.3 at %) as well as a small amount of Ca and F from electrolyte CaF_2. It demonstrates that the O in TiO_2 was completely removed. However, there are two distinct differences between the sample electrolyzed at 1800°C and that electrolyzed at 1750°C. First, the bright spots, which were rich in Ti, did not aggregate to large pieces of metallic Ti; second, some Ca and F were still contained. This was probably the consequence of intense fluctuation of molten CaF_2 during electrolysis, which resulted from large electrolytic current and high fluidity of electrolyte at high temperatures [24, 25].

Figure 2 shows the chronoamperometric plots of the electrolysis at three different temperatures. Each chronoamperometric plot had a current peak in the first 20–40 min and then decreased to a steady current plateau, which continued for 3-4 hr. The current plateau was almost flat, which indicated that the electrolytic process was stable. Then the electrolytic current dropped to the background current sharply in 5 min. The current behavior can be explained by a three-phase-interline (3PI) mechanism proposed by Deng et al. [26]. Briefly, the electrochemical reduction of insulating solid oxides can proceed at the place of 3PI, linking the conductor, oxide and electrolyte. As reaction persisting, the metal/oxide/electrolyte 3PI propagates, leading to metallization on the 3PI interface initially and then into the interior of oxide preforms until the reduction is completed [5, 6, 27]. Therefore, upon commencing the reduction, total length of the 3PI increased as the current raised. Once the current reached the maximum, the 3PI started to decrease to a steady status, leading to the formation of the current plateaus. Meanwhile, the reaction proceeded into the interior of the oxide until the oxide was completely electroreduced. Thus, the higher the temperature during the electrolysis, the higher the rate of the 3PI propagation, and the earlier the current peak appeared. In addition, the initial electrolytic current peak and current plateau became higher and the total electrolytic time was shortened as the temperature increased.

The design of the three electrodes, shown in Scheme 1(b) and the inset of Figure 3(a), was used for CV tests to demonstrate the electrochemical reduction process of TiO_2. Figure 3 shows the CV curves at 1750°C with and without TiO_2 slices as working electrode in molten CaF_2, respectively. Herein, the equilibrium potential of Ca^{2+}/Ca ($Ca^{2+} + 2e^- \leftrightarrow Ca$; Ca^{2+} and Ca are in standard status) is defined as the reference potential ($E^{\theta}_{Ca^{2+}/Ca} = 0$) as described in the other literature [3, 22, 28]. In Figure 3(a), with the potential becoming positive, oxidation currents increased rapidly to form a peak c1, which was the consequence of Ti dissolution [3]. If the working electrode was changed to a piece of Mo foil without TiO_2 slices, only two current plateaus of c2 and c3 would appear in both positive and negative potential directions, as shown in Figure 3(b). The difference between these two CV curves was a family of current peaks (a1–a5) shown in Figure 3(a). Apparently, these peaks were attributed to many electrochemical reactions. Moreover, we found that many types of titanium oxides existed in the Ti-O binary phase diagram due to different O ratios [23]. According to the O ratios in TiO_2, the reduction process can be divided into many steps, which were shown in (3)–(7):

$$4TiO_2 + 2e^- = Ti_4O_7 + O^{2-}, \quad (3)$$

$$3Ti_4O_7 + 2e^- = 4Ti_3O_5 + O^{2-}, \quad (4)$$

$$2Ti_3O_5 + 2e^- = 3Ti_2O_3 + O^{2-}, \quad (5)$$

$$Ti_2O_3 + 2e^- = 2TiO + O^{2-}, \quad (6)$$

$$TiO + 2e^- = Ti + O^{2-}. \quad (7)$$

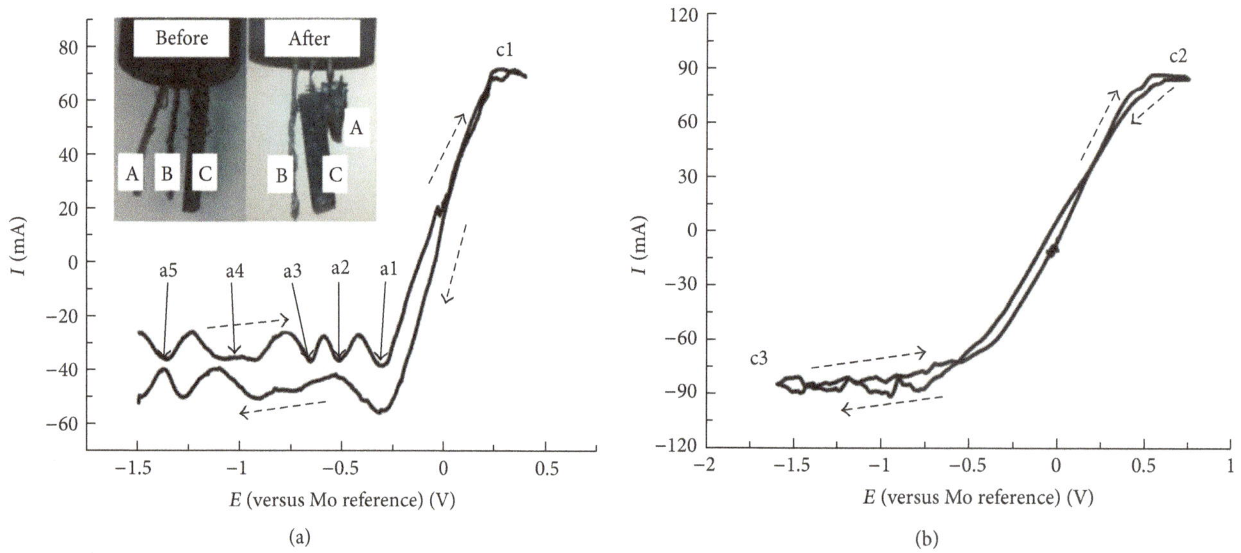

FIGURE 3: CV curves obtained at 1750°C. (a) A TiO_2 slice (from oxidizing Ti foil at 1300°C for 1.5 hr) wrapped with Mo wire as working electrode (the inset of (a) shows the design of three electrodes (working electrode (A), reference electrode (B), and counter electrode (C)) before and after CV test at 1750°C); (b) only Mo foil as working electrode. The scan rate was 20 mV s^{-1}.

According to the thermodynamic data [29], we can obtain the plots of the electrolytic potential against temperature (*E-T*), as shown in Figure 4. The theoretical reduction potentials of TiO_2, Ti_4O_7, T_3O_5, Ti_2O_3, and TiO at 1750°C were located at −1.30 V, −1.08 V, −0.96 V, −0.69 V, and −0.23 V, respectively. As what we have discussed above, the current peaks of the CV curve in Figure 3(a) have indicated a series of reactions during the electrochemical reduction process of TiO_2. Obviously, the peaks a5 (−1.28 V), a3 (−0.65 V), and a1 (−0.27 V) in Figure 3(a) were consistent with the theoretical potentials of electrochemical reduction of TiO_2, Ti_2O_3, and TiO, respectively. Therefore, it can be concluded that these three peaks resulted from reactions (3), (6), and (7). Besides, a broad peak (a4) ranged from −0.9 V to −1.13 V, indicating many reactions in this potential range. From the Ti-O binary phase diagram [23], there are a series of phases named Magnéli phases (Ti_nO_{2n-1} ($n \geq 4$)) with an O ratio between Ti_2O_3 and TiO_2. Definitely, the Ti_4O_7 is one of the Magnéli phases. As shown in Figure 4, the titanium oxide with a higher oxygen ratio will have a higher electrochemical reduction potential and vice versa. Therefore, the electrochemical reduction potentials of the Magnéli phases are among Ti_2O_3 and TiO_2. Besides, the reduction potential of the Ti_3O_5 should also be located between Ti_2O_3 and TiO_2. Thereby, the broad peak a4 was composed of reduction peaks of a series of Magnéli phases (Ti_4O_7 is the typical oxidation) and Ti_3O_5. Besides, one point should be highlighted here: the electrochemical reduction potentials of reactions (4) and (5) are very close to each other, as shown in Figure 4. This also indicates why the current peaks can be largely overlapped and merged to one broad peak (a4 in Figure 3(a)). People might question why there is another current peak a2 appeared at −0.51 V since no more Ti-O phases exist between TiO and Ti_2O_3 in the Ti-O binary phase diagram [23]. In fact, TiO can easily form $CaTiO_3$ or other higher perovskite phases

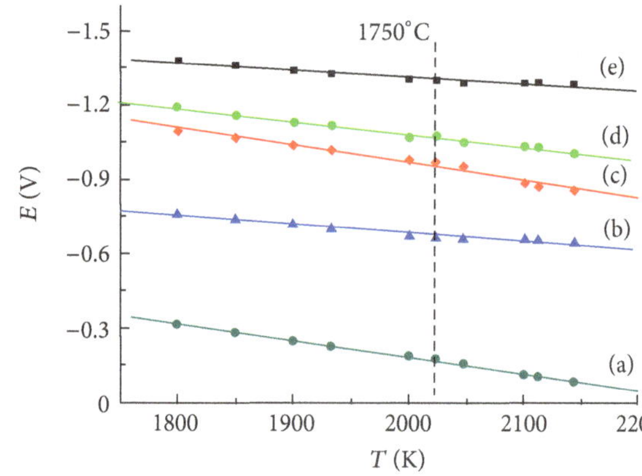

FIGURE 4: Plots of electrochemical reduction potentials of Ti oxides against temperatures: (a) $TiO + 2e^- = Ti + O^{2-}$; (b) $Ti_2O_3 + 2e^- = 2TiO + O^{2-}$; (c) $2Ti_3O_5 + 2e^- = 3Ti_2O_3 + O^{2-}$; (d) $3Ti_4O_7 + 2e^- = 4Ti_3O_5 + O^{2-}$; (e) $4TiO_2 + 2e^- = Ti_4O_7 + O^{2-}$. The vertical dashed line indicates the electrochemical reduction potentials of different Ti oxides at 1750°C.

($Ca_\delta TiO_\chi$ ($\chi/\delta \geq 2$)) with Ca^{2+} and O^{2-} in molten electrolyte as reaction (8) [16, 22]:

$$Ca^{2+} + TiO + 2O^{2-} - 2e^- = CaTiO_3 \quad (8)$$

Based on the thermodynamic data [29], we can infer that reduction potentials of this series of perovskite phases range from −0.36 V to −0.58 V. Therefore, some TiO will react with Ca^{2+} and O^{2-} at high temperatures to form $CaTiO_3$ or other higher perovskite phases, which will lead to the peak a2. The remaining TiO will continue the electrochemical reduction process to be reduced to metallic Ti, leading to the current peak a1 as discussed above.

4. Conclusions

Generally, TiO_2 was successfully electrochemically reduced to metallic Ti at high temperatures (1700–1800°C). The best result was shown by electrolysis at 1750°C; pure metallic Ti pieces were obtained. In addition, we have measured the procedure of the electrolytic reaction at 1750°C. It was found that the electrochemical reduction was separated into several steps. The oxygen in the TiO_2 was electroremoved as a step-by-step pathway ($TiO_2 \rightarrow Ti_4O_7 \rightarrow Ti_3O_5/Ti_2O_3 \rightarrow TiO \rightarrow Ti$) at different applied reduction potentials. This mechanism of electrochemical reduction can be helpful in controlling the production procedure of Ti in industry. Also, it can be used for producing Ti alloys by electrochemical reduction method at high temperatures.

Acknowledgments

The authors are grateful to the Science and Technology Ministry of China for the financial support to the national high technology project.

References

[1] J. L. Xu and Z. X. Qiu, *Brief Introduction to Direct Reduction of Titanium Dioxide to Titanium Technology*, Metallurgical Industry Press, Beijing, China, 3rd edition, 2004.

[2] W. J. Kroll, "The production of ductile titanium," *Transactions of the American Electrochemical Society*, vol. 78, no. 1, pp. 35–47, 1940.

[3] G. Z. Chen, D. J. Fray, and T. W. Farthing, "Direct electrochemical reduction of titanium dioxide to titanium in molten calcium chloride," *Nature*, vol. 407, no. 6802, pp. 361–364, 2000.

[4] I. Park, T. Abiko, and T. H. Okabe, "Production of titanium powder directly from TiO_2 in $CaCl_2$ through an Electronically Mediated Reaction (EMR)," *Journal of Physics and Chemistry of Solids*, vol. 66, no. 2–4, pp. 410–413, 2005.

[5] G. Z. Chen, E. Gordo, and D. J. Fray, "Direct electrolytic preparation of chromium powder," *Metallurgical and Materials Transactions B*, vol. 35, no. 2, pp. 223–233, 2004.

[6] T. Nohira, K. Yasuda, and Y. Ito, "Pinpoint and bulk electrochemical reduction of insulating silicon dioxide to silicon," *Nature Materials*, vol. 2, no. 6, pp. 397–401, 2003.

[7] K. Yasuda, T. Nohira, and Y. Ito, "Effect of electrolysis potential on reduction of solid silicon dioxide in molten $CaCl_2$," *Journal of Physics and Chemistry of Solids*, vol. 66, no. 2–4, pp. 443–447, 2005.

[8] X. Y. Yan and D. J. Fray, "Production of niobium powder by direct electrochemical reduction of solid Nb_2O_5 in a eutectic $CaCl_2$-NaCl melt," *Metallurgical and Materials Transactions B*, vol. 33, no. 5, pp. 685–693, 2002.

[9] X. Y. Yan and D. J. Fray, "Electrochemical studies on reduction of solid Nb_2O_5 in molten $CaCl_2$-NaCl eutectic," *Journal of The Electrochemical Society*, vol. 152, no. 1, pp. D12–D21, 2005.

[10] G. Qiu, M. Ma, D. Wang, X. Jin, X. Hu, and G. Z. Chen, "Metallic cavity electrodes for investigation of powders electrochemical reduction of NiO and Cr_2O_3 powders in molten $CaCl_2$," *Journal of the Electrochemical Society*, vol. 152, no. 10, pp. E328–E336, 2005.

[11] A. M. Abdelkader, A. Daher, R. A. Abdelkareem, and E. El-Kashif, "Preparation of zirconium metal by the electrochemical reduction of zirconium oxide," *Metallurgical and Materials Transactions B*, vol. 38, no. 1, pp. 35–44, 2007.

[12] A. J. Muirwood, R. C. Copcutt, G. Z. Chen, and D. J. Fray, "Electrochemical fabrication of nickel manganese Gallium alloy powder," *Advanced Engineering Materials*, vol. 5, no. 9, pp. 650–653, 2003.

[13] K. Dring, R. Bhagat, M. Jackson, R. Dashwood, and D. Inman, "Direct electrochemical production of Ti-10W alloys from mixed oxide preform precursors," *Journal of Alloys and Compounds*, vol. 419, no. 1-2, pp. 103–109, 2006.

[14] F. K. Meng and H. M. Lu, "Direct electrochemical preparation of NbSi alloys from mixed oxide preform precursors," *Advanced Engineering Materials*, vol. 11, no. 3, pp. 198–201, 2009.

[15] E. Gordo, G. Z. Chen, and D. J. Fray, "Toward optimisation of electrolytic reduction of solid chromium oxide to chromium powder in molten chloride salts," *Electrochimica Acta*, vol. 49, no. 13, pp. 2195–2208, 2004.

[16] K. Jiang, X. H. Hu, M. Ma et al., "'Perovskitization'-assisted electrochemical reduction of solid TiO_2 in molten $CaCl_2$," *Angewandte Chemie*, vol. 45, no. 3, pp. 428–432, 2006.

[17] W. Li, X. B. Jin, F. L. Huang, and G. Z. Chen, "Metal-to-oxide molar volume ratio: the overlooked barrier to solid-state electroreduction and a "Green" bypass through recyclable NH_4HCO_3," *Angewandte Chemie*, vol. 49, no. 18, pp. 3203–3206, 2010.

[18] F. Scholz, "Nobody can drink from closed bottles, or why it is so difficult to completely reduce solid TiO_2 to solid Ti," *ChemPhysChem*, vol. 11, no. 10, pp. 2078–2079, 2010.

[19] F. Cardarelli, "World intellectual property organization," Patent no. WO 03/046258 A2.

[20] F. K. Meng and H. M. Lu, "tudy on a new electrolysis technology of preparing high Nb bearing TiAl alloy from metal oxides," in *EPD Congress Proceedings*, pp. 711–717, 2009.

[21] G. Z. Chen and D. J. Fray, "Voltammetric studies of the oxygen-titanium binary system in molten calcium chloride," *Journal of the Electrochemical Society*, vol. 149, no. 11, pp. E455–E467, 2002.

[22] K. Dring, R. Dashwood, and D. Inman, "Voltammetry of titanium dioxide in molten calcium chloride at 900◦C," *Journal of the Electrochemical Society*, vol. 152, no. 3, pp. E104–E113, 2005.

[23] J. L. Murray and H. A. Wriedt, *Materials Properties Handbook: Titanium Alloys*, ASM International, Ohio, USA, 1994.

[24] G. Xie, *Theory and Application of Molten Salts*, Metallurgical Industry Press, Beijing, China, 1998.

[25] M. J. Zhang, *Molten Slats Electrochemistry Principle and Application*, Chemical Industry Press, Beijing, China, 2006.

[26] Y. Deng, D. Wang, W. Xiao, X. Jin, X. Hu, and G. Z. Chen, "Electrochemistry at conductor/insulator/electrolyte three-phase interlines: a thin layer model," *Journal of Physical Chemistry B*, vol. 109, no. 29, pp. 14043–14051, 2005.

[27] G. Z. Chen and D. J. Fray, "Understanding the electro-reduction of metal oxides in molten salts," *Light Metals*, pp. 881–886, 2004.

[28] D. T. L. Alexander, C. Schwandt, and D. J. Fray, "The electro-deoxidation of dense titanium dioxide precursors in molten calcium chloride giving a new reaction pathway," *Electrochimica Acta*, vol. 56, no. 9, pp. 3286–3295, 2011.

[29] D. L. Ye, *Handbook of Thermodynamic Data of Practical Inorganic Substances*, Metallurgical Industry Press, Beijing, China, 2nd edition, 2002.

7

Effects of Added Chloride Ion on Electrodeposition of Copper from a Simulated Acidic Sulfate Bath Containing Cobalt Ions

Bijayalaxmi Panda

Amity School of Engineering and Technology, Amity University, Noida, Uttar Pradesh 201303, India

Correspondence should be addressed to Bijayalaxmi Panda; blpanda@amity.edu

Academic Editors: M. Carboneras, M. Gjoka, and S. C. Wang

The effects of added chloride ion on copper electrodeposition was studied using Pb-Sb anode and a stainless steel cathode in an acidic sulfate bath containing added Co^{2+} ion. The presence of added chloride ion in the electrolyte solution containing 150 ppm of Co^{2+} ion was found to increase the anode and the cell potentials and decrease the cathode potential. Linear sweep voltammetry (LSV) was used to study the effects of added chloride ion on the anodic process during the electrodeposition of copper in the presence of added $Co^{2+} \approx 150$ ppm; the oxygen evolution potential is polarised by adding 10 ppm chloride ion at current densities (≥ 150 A/m^2), and further increase in chloride ion concentration increases the polarisation of oxygen evolution reaction more at higher current densities. X-ray diffraction (XRD) showed that added chloride ion and added Co^{2+} ion changed the preferred crystal orientations of the copper deposits differently. Scanning electron microscopy (SEM) indicated that the surface morphology of the copper deposited in the presence of added chloride ion and added Co^{2+} ion has well-defined grains.

1. Introduction

Copper is generally extracted through pyrometallurgical processes [1]. However several important factors such as non-availability/depletion of high-grade ores, increasing world demand, increasing process cost like labour cost, energy cost, and so forth, and emission of highly toxic and strongly acidic sulfur-oxide gases from smelter plants creating severe environmental pollution demanded an alternative technology to overcome these problems towards the end of nineteenth century. Thus in mid of 1980, hydrometallurgical processes involving leaching, solvent extraction, and electrowinning (L/SX/EW) were widely adopted for extraction of copper from secondary sources such as oxide ores, mixed sulfide and oxide ores, low grade sulfide ores, industrial wastes from metal plating, metal finishing, wastes from metallurgical industries, scrap copper, and alloys [1]. Although copper leaching and solvent extraction have achieved a state of advanced development, the commercial success of the process is dependent upon the ability to produce high-quality final product through the electrowinning process. The high power consumption associated with this process has been the subject of many investigations in the last few years [2–6]. The attempts made so far are [7]

(a) improvement in mass transport for cell operation at higher-current density without significant increase in energy requirement,

(b) selection of different routes for production of the copper metal,

(c) adoption of an alternative anode reaction,

(d) use of inorganic and organic depolarisers for decreasing the overpotential of oxygen evolution reaction at the anode and copper deposition reaction at the cathode,

(e) replacement of the Pb-Sb anode by a catalytic anode for decreasing the oxygen overpotential and many similar aspects.

In recent years, a more challenging problem is the efficient recovery of copper through electrodeposition process from the direct acidic leach solution and dilute industrial effluents with low power consumption [8, 9] eliminating SX process that involves alot of chemicals. The attempts described in (c)–(e) above are of particular interest as these that will be amenable to industrial implementation without any significant change in the standard plant practice. These attempts could bring significant decrease in cell voltage and power consumption through lessening of the anode potential or anode overpotential. Bivalent cobalt ion is a very useful addendum in copper electrodeposition [10–12] due to the following reasons.

(a) It considerably decreases the overpotential of oxygen evolution reaction.

(b) It significantly controls the lead corrosion of the lead-antimony anode and subsequently improves the cathode quality by reducing lead contamination.

(c) It can be used in conventional copper electrodeposition process without any modification to the existing plant cell.

Our earlier investigations [5, 12] reported the effects of cobalt ion and/or H_2SO_3 on copper electrodeposition from simulated acidic sulfate bath using Pb-Sb and/or graphite anode with significant decrease in power consumption.

In the present investigation, an attempt is made to see the effects of added Cl^- ion on the electrodeposition of copper from a simulated acidic copper bath containing added Co^{2+} ion. Small amounts of chloride ion alone are known to have an accelerating effect on the deposition of copper and reduces anode polarization [13]. Besides, Cl^- ion arises in the copper electrolyte from the makeup water. Several studies were undertaken to observe the effect of Cl^- ion on electrodeposition of copper [13–19] including the interaction of Cl^- ion with some entrained extractant residuals [20]. However, no literature appears to be available so far to our knowledge on the effect of Cl^- ion on electrodeposition of copper containing Co^{2+} ion. Pd/Sb is used as an anode material. A comparison of cell potential, anode potential, cathode potential, anode polarization characteristics, current efficiency, power consumption, deposit quality, deposit morphology, and the crystal orientation is reported in the absence and the presence of Cl^- during electrodeposition of copper in the presence of Co^{2+} ion.

2. Experimental Methods

2.1. Materials. Stock solutions of 40 g/L Cu^{2+}, 10 g/L Co^{2+} and 60 g/L sulfuric acid, and 10 g/L HCl were prepared separately using AnalaR grade reagents in doubly distilled water. The concentration of Cu^{2+} was measured by iodometric method, while H_2SO_4 was determined by acid-base titration, Co^{2+} by atomic absorption spectroscopy (Model 3100, PerkinElmer) and Cl^- by EDT (England ion meter).

2.2. Electrodeposition Experiments. The electrolysis cell consisted of a lidded 200 cm^3 double wall beaker. A stainless steel cathode (8.0 × 5.0 × 0.2 cm) and a lead-antimony (Sb = 6%) anode of the same dimensions were used. The interelectrode space was maintained at 3.0 cm for all the experiments. All electrodeposition experiments were carried out for 2 hours at room temperature (30 ± 1°C) using an electrolyte solution containing 20 g/L Cu^{2+} and 30 g/L H_2SO_4, and 150 ppm Co^{2+} ion. After electrolysis, the cathode was washed thoroughly with water followed by acetone and dried. The current efficiency (±0.3%) was calculated from the weight of copper gained by the cathode. The electrodeposited copper was analysed as >99.9% pure.

2.3. Polarisation Measurements. Linear sweep voltammetry (LSV) was used to examine the anodic polarisation behaviour, during copper electrodeposition containing added Cl^- and/or added Co^{2+}. A Pb-Sb (0.70 cm^2) electrode was used as the working anodic electrode. A platinum wire and a saturated calomel electrode (SCE) were used as the counter electrode and the reference electrode, respectively. The surface of the working electrode was freshly prepared before each experiment, initially rinsed with 1 M HCL followed by doubly distilled water. A scanning potentiostat (Model 362, EG&G Princeton Applied Research) was used for carrying out polarisation experiments between +1.2 V and +2.0 V. The linear voltammograms (V~I) were recorded by using an *X-Y* recorder (PAR Model RE0091, EG&G Princeton Applied Research) at a scan rate of 20 mV/sec during polarisation experiments.

2.4. Deposit Examination. An X-ray diffractometer (PW 1050, Philips) was used to determine the crystallographic orientations of the cathode copper deposits. Reproducible results were obtained using cathode-deposited sections and powders scraped from the cathode surface. The data matched those for copper powders reported in the literature (JCPDS, 1984). The deposit morphology of the electrodeposited copper samples was examined by SEM (SE 101B model, Philips).

3. Results and Discussion

3.1. Anode Potential. Effect of Added Chloride Ion Variation in the Presence of Added Cobalt Ion. The influence of $[Cl^-]_o$ on the anode voltage in the presence of $[Co^{2+}]_o \cong 150$ ppm in the electrolyte solution is shown in Figure 1. The anode voltage in the absence of Cl^- ion in the electrolyte containing only Co^{2+} (aq) (≅150 ppm) is found to be 1.58 V. Addition of 5 ppm of Cl^- ion to the same electrolyte increases the anode voltage to 1.63 V. No significant increase in the anode potentials is observed with further increase in $[Cl^-]_o$ up to 100 ppm.

3.2. Cathode Potential. Effect of Added Chloride Ion Variation in the Presence of Cobalt Ion. The cathode potential at zero concentration of Cl^- in the electrolyte containing Co^{2+} (aq) ≅ 150 ppm was found to be 0.25 V. The addition of $[Cl^-]_o \cong$ 5 ppm decreases the cathode potential from 0.25 to 0.19 V. Further increase in $[Cl^-]_o \cong 100$ ppm brings about no significant change in the cathode voltage in the presence of $[Co^{2+}.aq]_o \cong 150$ ppm as observed in Figure 2.

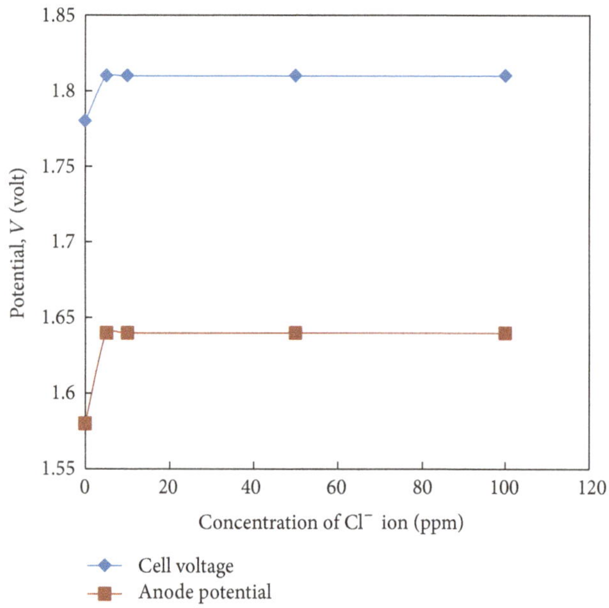

FIGURE 1: Effect of added $[Cl^-]_o$ on cell voltage and anode potential in the presence of added cobalt ion 150 ppm (Cu^{2+} = 20 g/L, H_2SO_4 = 30 g/L, T = 30 ± 1°C, CD = 150 A/m^2, t_d = 2 hr).

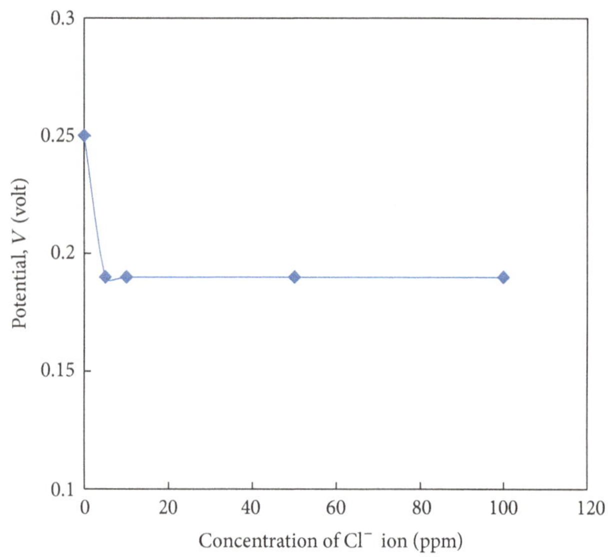

FIGURE 2: Effect of $[Cl^-]_o$ on cathode potential in the presence of added cobalt ion 150 ppm (Cu^{2+} = 20 g/L, H_2SO_4 = 30 g/L, T = 30 ± 1°C, CD = 150 A/m^2, t_d = 2 hr).

3.3. *Cell Voltage. Effect of Added Chloride Ion Variation in the Presence of Cobalt Ion.* Figure 1 shows the influence of $[Cl^-]_o$ on cell voltage in the presence of added Co^{2+} (aq) during electrodeposition of copper. The nature of the curve appears to be similar to that observed in the case of the anode potential. Addition of 5 ppm of Cl^- to the same electrolyte increases the cell voltage to 1.81 V. Further increase in $[Cl^-]_o$ in the range of 5–100 ppm brings about no significant increase in the cell voltage.

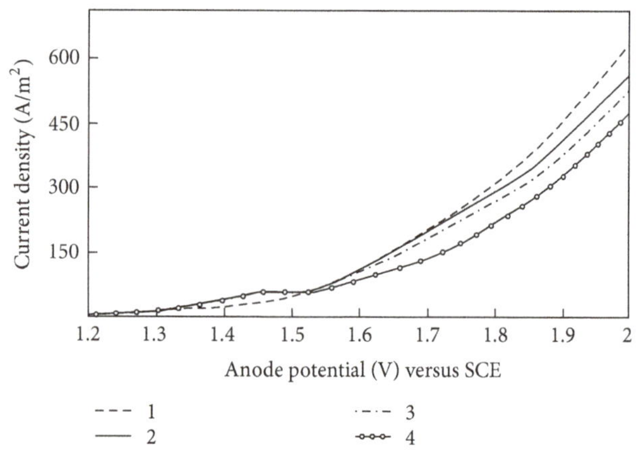

FIGURE 3: Anodic potentiodynamic curves in the presence of added chloride ion. Key: added $[Cl^-. aq]_o$ (1) nil; (2) 5 ppm; (3) 50 ppm; (4) 100 ppm (Cu^{2+} = 20 g/L, H_2SO_4 = 30 g/L, $[Co^{2+}]_o$ = 150 ppm T = 30 ± 1°C, scan rate 20 mV/Sec).

3.4. *Anode Polarisation.* The anodic potentiodynamic experiments conducted by varying the chloride ion concentration in the presence of $[Co^{2+}]_o \cong$ 150 ppm in the copper electrolyte solution were studied by LSV method, and the curves are shown in Figure 3. Figure 3 curve 1 shows current density versus anode potential curve in the presence of $[Co^{2+}]_o \cong$ 150 ppm. It was observed that the addition of $[Cl^-]_o \cong$ 5 ppm polarises significantly the anode potential of oxygen evolution reaction at higher current densities (Figure 3 curve 2); the region between 1.3 V and 1.5 V shows the probable occurrence of chlorine evolution that is not observed in curve 1. Further increase in $[Cl^-]_o$ to 50 ppm and 100 ppm (Figure 3 curves 3 and 4) increases the polarisation of oxygen evolution reaction.

3.5. *Current Efficiency and Power Consumption.* The effect of added chloride ion in the absence and the presence of $[Co^{2+}]_o$ on current efficiency and the corresponding power consumption are examined. It was found that the current efficiency remained ~98% for all additions of chloride ion and/or cobalt ion throughout the investigation, and the cathode remains smooth, bright, and compact. It is seen from Table 1 that the addition of $[Cl^-]_o$ in the range of 5–100 ppm in the presence of Co^{2+} (aq) shows increases in the power consumption by ≤35 kWh/ton of Cu than that observed in the presence of only $[Co^{2+}]_o \cong$ 150 ppm in the electrolyte solution.

3.6. *Crystallographic Orientations.* The electrodeposited copper samples produced from acidic copper bath electrolyte solution on stainless steel cathode in the presence of added Co^{2+} (aq) and/or Cl^- (aq) are examined by XRD to determine the preferred crystal orientations and the relative growth of copper on the preferred planes. Representative XRD traces are redrawn and shown in Figure 4. The XRD of all the copper deposits showed an fcc structure, with the 2θ positions of the peaks remaining constant. The XRD trace for the original

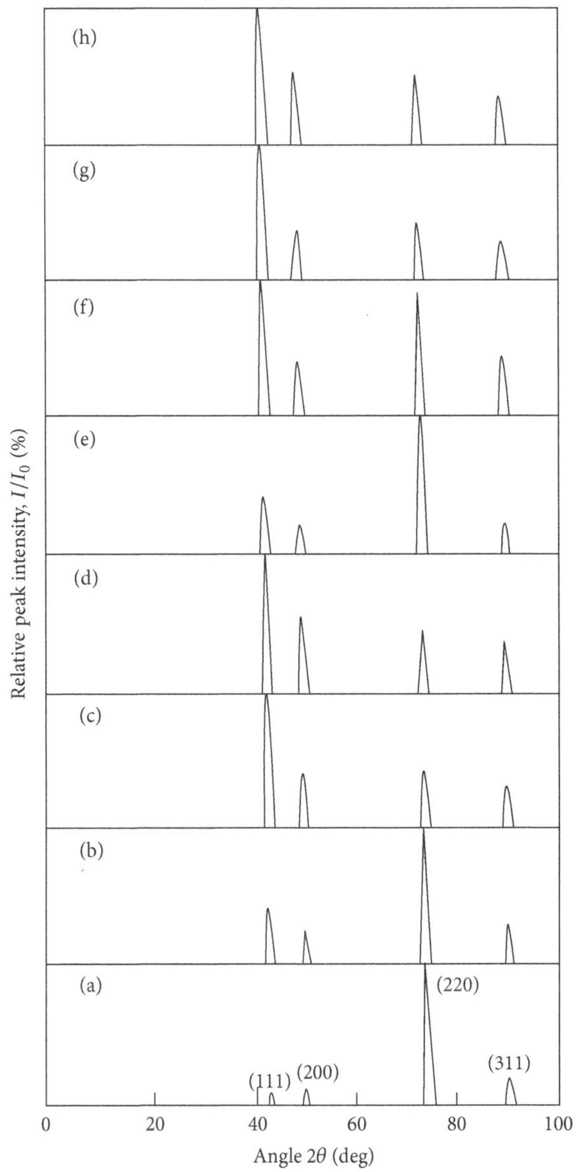

FIGURE 4: XRD patterns of electrodeposited copper samples. Key: added $[Co^{2+}]_o$ X ppm + $[Cl^-]_o$ Y ppm (a) Bl; (b) 0 + 5 ppm; (c) 0 + 100 ppm; (d) 150 ppm + 0; (e) 150 ppm + 5 ppm; (f) 150 ppm + 10 ppm; (g) 150 ppm + 50 ppm; (h) 150 ppm + 100 ppm. (Cu^{2+} = 20 g/L, H_2SO_4 = 30 g/L, $T = 30 \pm 1^\circ C$, CD = 150 A/m^2, t_d = 2 hr).

electrolyte (Figure 4(a)) in the absence of added Cl^- and Co^{2+} (aq) and in the presence of $[Co^{2+}]_o \cong 150$ ppm (Figure 4(d)) has been given here for comparison. The addition of only $[Cl^-]_o \cong 5$ ppm or 100 ppm changes the XRD pattern of copper deposit differently; while the presence of the former did not alter the most preferred (220) plane (in comparison to Figure 4(a)), the presence of the latter changes the most preferred (220) plane to (111) plane, and the growth along (200), (220), and (311) planes is more or less equally maintained (Figure 4(c)) and is similar in nature with that observed in the presence of only $[Co^{2+}]_o \cong 150$ ppm (Figure 4(d)). The

TABLE 1: Effect of chloride ion on power consumption in the presence of cobalt ion.

$[Co^{2+}$ (aq)$]_o$, ppm	$[Cl^-]_o$, ppm	E, kWh/ton of Cu
150	—	1564
150	5	1599
150	10	1599
150	50	1599
150	100	1599

(—) refers to nil.

addition of small $[Cl^-]_o \cong 5$ ppm to the electrolyte containing $[Co^{2+}]_o \cong 150$ ppm again retains the most preferred (220) plane (Figure 4(e)); the increase in $[Cl^-]_o$ to 10 ppm in the same electrolyte increases equal growth along (111) and (220) planes (Figure 4(f), further increase in the $[Cl^-]_o \cong 50$ ppm and 100 ppm in the presence of $[Co^{2+}]_o \cong 150$ ppm retained the (111) plane as the most preferred plane, and the growth along (200), (220), and (311) planes is more or less equally maintained (Figures 4(g) and 4(h)).

3.7. Surface Morphology. The effect of chloride ion on the surface morphology of the deposited copper samples in the absence and the presence of Co^{2+} (aq) is shown in Figures 5(a)–5(f). It is observed that polycrystalline pyramidal deposits are obtained in the presence of only $[Cl^-]_o$, ≅5 ppm and ≅100 ppm in the copper electrolyte solution (Figures 5(a) and 5(b)). The addition of $[Cl^-]_o$, 5 ppm to the electrolyte solution containing $[Co^{2+}]_o$, 150 ppm results in very small grains of copper deposits (Figure 5(c)). Increasing the $[Cl^-]_o$ to 100 ppm in the same electrolyte solution forms small size nodules and mosaic copper deposits (Figures 5(d)–5(f)).

4. Interpretation

It would be interesting to interpret the effects of Cl^- on electrodeposition of copper in the presence of Co^{2+} ion. The addition of only Co^{2+} ion to the copper electrolyte decreases the anode potential significantly but has no effect on the cathode potential as was found in our earlier investigation [12].

The decrease in anode potential in the presence of only Co^{2+} ion can be best explained by the following reactions. In the absence of Co^{2+} ion, oxygen is evolved from the electrolysis of water (1) taking place on PbO_2 surface on Pb-Sb anode which exhibits a high overpotential of ≥600 mV [1]:

$$2H_2O \longrightarrow O_2\uparrow + 4H^+ + 4e \qquad E_{SHE} = -1.229\,V \qquad (1)$$

$$Co^{2+} \longrightarrow Co^{3+} + e \qquad E_{SHE} = -1.84\,V \qquad (2)$$

$$Co^{3+} + 2H_2O \longrightarrow 4Co^{2+} + 4H^+ + O_2 \qquad (3)$$

While in the presence of only Co^{2+} ion, oxidation of Co^{2+} ion to Co^{3+} ion (2) allows the facile oxidation of H_2O in accordance with (3) and leads to lower oxygen overpotential [10, 21].

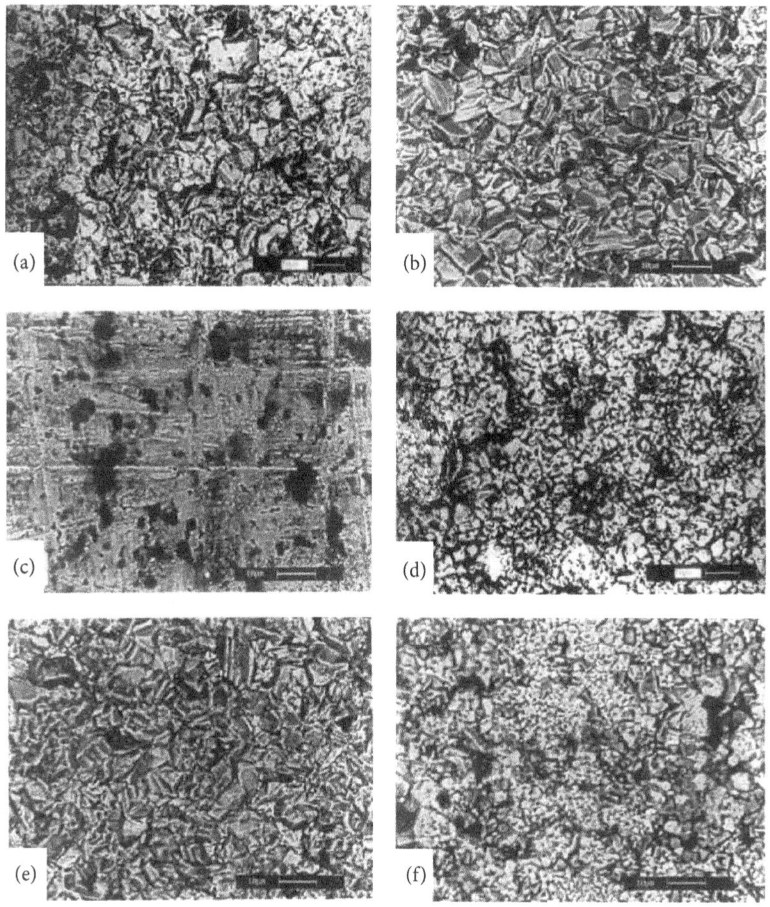

Figure 5: SEM photomicrographs of electrodeposited copper samples. Key: added $[Co^{2+}]_o$ X ppm + $[Cl^-]_o$ Y ppm (a) 0 + 5 ppm; (b) 0 + 100 ppm; (c) 150 ppm + 5 ppm; (d) 150 ppm + 10 ppm; (e) 150 ppm + 50 ppm; (f) 150 ppm + 100 ppm (Cu^{2+} = 20 g/L, H_2SO_4 = 30 g/L, $T = 30 \pm 1$°C, CD = 150 A/m^2, $t_d = 2$ hr).

An important result found in the present investigation is the increase in the anode potential when a small enough $[Cl^-]_o \cong$ 5–100 ppm is added to the electrolyte containing 150 ppm Co^{2+} (aq); nearly 50 mV increase in the anode potential is observed. The same is also indicated in the anodic potentiodynamic study; the addition of Cl^- ion to the copper electrolyte is found to polarise the facile oxidation of H_2O molecules in the presence of Co^{2+} ion. Thus, the increase in anode voltage due to the presence of Cl^- in the copper electrolyte containing Co^{2+} ion is most probably due to the partial discharge of Cl^- ion at the anode as given in (4) [21, 22] as well as the polarisation of the facile oxidation of H_2O molecules as described in (2) and (3) due to the Cl^- ions. Indication for this interpretation is the observed evolution of Cl_2 (g) at the anode during the experimental work:

$$2Cl^- \longrightarrow Cl_2 + 2e \qquad E^{\phi}_{SHE} = -1.35\,V \qquad (4)$$

The decrease in the cathode potential in the presence of Cl^- ion may be due to the ready discharge of the Cu^{2+}-Cl^- complex at the cathode [23].

5. Conclusions

The results observed in the present investigation can be summarised as follows.

(1) The addition of 5 ppm of chloride ion to the copper electrolyte solution in the presence of 150 ppm Co^{2+} ion increased the anode potential by ≅50 mV which is more or less maintained with further increase in Cl^- ion concentration up to 100 ppm. The increase in the chloride ion concentration (in the range of 0–100 ppm) in the electrolyte solution decreases the cathode potential in the presence of $Co^{2+} \cong 150$ ppm from 0.25 V to 0.19 V.

(2) The addition of ≅0–100 ppm of Cl^- to the electrolyte solution containing 150 ppm Co^{2+} increases the power consumption by 35 kWh/ton Cu.

(3) The increase in the chloride ion concentration in the absence and the presence of cobaltous ion changes the most preferred (220) hkl plane to (111) hkl plane.

(4) No significant change in the surface morphology of the copper deposits is observed. Smooth, bright deposits are obtained throughout the investigation.

Abbreviations

E_a: Anode potential
E_{cell}: Cell potential
E (kWh/ton Cu): Energy consumption
$[Cl]_o$: Initial concentration of Cl^- ion
$[Co^{2+}]_o$: Initial concentration of Co^{2+} ion
CD: Current density
T_d: Electrolysis duration
T: Temperature.

Acknowledgment

The author is thankful to Dr. R. K. Panda for his encouragement during the investigation and preparation of the paper.

References

[1] A. K. Biswas and W. G. Davenport, *Extractive Metallurgy of Copper*, Pergamon Press, Oxford, UK, 1976.

[2] K. K. Mishra and W. C. Cooper, "Anodes for electrowinning," in *Proceedings of the AIME Annual Meeting, Los Angeles*, D. J. Robinson and S. E. James, Eds., pp. 13–36, Metallurgical Society of AIME, Warrendale, Pa, USA, 1984.

[3] T. Subbaiah, P. Singh, G. Hefter, D. Muir, and R. P. Das, "Sulphurous acid as anodic depolarizer in copper electrowinning—part II," *Journal of Applied Electrochemistry*, vol. 30, no. 2, pp. 181–186, 2000.

[4] B. Panda and S. C. Das, "Electrowinning of copper from sulfate electrolyte in presence of sulfurous acid," *Hydrometallurgy*, vol. 59, no. 1, pp. 55–67, 2001.

[5] B. Panda, S. C. Das, and R. K. Panda, "Synergistic effects of added bivalent aqua cobalt ion, bivalent aqua iron ion and aqueous sulfurous acid on a graphite anode during electrodeposition of copper from a sulfate bath," *Hydrometallurgy*, vol. 72, no. 1-2, pp. 149–158, 2004.

[6] B. Panda, S. C. Das, and R. K. Panda, "Cathodic deposition of copper in the presence of aqueous sulfurous acid, bivalent aquacobalt ion, or both using a stainless steel cathode and a graphite anode," *Metallurgical and Materials Transactions B*, vol. 34, no. 6, pp. 813–819, 2003.

[7] W. C. Cooper, "Advances and future prospects in copper electrowinning," *Journal of Applied Electrochemistry*, vol. 15, pp. 789–805, 1985.

[8] D. Pletcher and F. C. Walsh, *Industrial Electrochemistry*, Chapman & Hall, London, UK, 1990.

[9] R. D. Armstrong, M. Todd, J. W. Atkinson, and K. Scott, "Selective electrodeposition of metals from simulated waste solutions," *Journal of Applied Electrochemistry*, vol. 26, no. 4, pp. 379–384, 1996.

[10] A. S. Gendron, A. V. Enel, and S. Abe, "Effect of cobalt added to electrolyteon corrosion rate of Pb-Sb anodes in copper electrowinning," *Canadian Metallurgical Quarterly*, vol. 14, pp. 59–62, 1975.

[11] C. Lupi and D. Pilone, "Energy saving in copper electrowinning," in *Aqueous Electrotechnology: Progress in Theory and Practice*, D. B. Dreisinger, Ed., TMS, Warrendale, Pa, USA, 1997.

[12] B. Panda, S. C. Das, and R. K. Panda, "Effect of added cobalt ion on electro-deposition of copper from sulfate bath using graphite and Pb-Sb anodes," *Hydrometallurgy*, vol. 95, no. 1-2, pp. 87–91, 2009.

[13] Z. D. Stanković, "The effect of $Cl(I)^-$ ions on kinetics and mechanism of anodic dissolution and cathodic deposition of copper," *Electrochimica Acta*, vol. 29, no. 3, pp. 407–409, 1984.

[14] I. J. Itzkovitch, V. A. Ettel, and A. S. Gendron, "Copper removal from thompson nickel anolyte by solvent extraction," *Canadian Mining and Metallurgical Bulletin*, vol. 67, no. 741, pp. 92–96, 1974.

[15] N. Pradhan, P. G. Krishna, and S. C. Das, "Influence of chloride ion on electrocrystallization of copper," *Plating and Surface Finishing*, vol. 83, no. 3, pp. 56–63, 1996.

[16] V. I. Lakshmanan, D. J. Mackinnon, and J. M. Brannen, "The effect of chloride ion in the electrowinning of copper," *Journal of Applied Electrochemistry*, vol. 7, no. 1, pp. 81–90, 1977.

[17] E. Ilgar and T. J. O'Keefe, "Surface roughening of electrowon copper in the presence of chloride ions," in *Aqueous Electrotechnology: Progress in Theory and Practice*, D. B. Dreisinger, Ed., The Minerals, Metals and Materials Society, 1997.

[18] T. C. Franklin, "Some mechanisms of action of additives in electrodeposition processes," *Surface and Coatings Technology*, vol. 30, no. 4, pp. 415–428, 1987.

[19] G. Fabricius and G. Sundholm, "The effect of additives on the electrodeposition of copper studied by the impedance technique," *Journal of Applied Electrochemistry*, vol. 14, no. 6, pp. 797–801, 1984.

[20] V. I. Lakshmanan, D. J. Mackinnon, and J. M. Brannen, "The effect of thiourea, LIX65N and chloride ion on the morphology of electrowon copper," *Journal of Applied Electrochemistry*, vol. 7, no. 2, pp. 127–133, 1977.

[21] "Max absorption (GAS): 162 NM shoulder (LOG E = 4.0); 175 NM (LOG E = 4.2); 187 NM (LOG E = 4.1)," in *CRC Handbook of Chemistry and Physics*, R. C. Weast and M. J. Astle, Eds., p. C-229, Boca Ranton, Fla, USA, 60th edition, 1979-1980.

[22] G. Milazza and S. Caroli, *Table of Standard Electrode Potentials, Project of the IUPAC Electrochemistry Commission*, John Wiley and Sons, 1978.

[23] A. J. Sukava and C. A. Winkler, "Depolarization by mercaptoacetic acid during electrodeposition of copper," *Canadian Journal of Chemistry*, vol. 34, pp. 128–132, 1956.

8

Drain Rate and Liquid Level Simulation in Blast Furnace Hearth Using Plant Data

Hemant Upadhyay[1] and T. K. Kundu[2]

[1] *Blast Furnace Department, Jindal Steel & Power Ltd., Raigarh, Chhattisgarh 496001, India*
[2] *Department of Metallurgical and Materials Engineering, Indian Institute of Technology Kharagpur, Kharagpur, West Bengal 721302, India*

Correspondence should be addressed to T. K. Kundu; tkkundu@metal.iitkgp.ernet.in

Academic Editors: M. Carboneras, P. Markovsky, and A. Pinkerton

Proper understanding and control of drainage of hot metal and slag from hearth are essential for a stable and efficient blast furnace operation. Various operational problems like irregular casting intervals, damage to lining, low blast intake, furnace pressurization, and so forth are normally encountered when liquid levels in the hearth exceed a critical limit where hearth coke and deadman start to float. Estimation of drain rate and liquid level in hearth needs to be simulated based on the operating parameters available as carrying out any direct measurement is extremely difficult due to the hostile conditions. Here, a mathematical model has been developed to simulate real-time liquid level and drainage behavior of the furnace hearth. Based on the computed drainage rate, production rate, and mass balance, the model is able to predict occurrence of slag-out time and cast close time which are in good agreement with the plant data.

1. Introduction

The iron-making blast furnace (BF) is a giant countercurrent heat exchanger and chemical reactor used for reduction of iron ore to molten iron. It is the most important commercial reactor for producing the majority of the world's primary steel. To improve blast furnace productivity, iron and steel industries are constructing larger furnace having more than 4000 m^3 internal capacity. Effective drainage of molten iron and slag from the hearth is the major problem occurring in larger-capacity and higher-productivity blast furnaces.

Proper drainage of blast furnace (BF) hearth is very important in order to maintain smooth hot metal production and stable burden descent. The condition of the hearth plays an important role, since it affects the reduction of the leftover FeO in slag [1], the dissolution of carbon in liquid iron, the distribution of various solutes between iron and slag, and the fresh hot metal flow from the dripping zone [2]. Inadequate drainage can make the blast furnace operation problematic when liquid levels exceed a critical limit characterized by very short distance between slag surface and raceways. Furthermore, at an elevated liquid level, the buoyancy force acting upon the coke column increases, and the deadman starts to float, causing sluggish or irregular descent of burden material. Therefore, it is very important for furnace operators to understand the mechanisms governing hearth drainage and access internal state of the liquids in the hearth. This knowledge is very important in order to take appropriate control actions to ensure a good drainage of liquids from the hearth and thereby guarantee a stable BF operation with minimum wear of the hearth linings [3]. Unfortunately, it is very difficult to carry out any direct measurement because of the hostile conditions in the blast furnace hearth, with high-temperature and chemically aggressive liquids. Hence, the estimation of drain rate and liquid level in hearth has to be simulated based on indirect measurements and the available operating parameters. Estimation of liquid levels and tap rates based on the assumption of a stationary or a floating deadman has been reported in the literature [4, 5]. Hearth liquid level estimation was carried out by strong correlation with the electromotive force measured between two electrodes mounted at the furnace shell and iron level in hearth [2]. A mathematical model for flow-rate simulation considering momentum conservation for a fluid flow at unsteady state and the effect of coke bed, pressure gradient, and drag interaction has been developed [6]. Measurements on liquid holdup and average residence time by high-temperature trickle-bed setup using a screw feeder and their modeling have been performed by Husslage et al. [7]. A two-dimensional model considering the simultaneous drainage of iron and slag from a packed bed has been developed by Tanzil and Pinczewski [8], but the estimates of slag residual volume showed large errors when applied to actual blast furnace. Half-cylindrical scale physical model and mathematical balance model have been developed by Nouchi et al., to show the effect of particle-free space on hearth drainage efficiency [9].

Although the above models provide reasonable estimates, they have some deficiencies. They either require a huge infrastructure or involve a lot of variables for metal and slag, whose values are not readily available, and they change significantly with process parameters. In order to solve this problem and provide the furnace operator with a tool to examine and control the tapping, a mathematical model is prepared to simulate the casting practice. Here, blast furnace hearth is considered as a pressurized vessel with a liquid supply from top and a drain (taphole) at bottom. The fluid flow from the taphole is considered to be dynamic and depends on production rate, slag rate, furnace pressure, quality of metal/slag/mud gun clay used, and the erosion behavior of the taphole. These parameters were studied over a large number of castings, and their relationship is established in terms of a tunable flow factor. A set of differential equations were used to represent the variation of flow rate using Bernoulli's principle, mass content, and fluid level in hearth. Based on these calculations, the slag-out time (t_s) and the tapping closure time (t_c) were predicted, and a real-time view of liquid levels in blast furnace hearth is provided.

2. General Description of the Blast Furnace Hearth Model

In blast furnace, the solid burden consisting of iron ore lumps (and/or pellet/sinter), coke, and flux (limestone and dolomite) is charged and from top using gas-leak-proof bell-less/double-bell charging system. Hot oxygen-enriched humidified-air blast blown through tuyeres (located just above hearth region) combines with carbon (of coke) at raceways and generates CO and heat. The hot CO reduces iron oxides (in the upper part of the furnace) as it ascends toward top, and molten iron and slag (molten oxides formed by gangue and flux) accumulate in the hearth in layers due to immiscibility and difference in specific gravity. The magnified view of the blast furnace hearth is shown in Figure 1, which is drained from a taphole.

The blast furnace hearth is never empty; it is always full of coke and contains varied quantities of metal and slag. Even after tapping, there is always some amount of residual slag in the hearth (Figure 1). Permeability and density of the

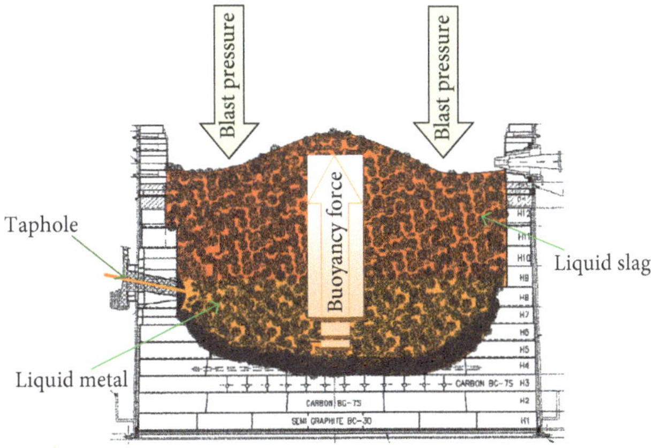

FIGURE 1: Cross-sectional view of blast furnace hearth.

coke grid are not uniform within the hearth; for example, deadman is the least permeable zone in the hearth. The fresh liquids generated are accommodated in the voids between the coke particles; hence, obtaining the representative average void fraction is necessary for hearth modeling. During the tapping, the cross-sectional area of the taphole does not remain constant as it continuously erodes. The rate and type of erosion depend on various factors like quality of clay used to close the taphole, furnace operating parameters, metal/slag quality, and extent of human intervention. As a consequence, the drain rate does not remain uniform. To proceed for any sort of modeling, it is of utmost importance to estimate the taphole erosion coefficient which may be a unique characteristic for each furnace.

The inner height of the BF hearth is generally considered to be from the top of the refractory lining to the tuyere center line, but for drainage calculation, the effective height is considered to be from the taphole center line to the tuyere center line. In order to avoid any mishap due to the rising liquid levels, there must be a defined safety level beyond which the liquid level should not rise. According to the general understanding, the slag being lighter remains on top of the metal layer, and it should come out only after the complete drainage of metal from the hearth. Contrarily, in actual practice, the interface of metal and slag does not remain completely flat, and, on approaching towards taphole, it takes an inclined profile due to which the slag comes out from the hearth before complete drainage of metal. At no point of time, slag comes out through taphole alone and even after drainage of metal layer as the instantaneous quantity of the generated (through dripping zone) metal comes out with the slag. All of these factors were kept in mind, and appropriate values were chosen while calculating the effective average drain rate.

3. Modeling Methodology

The delay time (t_h) between closing of last casting and opening of the current casting is obtained in the first step of the drainage calculation. The hot metal and slag production rates ($\overline{Q}_{\mathrm{mp}}, \overline{Q}_{\mathrm{sp}}$) are calculated based on burden, charge rate, burden and hot metal compositions and metal losses (assumed to be 1% here) as received from the process control system of BF-II, JSPL Raigarh Plant. The calculated production rate and the delay time are used to obtain the initial quantities of liquid metal (Q_{m0}) and slag (Q_{s0}) accumulated in hearth, prior to opening of taphole. The velocities with which the liquid metal (v_m) and slag (v_s) come out through the taphole are calculated using Bernoulli's principle in (1) for flow from pressurized vessel, considering blast pressure (P_b), outer pressure (P_{out}, atmospheric pressure), inner pressure (P_{in}, pressure at iron-slag interface, i.e., blast pressure plus the slag head pressure), persisting, and instantaneous vertical height of metal and slag levels from taphole (h_m, h_s) in hearth. Appropriate values for hot metal and slag densities are taken from the plant operating data (ρ_m= 6700 kg m^{-3}; ρ_s= 2800 kg m^{-3}). Consider the following:

$$v_m = \sqrt{2 \cdot \left(\frac{P_{\text{in}} - P_{\text{out}}}{\rho_m} + g \cdot h_m\right)},$$
$$v_s = \sqrt{2 \cdot \left(\frac{P_b - P_{\text{out}}}{\rho_s} + g \cdot h_s\right)}. \quad (1)$$

Using these velocities along with the cross-sectional area of the taphole (a_2) and with the flow factors (f_m,f_s) obtained by calibration with a large number of tapping results (as described later in this section), the drain rates of metal and slag ($\overline{Q}_{\text{md}}$, $\overline{Q}_{\text{sd}}$) at any point of time are derived as follows:

$$\frac{dQ_{\text{md}}}{dt} = -v_m \cdot a_2 \cdot f_m = \overline{Q}_{\text{md}},$$
$$\frac{dQ_{\text{sd}}}{dt} = -v_s \cdot a_2 \cdot f_s = \overline{Q}_{\text{sd}}. \quad (2)$$

As the taphole continuously erodes during the tapping, the effective taphole diameter (D_{th}) at any point of time (accordingly, the cross-sectional area of taphole) is correlated with the taphole erosion coefficient (K_{th} = 1.493 × 10^{-6} m s^{-1}) [6] and the drill diameter (D_{th0}):

$$D_{\text{th}} = D_{\text{th0}} + K_{\text{th}} (t - t_h). \quad (3)$$

These instantaneous drain rates in addition to the production rates give the rates of change of metal and slag volumes ($\overline{Q}_m$, $\overline{Q}_s$) in the hearth as follows:

$$\frac{dQ_m}{dt} = \frac{dQ_{\text{mp}}}{dt} + \frac{dQ_{\text{md}}}{dt} = \overline{Q}_m,$$
$$\frac{dQ_s}{dt} = \frac{dQ_{\text{sp}}}{dt} + \frac{dQ_{\text{sd}}}{dt} = \overline{Q}_s. \quad (4)$$

Having the initial liquid quantities and rate of change of liquid quantity in hearth, the remaining amounts of liquids (Q_m, Q_s) at any point of time are calculated according to the following equations:

$$Q_m = Q_{m0} + \int_{t_h}^{t} \overline{Q}_m dt,$$
$$Q_s = Q_{s0} + \int_{t_s}^{t} \overline{Q}_s dt. \quad (5)$$

Taking in to account the hearth geometry (diameter D_h = 8.4 m, accordingly, hearth area, a_1) and the coke grid void fraction (ε = 0.3), the liquid volume change rates are converted into liquid height change rates ($\overline{h}_m$, $\overline{h}_s$), which are used to find respective liquid height (h_m, h_s) at any point of time taking in to consideration the initial metal and slag heights (h_{m0}, h_{s0}) as follows:

$$\frac{dh_m}{dt} = \frac{\overline{Q}_m}{\varepsilon \cdot a_1} = \overline{h}_m, \quad (6)$$

$$\frac{dh_s}{dt} = \frac{\overline{Q}_s}{\varepsilon \cdot a_1} = \overline{h}_s, \quad (7)$$

$$h_m = h_{m0} + \int_{t_h}^{t} \overline{h}_m dt, \quad (8)$$

$$h_s = h_{s0} + \int_{t_h}^{t} \overline{h}_s dt. \quad (9)$$

On the basis of rate of change of height of liquid metal ($\overline{h}_m$) and slag ($\overline{h}_s$) using (6) and (7) and the predefined cutoff level, the slag-out time (t_s) and the cast closing time (t_c) are predicted. During the data collection for this work, it was found that the slag started coming out when there was still 150 mm of metal level in the hearth; hence, it is used as a cutoff level to predict the slag-out time (t_s). A flowchart for drain rate and liquid level calculation is provided in Figure 2. All of the calculation steps in the model are sequential and interrelated. The startup of calculations for time t draws the necessary values for variables from last step. These calculations are iterative in nature; thus, they are incorporated into a computer program.

The Bernoullis equation gives the fluid flow in the ideal case; so to derive the actual flow rates from the blast furnace hearth, v_m and v_s values need to be corrected in order to include the effect of friction, viscosity and pressure variations, and so forth. The friction depends on the quality of the mud gun clay used, the drilling practice, and the taphole length. In the same way, the viscosity depends on the liquid temperature and composition. Although the effect of furnace inner pressure has been taken into account, the load on the fluid surface, which is considered constant here, may vary in the long run. Now, many of the parameters, stated previously are not readily available in day-to-day operations; hence, the contributions of all of them are consolidated in terms of tunable parameters designated as "flow factors" (f_m and f_s) which are used to correct the fluid flow and calibrate the model in case of any deviation in the process parameters.

The model described above inherits the capability to re-calibrate and keep itself updated with the calculated parameters values. The differences in event times (t_s, t_c) are calculated from the model predicted and the corresponding actual values. The flow factors (f_m and f_s) values are optimized to eliminate these differences. The optimized flow factors values are saved and based on a predefined data pool and their average values are derived. Now, depending on the difference between the present and the optimized values of

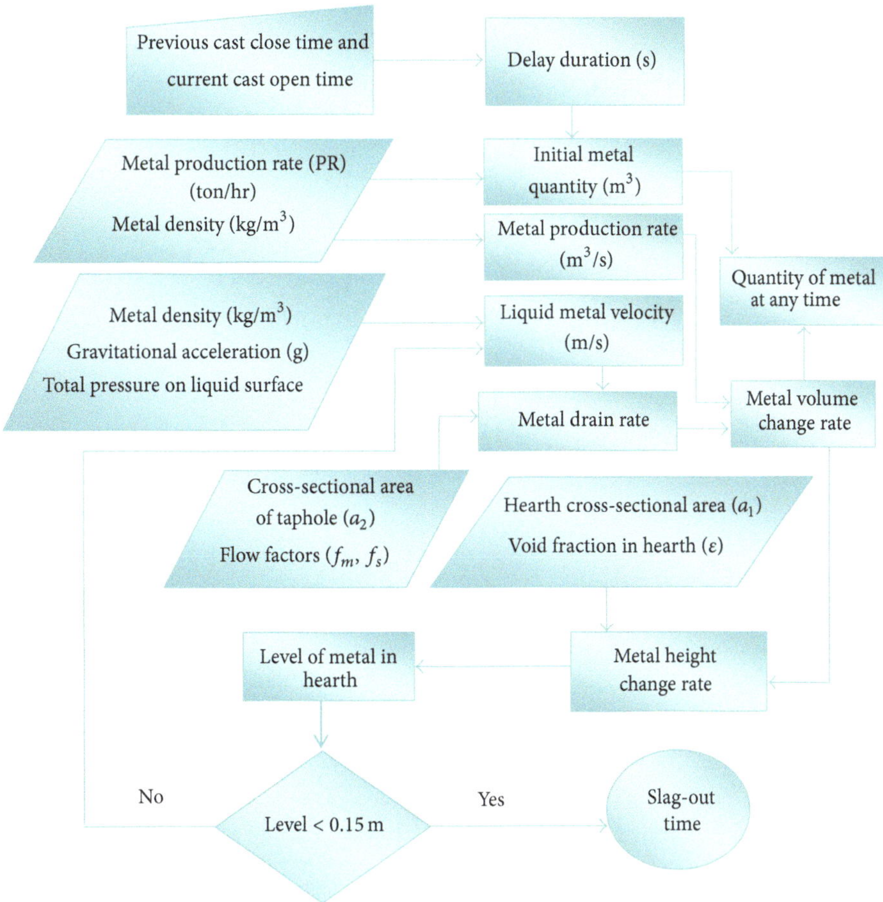

FIGURE 2: Flow-chart for drain rate and liquid level calculation to predict slag-out time from blast furnace hearth.

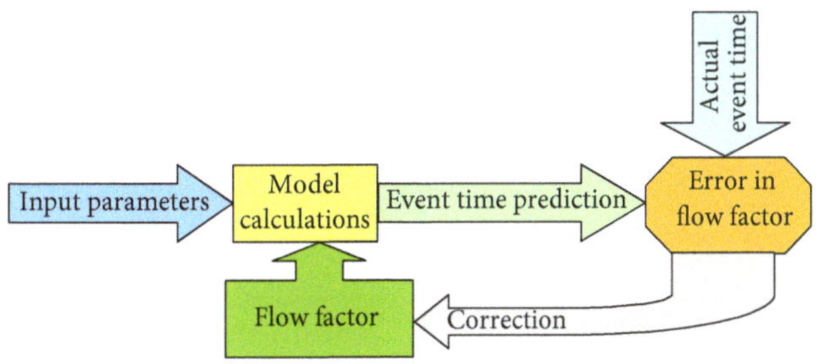

FIGURE 3: Flowchart for flow factor optimization.

flow factors, the correction factor is suggested as shown in Figure 3.

4. Results and Discussion

The results obtained by the model calculations are used for graphical representation of the liquid level variation in hearth as shown in Figure 4, which depicts the quantity of hot metal and slag present in the hearth at four distinct times. In the present study, it was observed that for the majority of cases the maximum liquid level in the hearth was 1.5 m above the taphole. Assuming the position of raceway bottom to be at about 1 m below tuyere center line, it still leaves a head space of 1.5 m between liquid level and tuyere. This modeled level matches well with the actual blast furnace practice, and, subsequently, smooth BF campaign is ensured as the blowing parameters never got disturbed due to raising liquid levels.

The volume flow rates and mass flow rates, when plotted individually for metal and slag against time duration as shown in Figures 5(a) and 5(b), reveal that there is a steep increase in volume flow rate for slag but equal steep decrease in slag mass flow rate after slag-out time (t_s = 64 min after tap open). This phenomenon can be attributed to the lower density of liquid

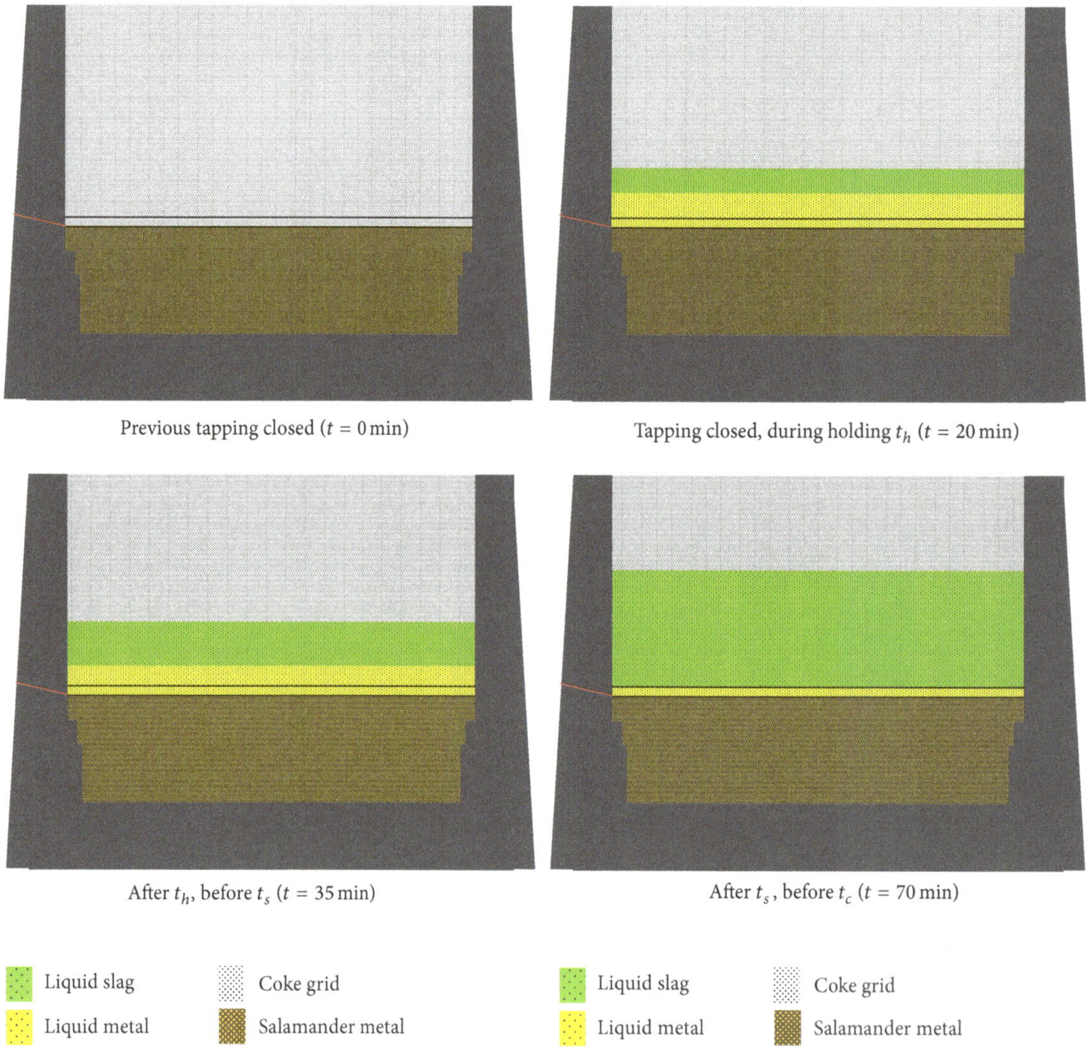

FIGURE 4: Model predicted liquid levels in the BF hearth at four distinct times.

slag than that of the metal. The same types of drainage trends are also observed in the plant under consideration. But for the combined plot for fluid flow, Figures 5(c) and 5(d) show an increase in both volumetric and mass flow rates during the slag flow part. It is due to the fact that whatever metal is produced it gets drained out with the slag which increases volume and mass of the tapped liquid. Calculated data plot for liquid metal and slag level variation in hearth against time, starting from the closing time of the last cast (i.e., including holding time t_h), is shown in Figure 6. It reveals that both liquid metal and slag level increase during the holding time (t_h = 20 min); after that, when taphole gets opened, the liquid metal only starts to come out from the hearth, and its level decreases, but the slag level continuously increases as no slag has drained until slag-out time (t_s = 84 min). During only the metal flow period, the volume of liquids drained remains less than the volume of liquids produced; hence, the total liquid level in hearth increases continuously with a reduced rate. At t_s, the total liquid level reaches at maximum due to accumulation of slag. After t_s, when slag starts coming out from the hearth, the total liquid level in hearth decreases, and, finally, at tapping close time (t_c = 138 min), there remains almost no liquid pool in hearth. By analyzing this trend, the casting practice can be tuned so that the total liquid level should not increase beyond the desired height at any point of time. It limits the buoyancy force exerted by the liquids and prevents the blast pressure from increasing due to excess liquid accumulation in hearth.

The model was tested on 227 castings of JSPL BF-II in October 2011, and it was found that the predicted slag-out time (t_s) and the cast closing time (t_c) are within a range of 10–15 minutes of the actual tapping data (Figure 7). The t_s was matching in 95% cases, whereas t_c was matching in 88% cases. The large deviations can be attributed to human interaction or some abnormality during the tapping. It is evident that, with further fine tuning, the present results and predictions can be utilized to optimize the blast furnace tapping operation for a stable casting behavior.

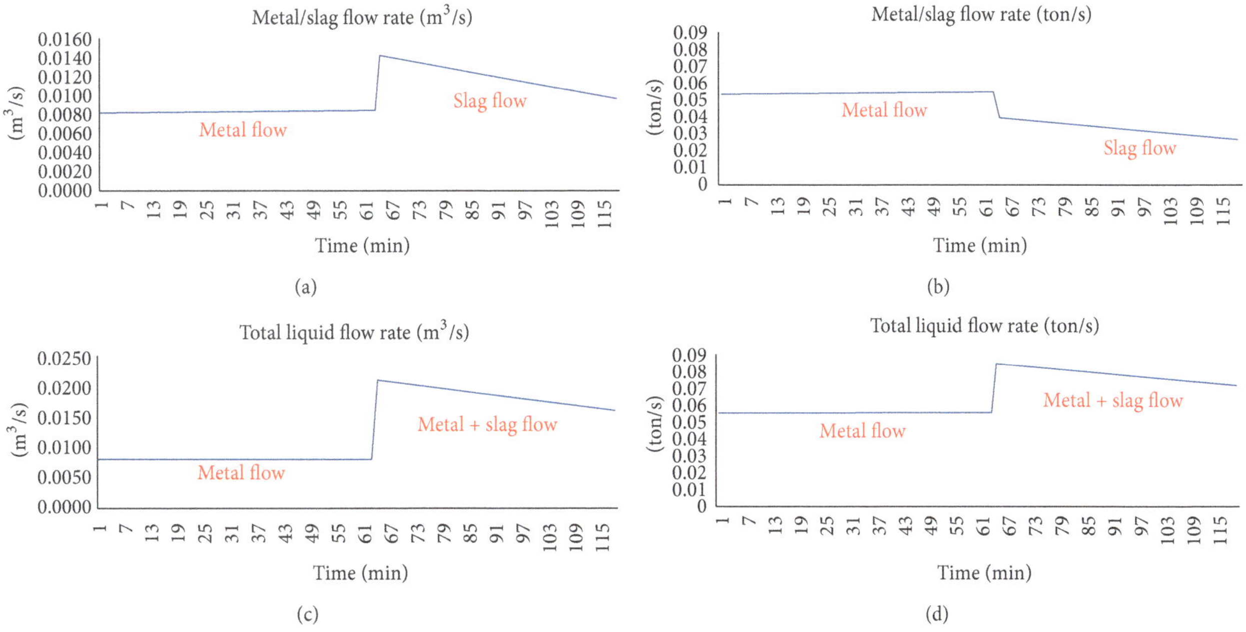

FIGURE 5: Metal and slag flow rates during tapping.

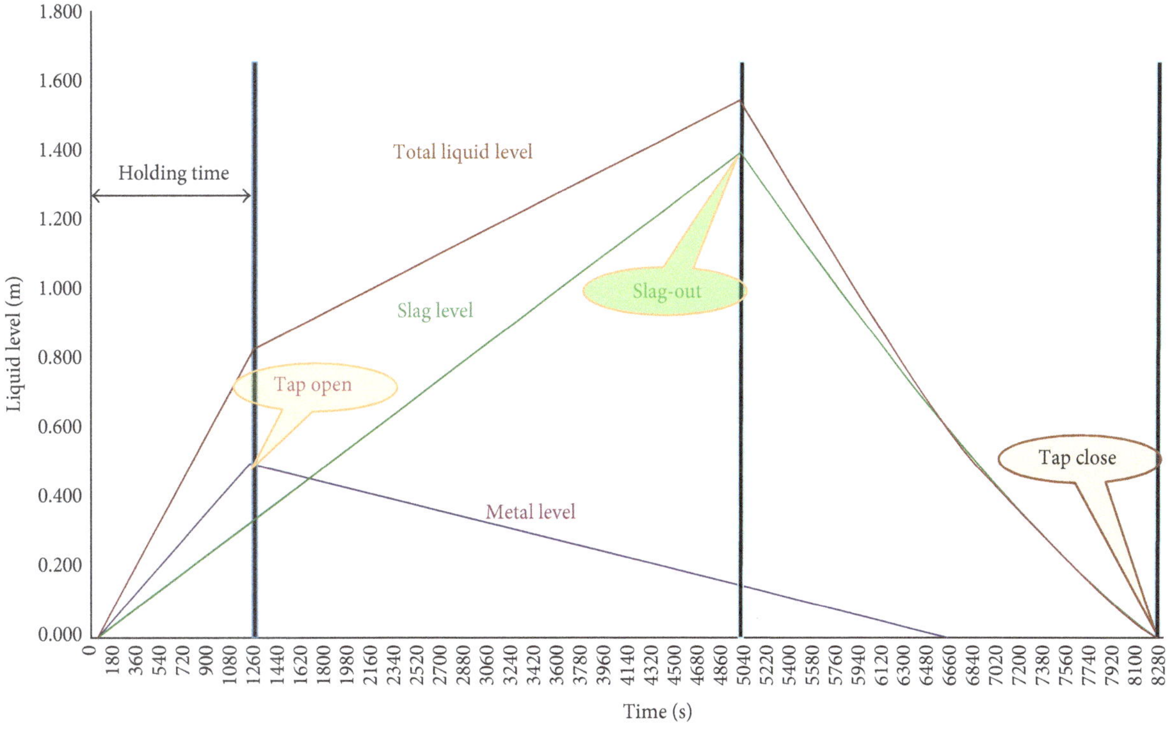

FIGURE 6: Variation in liquid levels in hearth during tapping.

5. Conclusion

A mathematical model on the drainage behavior has been developed which requires only the readily available data from an operating blast furnace. Effects of the unknown parameters, which vary from plant to plant, are taken into account by introducing flow factor concept and optimizing it. The developed model is very much useful as it provides the operator with a tool by which he/she can get a real-time view of the variation in the level of liquids in the hearth as well as the drainage behavior of the hearth. These data in turn help to control the parameters before they pose any threat to process stability. Our model predicts the slag-out time and the casting close time within acceptable range based on the

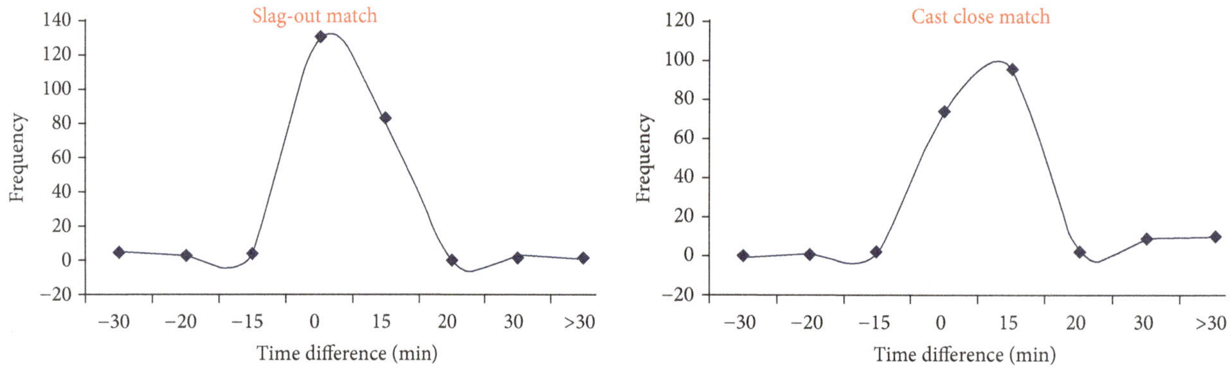

FIGURE 7: Comparison of the model predicted slag-out time and cast close time with the actual time.

trials conducted, and it can be used for process control and to judge irregularities in furnace operation.

References

[1] A. K. Biswas, *Principles of Blast Furnace Ironmaking*, SBA Publications, Kolkata, India, 1st edition, 2005.

[2] J. Brännbacka and H. Saxén, "Modeling the liquid levels in the blast furnace hearth," *ISIJ International*, vol. 41, no. 10, pp. 1131–1138, 2001.

[3] W. B. U. Tanzil, P. Zulli, J. M. Burgess, and W. V. Pinczewski, "Experimental model study of the physical mechanisms governing blast furnace hearth drainage," *Transactions of the Iron and Steel Institute of Japan*, vol. 24, no. 3, pp. 197–205, 1984.

[4] J. Brännbacka, J. Torrkulla, and H. Saxén, "Simple simulation model of blast furnace hearth," *Ironmaking and Steelmaking*, vol. 32, no. 6, pp. 479–486, 2005.

[5] T. Nouchi, M. Sato, K. Takeda, and T. Ariyama, "Effects of operation condition and casting strategy on drainage efficiency of the blast furnace hearth," *ISIJ International*, vol. 45, no. 10, pp. 1515–1520, 2005.

[6] K. Nishoka, T. Maeda, and M. Shimizu, "Effect of Various In-furnace conditions on blast furnace hearth drainage," *ISIJ International*, vol. 45, no. 10, pp. 1496–1505, 2005.

[7] W. M. Husslage, M. A. Reuter, R. H. Heerema, T. Bakker, and A. G. S. Steeghs, "Flow of molten slag and iron at 1500°C to 1600°C through packed coke beds," *Metallurgical and Materials Transactions B*, vol. 36, no. 6, pp. 765–776, 2005.

[8] W. B. U. Tanzil and W. V. Pinczewski, "Blast furnace hearth drainage: physical mechanisms," *Chemical Engineering Science*, vol. 42, no. 11, pp. 2557–2568, 1987.

[9] T. Nouchi, M. Yasui, and K. Takeda, "Effects of particle free space on hearth drainage efficiency," *ISIJ International*, vol. 43, no. 2, pp. 175–180, 2003.

CFD Model Study of a New Four-Port Submerged Entry Nozzle for Decreasing the Turbulence in Slab Casting Mold

Debasish Chatterjee

Department of Metallurgy & Materials Engineering, Indian Institute of Technology, Roorkee 247667, Uttarakhand, India

Correspondence should be addressed to Debasish Chatterjee; deb100dmt@gmail.com

Academic Editors: Y. Hao, Y.-d. Wang, and T. Yue

In recent years, the continuous casting process in which molten metal is constantly cast into semifinished shapes has been widely adopted in the steel and nonferrous metal producing industry, mainly due to its economic advantages associated with increased yields and the elimination of intermediate processing steps. Turbulent flow in submerged entrance nozzle and mold of continuous casting is associated with costly failure such as the entrainment of slag inclusions and the formation of surface defects. A number of previously studies have analyzed the flow. A CFD model study has done to decrease the turbulence within the mold by developing a 4-port submerged entry nozzle in comparison with conventional 2-port submerged entry nozzle. VOF (volume of fluid) simulation study has been done to find out the effect of turbulence within mold on slag metal interaction, level fluctuation, and risk of breakout. In all the cases the performance of new 4-port SEN over 2-port SEN has been compared by plotting graph.

1. Introduction

Steel maker around the world is emphasized to increase the production rate to cope with the global demand of steel in coming decade. Simultaneously the supply of quality steel is demanding by customer. But to meet this stringent demand is very difficult as production machine has certain drawbacks and limitations. One of these drawbacks is creation of turbulence within mold during casting of steel at high casting speed. The submerged entry nozzle (SEN) with two bifurcated ports is commonly applied in slab continuous casting. As the source of the flow, the SEN decides on the steel flow pattern in the mold. For unreasonable nozzle design, impingement of hot liquid metal with high momentum against the solidifying shell can contribute to shell thinning and costly breakout as well as large surface fluctuation [1]. Scientists have been interested in the fluid flow that is observed in the mold during continuous casting because the flow pattern affects the inner and the surface quality of the steel slab to a great extent. Figure 1(a) shows the basic phenomenon produced in the mold during the continuous casting. The flow pattern depends on the structure of the submerged entry nozzle (SEN) and other relative parameters. On the one hand, a calm and smooth mold level assists the uniform formation of the strand shell in circumferential direction and reduces the thermomechanical stresses in the strand shell. It also allows the homogeneous intake of casting powder into the lubrication gap between the strand shell and the mold wall and this, in turn, is essential for optimum lubrication and homogeneous heat transfer [2]. Several authors have studied fluid flow. Both physical and mathematical models have been developed to access the flow pattern and the surface waves. Turbulent flow and heat transfer coupled with solidification in slab continuous casting, particle image velocimetry (PIV) technique mold has been studied [1, 2]. Turbulent flow in SEN by large eddy simulation, turbulent flow in mold during ladle change operation, decreasing turbulent flow by use of EMBR, effect of submergence depth of SEN on mold turbulence, and so forth have been studied by many researchers in past [1–8]. The current research is to decrease the turbulence in the mold to increase the product quality and indirectly increasing the yield of caster without hampering the normal casting process.

2. Present Work

Turbulent flow in submerged entrance nozzle and mold of continuous casting is associated with costly failure such as the entrainment of slag inclusions and the formation of surface

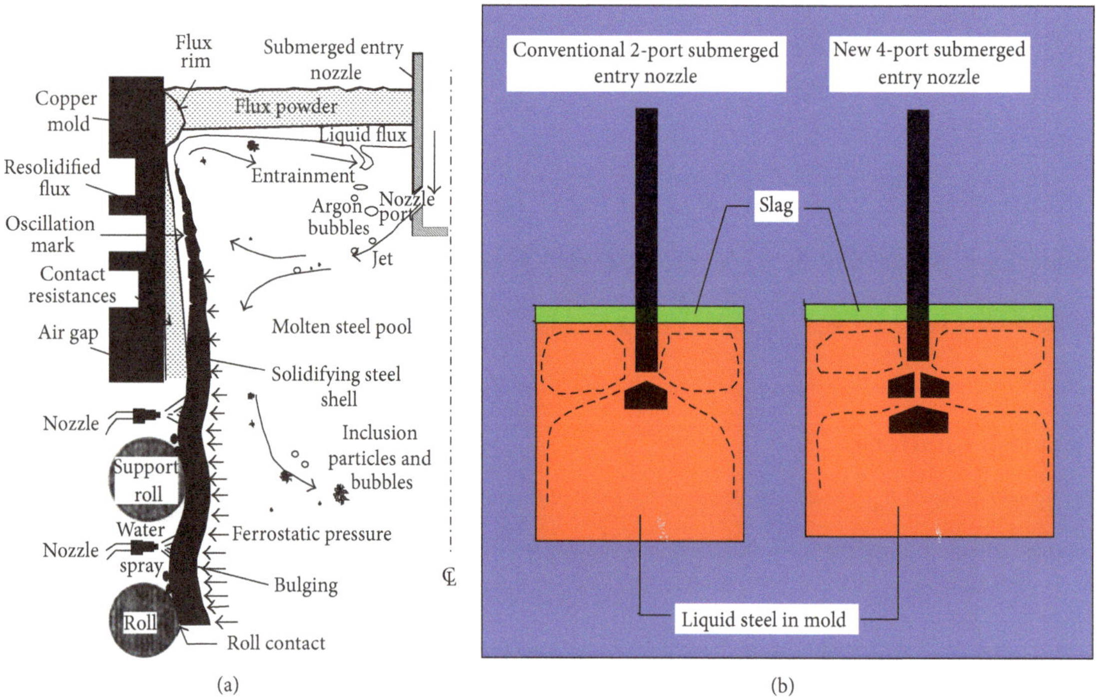

FIGURE 1: (a) Schematic of flow phenomena in mold of continuous casting [2] and (b) schematic of 2-port and new 4-port SEN.

defects [3]. The mold is one of the most important components of a continuous caster. It is known that the quality of metal castings is affected by a number of parameters directly related to the turbulent transport processes taking place within the mold. An understanding of transport processes in the liquid metal pool is of particular importance because it can be used in controlling the growth rate of the solidifying shell, thus affecting the microstructure of the final casting. The transport processes also control the formation of various defects, which can hamper subsequent processing. Examples of defects formed in this region are macrosegregation, surface irregularities, such as cold folding and oscillation marks, and more serious defects such as longitudinal or transverse cracks. A more catastrophic problem, which may occur, is "breakout," where the shell ruptures and molten metal is discharged in an uncontrolled manner [4].

The present work has developed a new 4-port submerged entry nozzle over conventional 2-port submerged entry nozzle which is schematically shown in Figure 1(b). The objective is to lower the turbulence within the mold. The conventional bifurcated SEN has 2 holes symmetrically opposite position embedded in a vertical cylindrical nozzle. The 4-port nozzle has 4 holes symmetrically opposite position embedded in the upper and lower part of a vertically cylindrical nozzle. The two parts are connected each other by a vertical cylindrical hole which has less diameter than the diameter of the main central nozzle. When the same volume of fluid will eject from 4 holes (each hole diameter of 4-port SEN = each hole diameter of 2-port SEN), it will experienced a less velocity compare to 2-port SEN. In the present work computational fluid dynamics software Fluent has been used to study the turbulence phenomena of 4-port nozzle and it has been compared with 2-port SEN.

3. Results and Discussion

3.1. Velocity Profile. The contour of velocity has been plotted in Figures 2(a) and 2(b) for 2-port and 4-port submerged entry nozzle, respectively. A 1000 mm wide, 220 mm thick, and 900 mm long mold was used for simulation. The submerged entry nozzle having internal diameter 60 mm and port of 40 mm diameter was used. It was immersed 150 mm in the liquid. The simulating liquid was water whose kinematic viscosity is equal to liquid steel. The casting speed was taken 1.6 m/min. the velocity of flowing water through the submerged entry nozzle was 2 m/sec. When the fluid impinges on the dome of the SEN, it flows through bifurcated nozzle for 2-port condition. The fluid is hitting the mold side wall and creating an upper and downward motion. It is creating turbulence within the mold. The boundary condition used for this simulation was velocity inlet at the upper entry of SEN. The lower part of the mold was outflow boundary condition as amount of liquid through the SEN to the mold will be equal to exact amount of metal solidified as going to out from old as semifinished product. The other side of the mold used as wall boundary condition. The velocity has been plotted along X and Z direction in Figures 2(c) and 2(d). It has been found out that the velocity of flowing fluid for 2-port SEN has high value as compared to 4-port SEN. It means that 4-port nozzle will provide less turbulence in mold compare to 2-port nozzle.

3.2. Slag Metal Interaction in Mold. The VOF (volume of fluid) simulation has been done to simulate the slag metal interaction within the mold. Here 4 phases have been used. The phases are air (primary phase), slag (ice water mixture), steel (water has been used as simulating material), and tracer (water). At the onset the slag was stagnant. After flowing

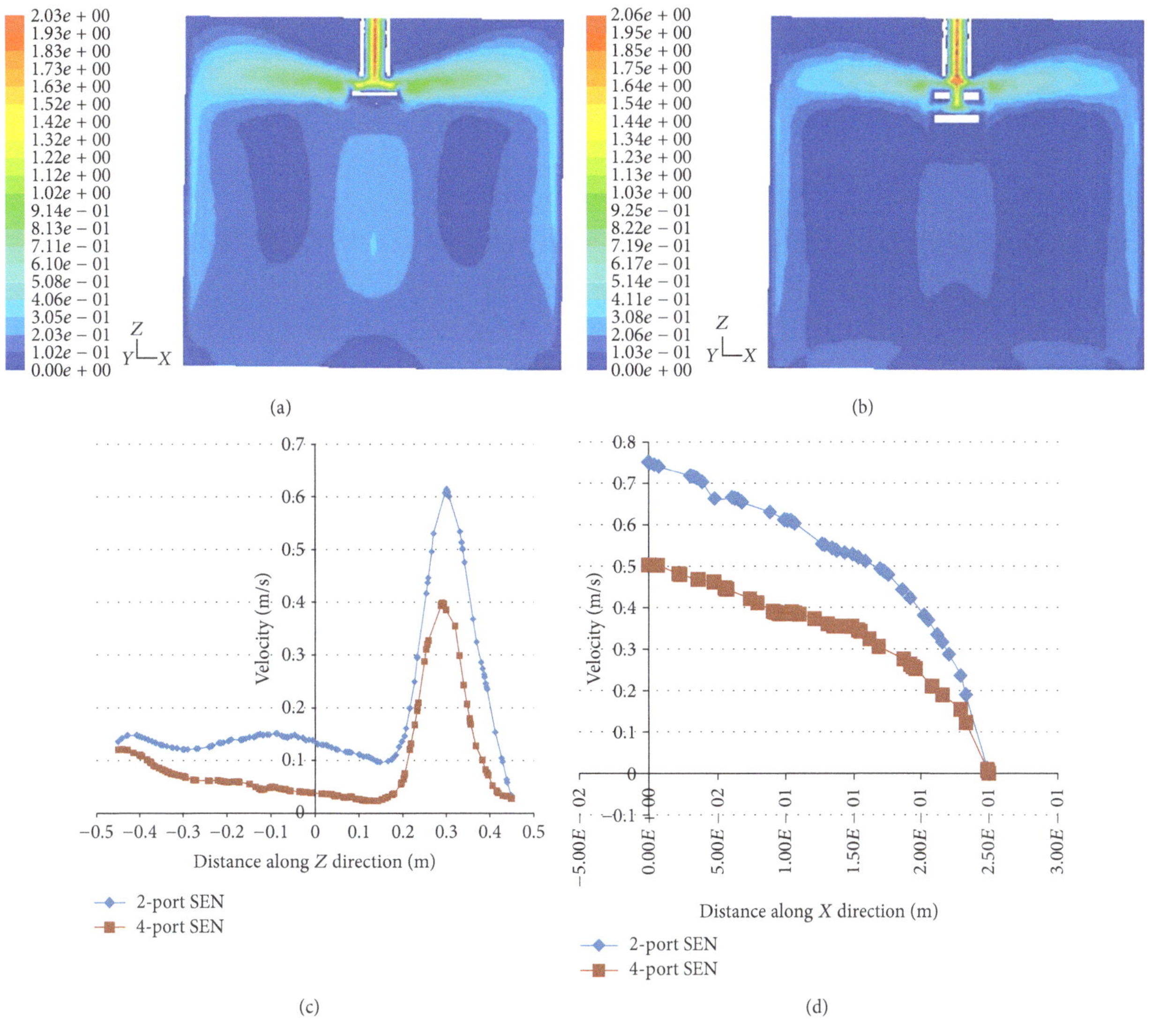

FIGURE 2: (a) Velocity contour of 2-ports SEN, (b) velocity contour of 4-port SEN, and (c) velocity plot along Z direction, and (d) velocity plot along X direction.

fluid through the mold upper slag phase shape was changed. For 2-port SEN (submerged entry nozzle), the shape was much distorted as compared to 4-port SEN. It has been demonstrated in Figures 3(a), 3(b), 3(e), and 3(f).

Slag eye has great effect on the quality of cast steel. As mold is the last metallurgical reactor, slag has significant role to supply quality of steel to customer. Due to opening of slag eye reoxidation of the melt takes place within the mold which ultimately causes severe sliver defects after hot rolling. Slag act as a layer which capture the inclusion generated by previous subsequent processes. But due to turbulence and overflow, slag cannot remove the inclusion from the upcoming stream of steel. Instead it increases the number of large size (>50 micro meter) inclusions within the mold. As the slag eye opened for 2-port nozzle, so it has detrimental effect on the steel. But compared to 4-port nozzle, no opening of slag eye was observed in the computational simulation results. It has been shown in the Figures 3(c) and 3(g).

Figures 3(d) and 3(h) show volume fraction of the incoming fluid within the mold after certain time elapsed. Not only that it showed the tracking of fluid within the mold. The fluid which ejected from the submerged entry nozzle impacted with mold side wall with a high speed. It creates an upper and lower flow of thin layer of fluid along the side wall of the mold. For the case of 2-port nozzle as compared to 4-port nozzle, the impinging velocity was high which is evidence from the large amount of fluid was climbing along the side wall of the mold. When this upward flow fluid interacts with the slag layer, an opening of slag eye was formed. Beside this, the mold that contains 2-port nozzle shows the presence of scattered volume of tracer fluid in the mold. It is not observed for the case of mold containing 4-port nozzle. It signifies that 2-port SEN creates more turbulence in the mold as compared to 4-port SEN.

3.3. Surface Fluctuation. Higher casting speed of liquid steel has been an important requirement in the steel-making

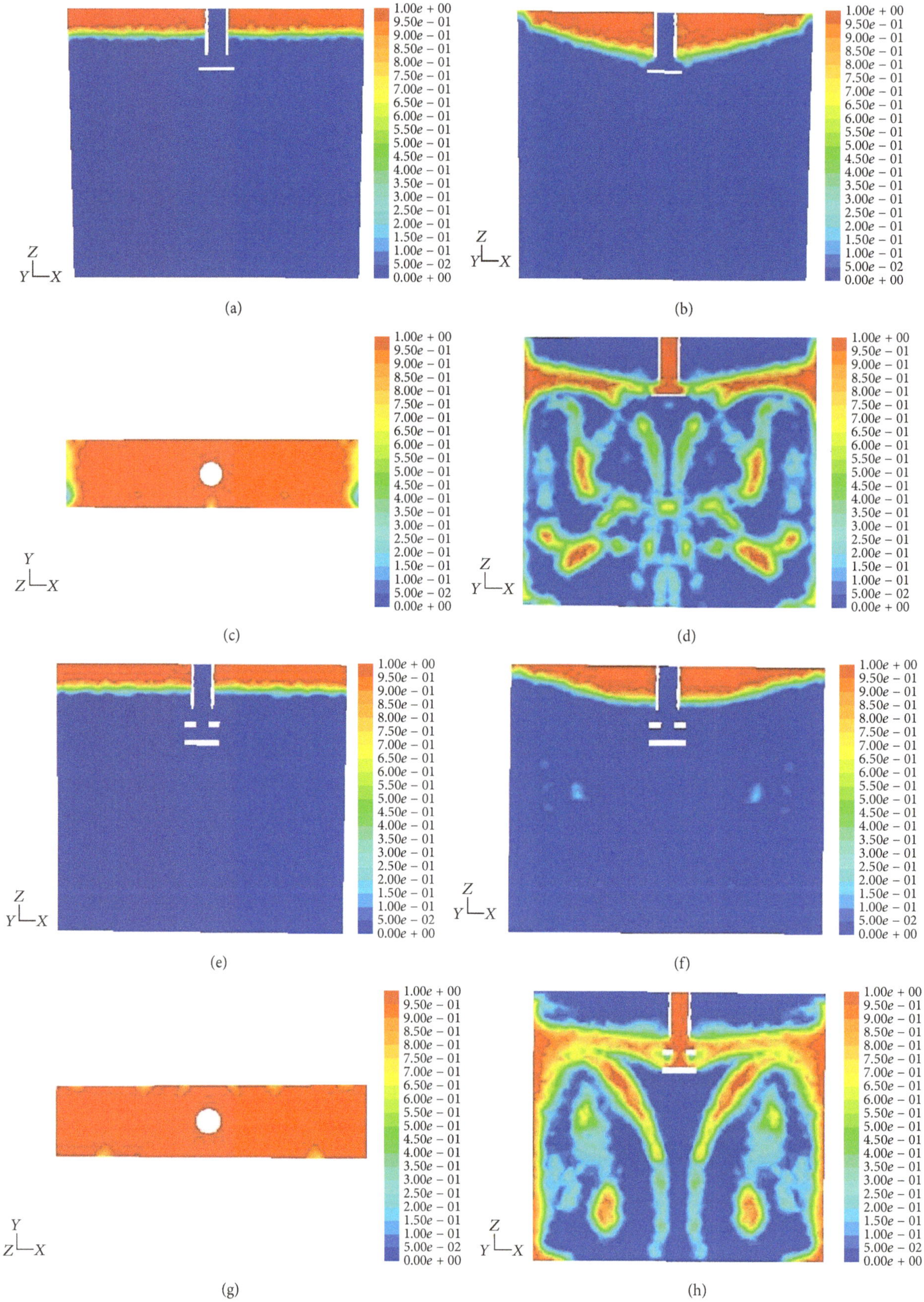

FIGURE 3: 2-port SEN: (a) slag volume before flow, (b) after flow, (c) slag eye, and (d) volume fraction of incoming fluid. 4-port SEN: (e) slag volume before flow, (f) after flow, (g) no slag eye, and (h) volume fraction of incoming fluid.

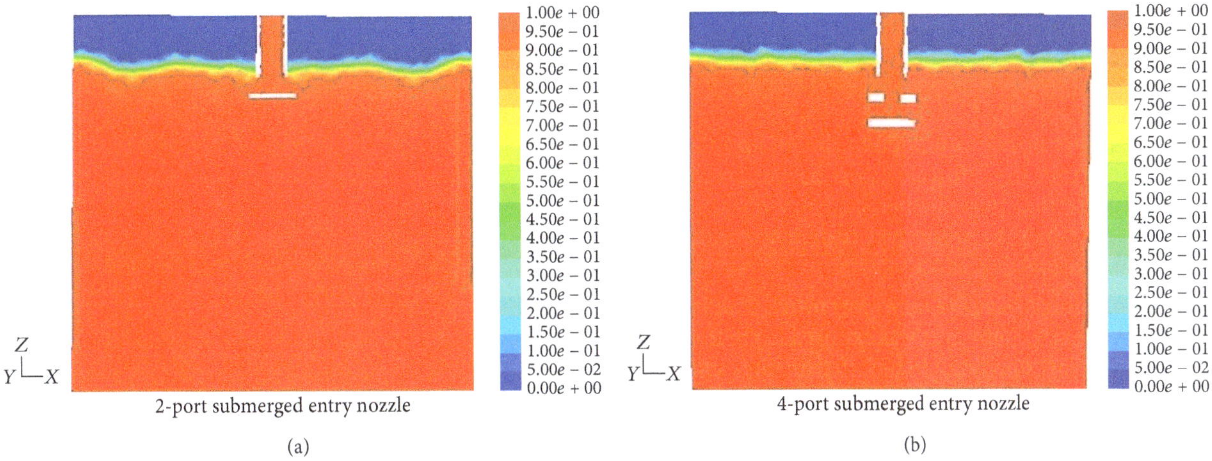

FIGURE 4: (a) Level fluctuation of 2-port SEN and (b) level fluctuation of 4-port SEN.

industry. In a continuous slab caster, a higher casting speed may lead to several problems such as higher meniscus turbulence and frequent vortex formation, due to which there is entrainment of slag into the liquid steel. This may lead to a compromise with the quality of steel produced. Thus, modeling and CFD analysis of free-surface fluctuation become necessary for design of submerged entry nozzles [5]. There is a close correlation between the fluctuation of the mold steel level and the occurrence of surface defects of final products: it is considered that when the mold steel level fluctuates significantly, casting powder and other impurities floating on the surface of molten steel are entrapped in steel, and they appear in the form of surface defects of steel sheet products during rolling. To prevent this, mold level control is designed so as to minimize the fluctuation of the mold steel level.

A new 4-port submerged entry nozzle has been designed over conventional 2-port submerged entry nozzle. The VOF (volume of fluid) simulation has been done to capture the level fluctuation within the mold. It is shown in Figures 4(a) and 4(b). Here only 2 phases have been simulated. One phase is air which is primary phase and another phase is water which is secondary phase. After certain time interval (6-7 sec), it was observed from simulation plot that the surface of the 2-port mold was much more zigzag. But after the same time interval, 4-port SEN mold surface was quiescent and no distortion on the surface was observed. The upper blue colour was air phase and the red colour was water phase. So it is concluded that the 4-port nozzle will help to produce much cleaner steel as compared to 2-port nozzle. The entrainment of slag within the melt in the mold will be less for 4-port submerged entry nozzle as compared to 2-port submerged entry nozzle.

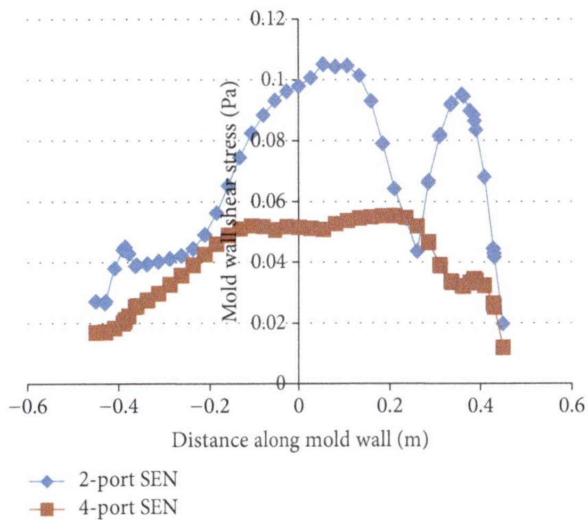

FIGURE 5: Typical shear stress curve along the mold wall for 2-port and 4-port nozzle.

3.4. Risk of Breakout. In the continuous casing production, breakout accident is the most typical and greatest damage industrial accident, especially, sticking type breakout. It is reported that the loss of once reached up to several million money. Breakout phenomenon is caused by various factors that deviate from the steady state and interaction; its mechanism is more complex. To prevent such accidents, people have done a lot of experimental research and take appropriate measures on equipment and technology; however, breakout accident still happened some times. Due to breakout damage of equipment takes place. Not only that liquid steel loss from tundish as well as from ladle. Beside this, it takes several hours to restart the continuous casting machine which ultimately hamper the production and decrease the yield of the caster.

A typical shear stress curve has been plotted along the side wall of mold for 2-port and 4-port nozzle which is shown in Figure 5. The shear stress is generated due to interaction of high speed incoming fluid from submerged entry nozzle to the side wall of the mold. The shear stress is much more for case of 2-port nozzle as compared to 4-port submerged entry nozzle. This signifies that the susceptibility of breakout for 2-port nozzle will be more than 4-port nozzle as it will thin the solidified layer much more as compared to 4-port submerged entry nozzle.

4. Conclusions

The following conclusions can be drawn from the above study of the simulation.

(i) The 4-port nozzle is unique in design and the fluid ejected from the bifurcated port has less velocity as compared to 2-port submerged entry nozzle. It will indirectly help to create less turbulence within the mold.

(ii) The opening of slag eye for 4-port nozzle is nil. But for 2-port nozzle, some area was exposed to environment due to slag opening. It has detrimental effect on product quality due to generation of more inclusion within the cast structure.

(iii) The surface fluctuation is more for 2-port SEN where surface fluctuation is almost nil for 4-port submerged entry nozzle. It means that chances of entrainment of slag particles to the melt are less for 4-port SEN.

(iv) Shell thinning will be less for 4-port nozzle. So chances of breakout will be less for 4-port SEN.

References

[1] D. F. Wu and S. S. Cheng, "Effect of SEN design on surface fluctuation and solidifying shell in slab mold and its optimization," *Acta Metallurgica Sinica (English Letters)*, vol. 21, no. 5, pp. 341–350, 2008.

[2] X.-G. Zhang, W.-X. Zhang, J.-Z. Jin, and J. W. Evans, "Flow of steel in mold region during continuous casting," *Journal of Iron and Steel Research International*, vol. 14, no. 2, pp. 30–41, 2007.

[3] C.-B. Ji, J.-S. Li, S.-F. Yang, and L.-Y. Sun, "Large eddy simulation of turbulent fluid flow in liquid metal of continuous casting," *Journal of Iron and Steel Research, International*, vol. 20, no. 1, pp. 34–39, 2012.

[4] X. K. Lan and J. M. Khodadadi, "Fluid flow, heat transfer and solidification in the mold of continuous casters during ladle change," *International Journal of Heat and Mass Transfer*, vol. 44, no. 5, pp. 953–965, 2001.

[5] R. Kumar Das and S. Kumar Dash, "Free-surface fluctuations for design of submerged-entry nozzle," *International Journal on Theoretical and Applied Research in Mechanical Engineering*, vol. 1, no. 2, pp. 109–114, 2012.

[6] Y. Haiqi, W. Baofeng, L. Huiqin, and L. Jianchao, "Influence of electromagnetic brake on flow field of liquid steel in the slab continuous casting mold," *Journal of Materials Processing Technology*, vol. 202, no. 1–3, pp. 179–187, 2008.

[7] W. Zhang, J. Gao, P. K. Rohatgi, H. Zhao, and Y. Li, "Effect of the depth of the submerged entry nozzle in the mold on heat, flow and solution transport in double-stream-pouring continuous casting," *Journal of Materials Processing Technology*, vol. 209, no. 15-16, pp. 5536–5544, 2009.

[8] X. Miao, K. Timmel, D. Lucas, Z. Ren, S. Eckert, and G. Gerbeth, "Effect of an electromagnetic brake on the turbulent melt flow in a continuous-casting mold," *Metallurgical and Materials Transactions B*, vol. 43, pp. 954–972, 2012.

10

Alloy Designation, Processing, and Use of AA6XXX Series Aluminium Alloys

Prantik Mukhopadhyay

Electron Microscope Group, Defence Metallurgical Research Laboratory, Kanchanbagh, Hyderabad 500058, India

Correspondence should be addressed to Prantik Mukhopadhyay, prantikmukherje@yahoo.com

Academic Editors: R. Hebert and Y. Yamabe-Mitarai

The strength-to-weight ratio offered by AA6XXX alloys and their enhanced mechanical properties have become crucial criteria for their use in light weight military vehicles, rockets, missiles, aircrafts, and cars, used for both defence and civil purpose. The focus of this review paper is to put together the latest knowledge available from various sources on alloy design, industrial processing, development of properties, and potential use of AA6XXX alloys. The direct chill (DC) cast AA6XXX wrought alloys which are subsequently processed by fabrication process like hot working, cold working, process annealing, and age hardening heat treatments are the foci of this review though designation section also contains the designations of cast alloys to provide the reader a broad overview on designation. World-wide accepted designations are briefly tabled along with their alloying elements. The effects of the alloying elements which are generally used for AA6XXX wrought alloys are discussed incorporating their interactions during wrought AA6XXX alloy fabrication. The significance of the alloying and also the processing to develop the certain properties and the underlying strengthening mechanisms are discussed. The frequent and versatile uses of these AA6XXX alloys for the structural applications both in defence and civil purpose are put forth.

1. Introduction

The properties of AA6XXX wrought alloys like high strength-to-weight ratio, plasticity, capacity for crucial shape forming along with their ease for joining and good corrosion resistance have become increasingly focused for versatile applications such as design of armour structures, rocket, missile casing, light-weight defence vehicle, cars, and marine structures [1–5]. Varieties of AA6XXX wrought alloys have been developed by manipulating their compositions and tailoring the fabrication technology for the above-mentioned applications [2–10]. The requirements for producing high-quality AA6XXX wrought alloys are inevitably associated with precision alloy design, innovative technology development in processing and enhancement of mechanical properties [1, 2]. The genesis of these upgraded Taylor-made properties demands the knowledge of underlying mechanisms which govern the strengthening phenomenon [3–10]. The focus of this review paper is to put together the latest knowledge available from various sources on alloy designation, industrial processing, development of properties, and potential use of AA6XXX wrought alloys.

2. Designation

Aluminium offers weight saving because of its light weight. The softness of this metal can restrict its use for the engineering purposes. Hence, the strengthening of this metal is required for its use in cars, light-weight armour vehicles, rockets, missiles, and aircraft structures in civil and defence sectors, which demand high strength-to-weight ratio. The strengthening of Al is principally done by alloying it with elements like Cu, Zn, Mn, Mg, Si, and Li and processing its alloys. Generally, a four-digit numerical designation is used to specify aluminium and its alloys (Table 1), where first digit denotes the principal alloying element except AA1XXX series, which stands for purity of aluminium. The second digit stands for the changes to impurity limits. The minimum %Al (for AA1XXX series) and different aluminium alloys in

Table 1: AA Designation of wrought aluminium and its alloys [1, 2].

Alloying elements	Series designation
Pure aluminium	AA1XXX
Copper	AA2XXX
Manganese	AA3XXX
Silicon	AA4XXX
Magnesium	AA5XXX
Magnesium and silicon	AA6XXX
Zinc	AA7XXX
Lithium	AA8XXX

Table 2: AA Designation of cast aluminium and its alloys [1, 2].

Alloying elements	Series designation
Pure aluminium	AA1XX.X
Copper	AA2XX.X
Silicon, with copper and/or magnesium	AA3XX.X
Silicon	AA4XX.X
Magnesium	AA5XX.X
Zinc	AA7XX.X
Lithium	AA8XX.X

the group (for other series) are represented by the third and fourth digits [1, 2].

The castings and foundry ingots of Al and its alloys are designated (Table 2) by a four-digit number with the incorporation of a decimal point. Number 1 in the first digit stands for aluminium with high purity. The second and third digits are used for minimum Al percentage. The fourth digit is used after decimal point and it is 1 for cast ingot and 0 for casting.

2.1. Effect of Alloying Elements in AA6XXX Alloys. Pure aluminium is soft and formable. Aluminium is frequently alloyed in order to increase strength. The alloying also affects the weldability, corrosion resistance, stress corrosion cracking, conductivity, and density. AA6XXX aluminium alloys can be strengthened by the following major strengthening mechanisms along with the solid solution strengthening:

(i) grain-boundary strengthening [3, 4],

(ii) precipitation or age hardening [3, 4],

(iii) work hardening [3, 4].

The AA6XXX alloy design [1, 2, 5] aims at significant enhancement of properties. The properties depend on microstructure of metals and alloys. Hence, the final goal of the alloy design is to design a perfect microstructure so that the principal goal such as strength, ductility, and toughness along with other properties like corrosion resistance, stress corrosion cracking resistance, and fatigue resistance are enhanced. The role of composition (Table 3) of metals and alloys is not insignificant because it provide the suitable structure with enhanced properties.

The major alloying elements of AA6XXX series alloys are Si and Mg. The addition of these two alloying elements is done in the proper ratio to form Mg_2Si valence compound. Mg/Si ratio of 1.73 is required for the formation of magnesium silicide (Mg_2Si). Generally, AA6XXX alloys, produced by DC casting techniques, contain either extra Si or extra Mg. Presence of excess Si produces higher strength and higher formability but incorporates the intergranular corrosion tendency while excess Mg increases the corrosion resistance but reduces formability and strength. The formation of Mg_2Si precipitates give rise to the simple eutectic system with aluminium. At elevated temperature, the solute element dissolves in the solid solution but because of the decrease in solubility at lower temperature forms age-hardenable Mg_2Si precipitates.

Copper has appreciable solubility and strengthening effect. The addition of Cu, in variable concentrations, produces substantial solid solution and precipitation strengthening. In presence of magnesium and silicon, copper produces age-hardening effects at room temperature. The weldability and corrosion resistance are decreased and the weight of the alloy is increased with the copper addition.

During the extraction processing of aluminium from alumina trace elements like Fe, Na, Li, Ca, Ti, V, Cr, and compounds such as oxides, nitrides, borides, carbides, and H can go into the structure. All traces are not harmful for the properties of Al and its alloys. Elements like Cr, Ti, Li play crucial role in properties enhancement and sometime additions of those elements are deliberate. The deliberate additions of trace elements like Cr, Mn, Zr, and Sc can restrict the softening mechanisms during elevated temperature deformation. The fine grain structure ensures strengthening by grain refinement and work hardening. The trace elements show significant role (Table 4) in altering the physical, mechanical, and corrosion behavior of Al and its alloys.

The presence of trace elements Fe in AA6XXX alloys, as $FeAl_3$, Fe_2SiAl_8, $FeMg_3Si_6Al_8$, and $FeAl_6$ can be harmful for the properties of these alloys. Iron hinders ductility and toughness by the formation of coarse constituent with Al and other alloying elements such as Cu, Ni, Mn, and Si. Iron reduces the strength and decreases the corrosion resistance and fatigue resistance. Additions of Cr and Mn alloying elements are done as the Fe corrector of AA6XXX alloys to correct the shape and size of the iron-bearing precipitates, while restricted addition of Zn can cause extra strengthening without much loss in corrosion resistance. The presence of Manganese not only raises the recrystallization temperature but also shows strengthening by both the dispersion hardening and the age hardening. It also enhances the corrosion resistance.

Alloying elements B, Ti in some AA6XXX alloys cause the control of grain size and produce grain boundary strengthening of those alloys. The machinability can be higher by the addition of Bi and Pb in AA6XXX alloys. The solid solution strengthening to some extent can be caused by lattice parameter control in commercial alloys due to presence of Mg and Si in solid solution.

TABLE 3: Composition of AA6XXX alloys [6–8].

Designation	Si, %	Mg, %	Cu, % maximum	Mn, % maximum	Cr, % maximum	Others, %
AA6003	0.35–1.0	0.8–1.5	0.10	0.8	0.35	—
AA6005	0.6–1.0	0.4–0.6	0.10	0.10	0.10	—
AA6053	0.44–1	1.1–1.4	0.10	—	0.15–0.35	—
AA6061	0.4–0.8	0.8–1.2	0.15–0.40	0.15	0.04–0.35	—
AA6063	0.2–0.6	0.45–1.0	0.10.	0.10	0.10	—
AA6066	0.9–1.8	0.8–1.4	0.7–1.2	0.6–1.1	0.40	—
AA6070	1.0–1.7	0.50–1.2	0.15–0.40	0.4–1.0	0.10	—
AA6101	0.3–0.7	0.35–0.8	0.10	0.03	0.03	B 0.06% maximum.
AA6105	0.6–1.0	0.45–0.8	0.10	0.10	0.10	—
AA6151	0.6–1.2	0.45–0.8	0.35	0.20	0.15–0.35	—
AA6162	0.4–0.8	0.7–1.1	0.20	0.10	0.10	—
AA6201	0.5–1.0	0.6–1.0	0.10	0.03	0.03	B 0.06% maximum.
AA6253	0.4–1.05	1.0–1.5	0.10	—	0.04–0.35	Zn 1.6–2.4%
AA6262	0.4–0.8	0.8–1.2	0.15–0.40	0.15	0.04–0.14	Pb and Bi 0.4–0.7% each
AA6351	0.7–1.3	0.4–0.8	0.10	0.4–0.8	—	—
AA6463	0.2–0.6	0.4–1.0	0.20	0.05	—	—
	0.2–0.5	0.4–0.8	0.15–0.40	0.10	—	—

TABLE 4: Trace elements in Al and its alloys [2].

Trace elements	Concentration (wt %)	Properties
H	0.0001	Presence of S, Mg and Li increase the absorption of H while Si, Sn, Cu, and Be reduce it. Formation of gas porosity after solidification.
Fe and Si	0.1–1.0	Formation of Al_3Fe and AlFeSi second phase compounds, which reduce the fatigue and fracture properties of Al alloys.
Na, Li and Ca	<0.01	These elements enhance the Al-Si eutectic structure resulting in higher strength properties and pressure tightness effect. Rolling might be difficult because of formation of edge cracking tendency. Li increases the strength at elevated temperatures, maintaining low density.
Ti, V, Cr, Mn	0.001–0.05	Cause grain refinement during solidification, retard recrystallization during deformation. Presence of Mn also acts as Fe-corrector correcting the shape of Fe bearing precipitates.
Ga	0.001–0.02	The corrosion behavior, etching, and brightening of some alloys are altered.

3. Processing of AA6XXX Alloys

Processing of Al and its alloys plays very crucial effect on the property determination. As earlier explained properties of Al alloys principally depend on their microstructure and their constituents such as average grain size, grain size distribution, precipitate volume fraction, and number fraction and the crystallographic orientation aspects like texture of Al and its alloys [6–10]. The composition is designed to render certain structure and texture in Al alloys after solidification but such structure and texture can further be developed by incorporating different processing methods like metal working fabrications such as rolling, forging, extrusion, wire drawing, forming, annealing, and age hardening. The manipulation of process parameters in metal working, subsequent annealing, and aging treatment shall be able to produce diverse structures with same composition. That manifests the significance of work hardening and microprocesses such as recovery, recrystallization, and grain growth to design the desired properties of Al alloys.

Hence, composition and processing both are inseparable tools for developing the desired structure with enhanced mechanical properties. The details of processing of Al alloys are reflected by the following temper designation (Table 5) letters, which follows the alloy designation [2, 11, 12]. This temper designation is used for all forms of wrought and cast Al alloys except ingot and it is based on the sequence of processing used to produce the following tempers such as fabrication, work hardening, annealing, solution, and stabilization treatment.

The temper designation H, which is suitable for wrought products, is not only involved with the work hardening process but also involved with two other elevated temperature treatments, namely, subsequent partial recrystallization annealing and stabilization heat treatments for Al alloys, which are liable to soften by over ageing process. The subdivisions of work hardening (H) are following (Table 6).

Apart from the heat-treatment process involved in H temper, separate heat-treatment process for cooling from elevated temperature, natural, artificial, step aging process,

Table 5: Basic temper designation of Al alloys [2, 11, 12].

Temper	Treatment	Use
F	Fabricated	The products fabricated by casting, hot working, cold working processes without control over elevated temperature treatment or work hardening is used.
O	Annealed	The castings which are annealed to enhance ductility and dimensional stability. The wrought semis which are annealed to obtain lowest strength temper.
H	Strain hardened	The wrought products only which are work hardened with or without supplementary elevated temperature treatments to get desired ductility at lower strength. The H is always followed by two or greater than two digits.
W	Solution treated	The alloys which age at room temperature after solution heat-treatment to produce unstable temper. This designation is proper only when the time of the aging is indicated.
T	Stable tempers	The products which show stable tempers after heat treatment with or without supplementary work-hardening. The T is always followed by one or several digits.

Table 6: Subdivision of H temper [2, 11, 12].

Subdivisions	Treatment	Use
H1	Work hardened	Only work hardened products without elevated temperature treatment. The H1 is followed by digits, which show the degree of work hardening.
H2	Work hardened and partially annealed	Work hardened products are subsequently partially annealed associated with reduction of strength. The H2 is followed by digits, which show the residual work hardening after partial annealing.
H3	Work hardened and stabilized	The properties of work hardened products are stabilized either by low temperature treatment or by the heat introduced by fabrication. The stabilization of properties is crucial to restrict the age softening of these products. The H3 is followed by digits, which show the remaining work hardening after stabilization.

the combination of stress relieving, and step aging is designated by T temper. The digit following the letter T indicates the type of heat treatment used in Al alloys. The description of T temper in terms of treatment and use is shown in Table 7.

3.1. Homogenization. The first processing step after melting and DC casting is known as homogenization of AA6XXX alloys. The homogenization treatment of castings and ingots is performed at elevated temperature, (460°C–550°C) which reduces the micro segregation in the castings and ingots, and this can produce compositional homogeneity [13–16]. This homogenization treatment of microstructure plays significant role on the subsequent processing stages such as hot working, cold working, and annealing. The homogenization treatment produces the uniform composition and is feasible when the diffusion distances are short and dendrite cell size and arm spacing are not big. The homogenization heat treatment contains the following phenomena in Al alloys [13–16]:

(i) uniformity in composition,

(ii) precipitation of supersaturated elements,

(iii) reduction of microsegregation,

(iv) dissolution of unstable phases or precipitates,

(v) size enlargement of stable intermetallics,

(vi) grain growth.

The nonuniform solid solubility and diffusion rate in the Al matrix of the alloying and trace elements make the effect of homogenization treatment diverse [17–20]. Alloying elements Cr, Mn, V, and Zr show very low diffusion rate, Fe shows low solubility and low diffusion while Cu, Mg, Si, and Zn show high solubility and high diffusion rates. Alloying elements like Fe, Cr, Mn, V, and Zr if present in the Al alloy hinder the homogenization process and make difficult to dissolve inclusion during homogenization. In the Zr- and Sc-containing Al alloys, fine dispersion of Al_3Zr, Al-Sc, Al-Mg-Sc is formed which has beneficial effect on the properties of Al alloys refining the grain size. In Fe, bearing AA6XXX alloys the transformation of metastable phase ($FeAl_3$) to stable phase (α-Al(Fe, Mn)Si) [16, 17] is as

$$FeAl_3 \longrightarrow \beta\text{-AlFeSi} \longrightarrow \alpha\text{-Al(Fe, Mn)Si}. \qquad (1)$$

The homogenization treatment can increase the size of Mn-containing dispersoids [2], which can later contribute in recrystallization by self-annealing of AA6XXX alloys just after hot rolling. The self-annealing is enhanced by the presence of large Mn-containing dispersoids, which provide the particle stimulated nucleation at elevated temperature (350°C) just after hot working in AA6XXX alloys. The self-annealing of AA6XXX alloys just after the hot rolling eliminates the possibility of surface-roughening phenomenon, that is, roping during subsequent forming process. Conventionally, the chances of roping are prevented by recrystallization annealing (continuous or batch annealing) treatment before or between cold working process but that increase the cost of processing and also the ability to solution treatment of the final sheet. The two-stage homogenization treatment (first stage at 550°C and subsequently second stage at 460°C)

Table 7: Designation of elevated temperature treatments [2, 11, 12].

Temper	Treatment	Use
T1	The elevated temperature fabrication process followed by stable temper aging.	The products are not worked at ambient temperature following elevated temperature fabrication. The product properties cannot recognize the ambient temperature straightening and flattening effects.
T2	The elevated temperature fabrication process followed by ambient temperature working and stable temper aging.	The strength of the products is enhanced by ambient temperature working and the product properties can recognize the flattening or straightening effects.
T3	Solution heat treated, cold worked, and aged to a substantially stable condition.	The strength of the products is enhanced by ambient temperature working after solution treatment and the product properties can recognize the flattening or straightening effects.
T4	Solution heat treated and aged to a substantially stable condition.	The products are not worked at ambient temperature following solution heat treatment. The product properties cannot recognize the ambient temperature straightening and flattening effects.
T5	Cooled from an elevated temperature shaping process and artificially aged.	The products are not worked at ambient temperature following elevated temperature fabrication. The product properties cannot recognize the ambient temperature straightening and flattening effects.
T6	Solution heat treated and then artificially aged.	The products are not cold worked after solution heat-treatment or in which the effect of cold work in flattening or straightening may not be recognized in mechanical property limit.
T7	Solution heat treated and overaged/stabilized.	The artificial aging of the wrought product after the solution heat-treatment to beyond the point of highest strength.
T8	Solution heat treated, ambient temperature worked, and then artificially aged.	The strength of the products is enhanced by ambient temperature working after solution treatment and the product properties can recognize the flattening or straightening effects.
T9	Solution heat treated, artificially aged, and then work hardened.	The work hardening to enhance strength after artificial aging.
T10	Cooled from an elevated temperature shaping process, cold worked and then artificially aged.	The products which are cold worked to improve strength or in which the effect of cold work in flattening or straightening is recognized in property limits.
Tx51	Stress relieving before ageing.	
Tx52	Stress relieving by compressing.	
Tx54	Stress relieving by combined stretching and compressing.	
T72	Corrosion resistance is enhanced by overaging.	
T73	Corrosion resistance is enhanced by step aging.	
T76	Step aging leading to enhancement of exfoliation corrosion behavior	
T73x51 T76x51	Combined stress relieving/step ageing	

can be prescribed for AA6XXX alloys to increase the size of Mn-containing dispersoids and to bring its effect on re-crys-tal-li-za-tion during self-annealing of AA6XXX [2, 13–20]. The presence of Mn, Cr, Zr, and Sc can hinder the recrystallization during self-annealing by pinning grain boundary [21]. The concentration of those elements should be kept low when the self-annealing is aimed just after hot working in AA6XXX alloys.

The removal of trace element H from castings and ingots can be possible by proper selection of homogenization temperature and time. The slow heating to an elevated temperature and presence of fluoride in furnace atmosphere during homogenization gradually removes H from Al alloys [2]. The surface oxidation and localized melting during homogenization should be avoided by selection of proper furnace atmosphere and slow heating.

3.2. Hot Working. Hot working processes such as rolling and extrusion are the first metal working processing of the DC cast ingot of AA6XXX alloys which converts DC cast ingots into wrought product [3]. The plastic deformation of AA6XXX takes place during metal working process, which is classified according to the type of force associated with. Rolling and forging are direct compression type, the extrusion and deep drawing are of indirect compression category while the stretch forming is tension-type metal working. The metal working is classified depending upon the temperature of plastic deformation in two broad categories, namely, hot working and cold working.

Hot working by rolling and forging of AA6XXX alloys is done at elevated temperature (>0.5 T_0, while T_0 if the melting point of AA6XXX alloys) with high strain rate. The hot working starting temperature is around 460°C while the hot working finish temperature is around 330°C [15, 22–24]. The temperature of work piece in hot working depends on [3]:

(i) initial temperature of tools and work piece,

(ii) heat generation due to plastic deformation,

(iii) heat generation by friction,

(iv) heat transfer between the deforming alloy and surroundings.

Strain during the hot working of AA6XXX alloys by rolling and forging process is large compared with creep and tension test and the strain rate is in the range of ($0.5\,s^{-1}$–$500\,s^{-1}$) [3]. The conventional hot rolling of AA6XXX alloys is done in two stages, first by breakdown mill, which produces the transfer slab of 30–50 mm thick and later by tandem mill (3–5 stand), which produces around 6 mm thick sheet of AA6XXX alloy [15]. Deformation capability of work piece is enhanced at elevated temperature and alloy work piece shows high plasticity enabling the large deformation reductions without consumption of high energy. Salient features of hot working are the following [3, 15]:

(i) complete elimination of cast dendritic structure,

(ii) no existence of elongated grains produced by rolling and extrusion,

(iii) virtually no strain hardening and flow stress becomes constant,

(iv) large plastic deformation capability,

(v) Replacement of deformation structure by microprocess such as simultaneous dynamic recovery and rapid recrystallization and formation of strain free structure [10, 15].

The microprocess such as dynamic recovery phenomenon can be widely observed during hot working by vacancy migration induced dislocation movement by climb process and cross slip (cross slip is valid only for screw dislocations) at elevated temperature. The initial hot deformation produces dislocations, which accumulate in tangles and form subboundaries by dynamic recovery process [10]. The resulting subgrain structures are stable in size during steady-state deformation, that is, at constant stress, strain rate, and temperature of hot working process. During dynamic recovery, size of the subgrains, increases with the process temperature and decreases with strain rate [3]. Stacking fault energy plays the crucial role in controlling the dynamic recovery. Generally, alloys with high stacking fault energy are favored for dynamic recovery for both the rate controlling recovery processes cross slip or climb of dislocation. The presence of alloying elements in solid solution reduces the stacking fault energy and increases the separation of partial screw dislocation, which will be hard to recover dynamically during hot working process by cross-slip. The recovery by dislocation climb is also hindered by lowering of stacking fault energy. (In this context either the theory of constriction of the fault, which needs elastic energy, depending on the width of fault between the partial jogs, or the theory of advancement of extended jogs by absorption of multiple vacancies, which is determined by the fault width is applicable.) Recrystallization, occurring during or just after the hot deformation, produces the strain-free grains surrounded by high-angle grain boundaries. Initiation of dynamic recrystallization depends on the critical strain rate and strain which occurs during hot rolling. The critical sized subgrains surrounded by favorable misorientations (for growth) present in the deformed and recovered substructure can produce the nuclei for recrystallization, which follow the subsequent growth to replace the deformed substructure and result in formation of strain-free grains discontinuously [15, 25–30]. The addition of alloying elements such as Ti, B, Zr, and Sc usually retard the grain-boundary migration and hinder the recrystallization during and just after hot working process. Presence of precipitates in AA6XXX alloys can also pin the dislocation, cell wall, and grain-boundary movement during the microprocesses such as deformation, recovery, and recrystallization. The contribution of precipitates on these phenomena depends on the precipitate size, interprecipitate distance, or the derived volume fraction of precipitates [10, 31]. The cell size and the subgrain size are reduced due to presence of precipitates by the pinning process and by the formation of deformation zone (with reduced cell size) around the precipitates. Larger precipitates or dispersoid, produced by two-stage homogenization treatment during the processing of AA6XXX alloys, generally give rise to particle, stimulated nucleation which helps in nucleation of recrystallization grains during self-annealing of AA6XXX alloys while smaller precipitates restrict the grain boundary migration during recrystallization.

(vi) The solidification texture of AA6XXX alloy is converted to hot rolling texture (Table 8), which has been well studied and contains the recrystallization texture cube orientation (intensity of cube orientation in hot-rolled AA6XXX is much lower than that of hot-rolled AA1XXX alloy) and remnants of deformation texture orientation such as S, brass, and copper in the partially recovered aluminium alloys [30]. Presence of large precipitates randomizes the hot-rolled texture and decreases the cube texture intensity. However, the formation of texture during hot rolling strongly depends on the composition of this alloy and process parameters like strain, strain rate, and hot working temperature.

Table 8: Volume fraction of crystallographic textures of AA6XXX alloys [33–35].

Products	cube	Goss	Bs	Cu	S
Reroll after self-annealing	30.00	3.10	4.50	3.30	14.30
Conventional-reroll	36.00	2.40	2.70	2.30	14.40
Batch annealed-reroll	36.60	1.80	3.00	2.60	14.60
Batch annealed-final gauge (self-annealed during HR)	10.00	2.20	5.60	6.60	14.40
T4 Conventional-final gauge	11.40	3.30	4.00	1.60	7.60
T4 Random	3.50	3.50	6.60	6.60	14.00

During hot working, large strain is involved. Hence, the strain hardening is not desirable. Lower limit of hot working temperature for AA6XXX alloys can be based on the lowest temperature at which rapid recrystallization occurs to avoid strain hardening [3]. Recrystallization temperature depends on the degree of deformation, and for large degree of deformation, the hot working temperature reduces. Upper limit of hot working temperature is determined by either heavy oxidation or localized melting due to grain-boundary segregation of solute elements. The multiple pass hot working is widely used process for AA6XXX Al alloys. In multiple pass processing, the temperature in all passes is kept above minimum temperature for recrystallization. And the temperature of last pass is kept low to enhance the formation of fine grains, which can also be ensured by providing slightly larger deformation (45% thickness reduction) at the last pass. The hot working of AA6XXX alloys has the following advantages [3]:

(i) the higher plastic flow at elevated temperature makes plastic deformation easier,

(ii) accelerated diffusion during hot working reduces the cast inhomogeneities in composition,

(iii) blow holes and pin holes of cast ingot eliminated by welding of cavities,

(iv) cast dendritic inhomogeneous structure is replaced by equiaxed strain-free grains, which provide higher ductility and toughness.

Magnesium, as the alloying element, has significant effect on the hot workability of AA6XXX alloys, however, Si, Cu, and Fe in this regard are not very effective. Alloying elements, namely Cr, Mn, and specially Zr tend to enhance the press effect and provide higher strength [2]. The hot working of Mg_2Si containing slab spheroidizes the Mg_2Si and induces coagulation, which reduces the number of Mg_2Si precipitates [2].

3.3. Cold Working. During hot working at elevated temperature, oxidation takes place, also the expansion and contraction during elevated temperature treatment result in poor surface finish. In order to enhance the surface condition and strength, wrought AA6XXX alloys are cold worked after hot working [3]. Aiming at the multiplication of dislocation density the cold working renders work hardening. The strength exhibited AA6XXX Al alloy after cold working is much higher than shown after hot working, which is very much desired for structural applications in automobile, aerospace, and marine sector. Plastic deformation of AA6XXX alloy is performed at room temperature during cold working, where these alloys show less plasticity thereby restricting large reductions during cold working in a single pass. In multipass cold rolling mill, used for cold rolling of AA6XXX Al alloy, total cold deformation is done by several rolling stands. The hardness and strength of the rerolled (cold rolled) sheet gradually increase from the entry stand to exit rolling stand. The optical microstructure study (up to 2000X) after cold rolling reveals the elongated grains with high aspect ratio. Grains elongate along the cold working direction, whereas along normal direction the grains squeeze. With the higher percentage of cold deformation, the grains appear as long bands parallel to the working direction. At higher magnification, many other features of the deformed substructure such as the grain subdivision, cell structure, and dislocation can be revealed by electron microscope (TEM) [32]. The cell structure study of cold worked Al alloys has already been well studied. Cell structure is composed of cell wall and cell interior. The tight-packed dislocations generally form the cell wall. Dislocation density of cell interior is less than that at the cell wall. Work hardening during cold working depends on the cell size, the fraction of cell wall, and the fraction of cell interior. Recent research shows that (Figure 1) deformation-induced cell size in AA6XXX alloys varies with strain rate during dynamic loading conditions [32]. The subdivision of grain and grain rotation during the cold deformation produce deformation texture. The hot-rolled texture is usually replaced by deformation texture, which comprises with the S, copper, and brass texture orientations. Cold deformation gradually reduces the cube texture intensity of hot deformed Al alloy. The cube-oriented grains rotate causing change in orientation. Copper and brass type of shear of grains produce the deformation texture. The deformed substructure, which can be observed under high magnification, would be heterogeneous [10, 25]. The presence of transition and shear band in the deformed structure produces the deformation inhomogeneity.

3.4. Annealing. Annealing process is conventionally performed before cold working or between cold working (intermediate/process annealing). Annealing is of two types batch and continuous annealing. AA6XXX alloy hot-rolled sheets are reheated in furnace during batch annealing process in the temperature regime 380°C–460°C (Table 9), while in continuous annealing, slightly elevated temperature around

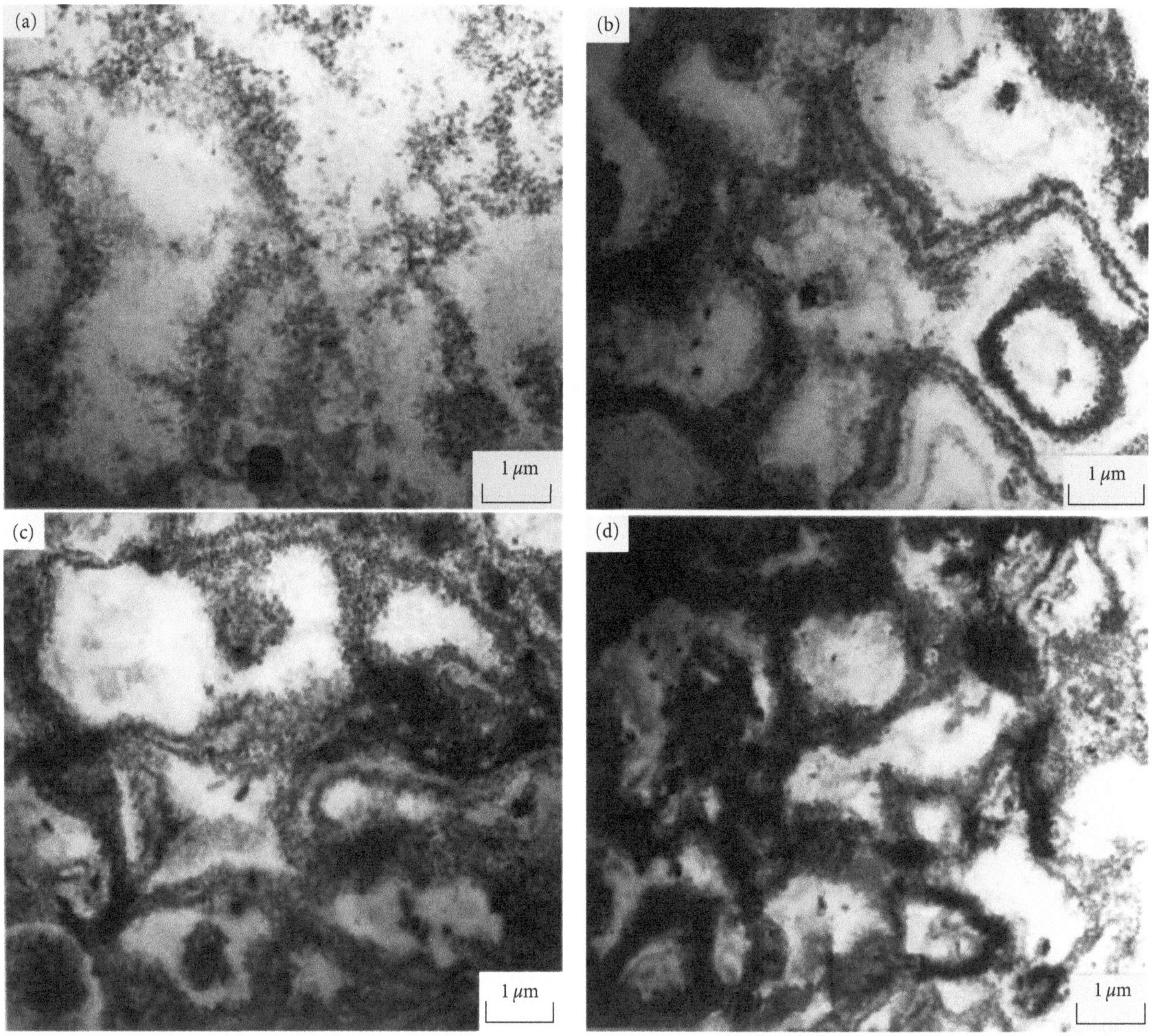

FIGURE 1: Dislocation substructure [32] of AA6061-T6 aluminium alloy at strain rates (a) 10^{-3} s^{-1}, (b) 2×10^{-3} s^{-1}, (c) 3×10^{-3} s^{-1}, and (d) 4×10^{-3} s^{-1} at true strain 0.25. The cell structure refined with strain rate.

510°C is used for aluminium alloy sheets [26]. In AA6XXX, softening of the sheet occurs during annealing which is intentionally designed to induce plastic deformation capability during cold rolling. Subsequent formation of strain-free equiaxed grains induces softness due to absence of dislocations. The last step for cold rolling-intermediate annealing process can be annealing when soft final gauge is required while last step can be cold rolling when high strength of final gauge is required. Strength and hardness of the final gauge produced can be varied by manipulating the combined process cold rolling and intermediate annealing.

The strain-free grains and the crystallographic texture, which are formed during annealing, determine the subsequent final forming character of AA6XXX alloy sheets. cube texture orientation, after annealing, is the characteristic of aluminium [27]. Presence of large precipitates randomizes the cube texture in aluminium alloys [25, 30]. The large precipitates form the deformation zone during cold working in the surrounding matrix and fine subgrain structure can be formed in the deformation zone by static recovery process. The presence of solutes in the solid solution and fine precipitates (of Zr, B, Sc, Ti) can hinder subgrain formation by pinning of dislocation during static recovery, which can result in no or delayed subgrain structure formation during annealing [2]. Recrystallization can occur by nucleation and growth (discontinuous recrystallization) with the driving force of very high dislocation density, or it can occur continuously without any high-angle grain-boundary migration by extended recovery of dislocation at the low-angle grain boundary which transform the low-angle grain boundary to high angle grain boundary (recrystallization *in situ*). The cube oriented grains generally determine the recrystallization texture in discontinuous recrystallization with the suppression of rolling texture orientations while rolling texture orientations become prominent after recrystallization *in situ*. Both discontinuous and continuous recrystallizations can contribute to the formation of strain-free grain after annealing in the commercial AA6XXX alloys, which have

TABLE 9: Annealing, solutionizing, and aging: processing data [7].

Designation	A. T (°C)	S. T (°C)	Ag. T (°C)
AA6053	413	520	171–178
AA6061	413	528	160–177
AA6063	413	521	176–183
AA6066	413	532	176–180
AA6070	413	546	160
AA6463	413	521	177

heterogeneity in solute level and precipitate distribution. The sources of nucleation of strain-free recrystallized grains (N_R) in aluminium alloys are stable cube grains (N_C) after deformation, deformation zones around the precipitate (N_Z), deformation heterogeneities (N_{HD}) such as transition and shear bands, and also the deformed grain boundary (N_G) [25–30]. The N represents the number of nuclei per unit volume:

$$N_R = N_G + N_C + N_{HD} + N_Z. \tag{2}$$

Whatever be the mechanism, the presence of large precipitate in AA6XXX alloy induces the particle-stimulated nucleation and randomize the cube texture after annealing recrystallization [25–30]. Extended recovery can result in the elongated band-shaped grains after recrystallization and stabilize the deformation texture orientation like S, brass, and copper [28–30]. The conventional annealing is not only done to facilitate the cold rolling, but also to eliminate the roping defect after forming of final gauge. Self-annealing just after hot rolling of AA6XXX alloys in the temperature regime 270–300°C can eliminate the surface roughening by roping of AA6XXX alloy, without extra processing such as conventional annealing.

3.5. Aging Treatment. The elevated temperature treatments of AA6XXX alloys are done to regulate the properties by the formation of age-hardenable precipitates. This treatment involves solutionizing at elevated temperature around 520–550°C and then quenching to room temperature followed by room temperature aging or elevated temperature aging (artificial aging) in the temperature range 160–180°C [2, 7, 36–39] (Table 9). Quenching to room temperature from the solutionizing temperature produces the supersaturated solid solution (SSS). Slow cooling is avoided after SSS to prevent the separation of equilibrium precipitate from solid solution. The aging at room temperature or at elevated temperature (below the GP zone solvous) produces the nonequilibrium metastable or transition phases such as Guinier-Preston (GP) zones and hexagonal β'' and $\beta'(Mg_2Si)$, which are coherent to the matrix and strengthening occurs. These β'' and $\beta'(Mg_2Si)$ are coherent unlike incoherent equilibrium fcc Mg_2Si and they are the source of precipitation hardening or age hardening of AA6XXX alloys. The age hardenable precipitation formation sequence in AA6XXX alloys is the following [2, 7, 36–38]:

$$\begin{aligned} &\text{SSS} \longrightarrow \text{GP zones} \longrightarrow \beta'' \longrightarrow \text{hexagonal } \beta'\left(\text{Mg}_2\text{Si}\right) \\ &\longrightarrow \text{fcc } \beta\left(\text{Mg}_2\text{Si}\right). \end{aligned} \tag{3}$$

Recent observation reveals the existence of other precipitate phase like hexagonal Q' and Q phases in Al-Mg-Si-Cu alloys [38]. The lath features of these two phases distinguish Q' and Q phases from needle-shaped β' precipitate [38].

The AA6XXX alloys used in automobile industries are subjected to another elevated temperature aging treatment known as reversion and reaging (RRA) where the artificially aged AA6XXX alloys are heated to above GP solvus and then quickly aged to temperature below GP solvus. This RRA technology offers fine grain structures at the surface which are wanted for forming processes used in automobile industries.

In many AA6XXX alloys, the following sequence of processing is performed: solution treatment, preaging, deformation and final aging. The pre-aging can play crucial role to determine final properties of many AA6XXX alloys. In this process, the aging of AA6XXX alloy containing less dislocations (which is not in deformed state) can be performed at slightly lower temperature and shorter time than the normal aging. The presence of less dislocations hinders the heterogeneous nucleation of precipitates at the dislocations in the AA6XXX alloy matrix and that can lower the chance heterogeneous nucleation after deformation during final aging. The uniform distribution of precipitates, after preaging, gives rise to uniform substructure after deformation replacing the occurrence of deformation heterogeneity.

The resistance to stress corrosion cracking is enhanced by second aging treatment but the strength is slightly decreased because of overaging during second aging treatment. Repeated treatment at elevated temperatures might contribute in the loss of magnesium but the effect is not very significant. The quenching rate plays significant role for those AA6XXX alloys which contain high solute concentration of magnesium, silicon, copper, manganese, and chromium. Room-temperature aging contributes to only limited hardening in AA6XXX alloys [2]. Cold working is sometimes performed after artificial aging to increase strength (35% in ultimate tensile strength and 55% yield strength). The ductility (elongation up to 60–80%) is reduced by cold working after aging. The alloying elements such as Zn and Fe do not contribute much in the aging process but Cu accelerates the aging while Mn and Cr can reduce the grain-boundary precipitations preventing embrittlement and intergranular corrosion.

4. Strengthening Methods of AA6XXX Alloys

4.1. Grain-Boundary Strengthening. The grain boundaries represent the misorientation surfaces between adjacent grains in polycrystalline AA6XXX and they are associated with high energy [3]. Fine-grain structures of AA6XXX alloys are wanted to exhibit high strength at the same time higher toughness. The increase in the fraction of grain-boundary surfaces, associated with fine grain structure, can be the source of strengthening. There are several mechanisms of strengthening in fine grain polycrystalline AA6XXX alloys [40–49].

(i) Activation of higher number of slip systems and mutual interference of slip systems near grain-boundary region than in the grain center, which makes effect of grain boundary felt in the grain center in fine grain structure.

(ii) Contribution of texture orientation dependent factor, that is, Taylor factor in the texture strengthening. Individual grains in the polycrystalline AA6XXX alloy will have different work hardening rate because of its own texture oriented deformation capability, which is reflected by the Taylor factor. In fcc AA6XXX alloys the average value of that factor is of the order of 3.10 and that textural work hardening (equivalent to T^2, where T is the Taylor factor) can be 10 times higher than for AA6XXX single crystals [3].

(iii) Misoriented regions between grains or the grains-boundaries can also act as the barrier against the dislocation movement during deformation, which creates the large (>50 dislocations) dislocation pile-up and cause strengthening. The dependence of yield strength on grain size of AA6XXX alloys can be represented by well-known Hall-Petch Equation, which is generally explained by the dislocation pile-up theory.

Hence, the fine grain structure of AA6XXX alloys has certain advantages in strengthening. That fine grain structure of AA6XXX alloys can be ensured during processing starting from solidification of casting to elevated temperature processing such as annealing. Grain refinement of AA6XXX alloy microstructure involves the preferential nucleation of fine-equiaxed grains at the place of solidification dendrites [50–55]. Higher solidification rates in DC casting than conventional sand casting can reduce the dendritic cell size or arm spacing but best grain refinement occurs because of addition of insignificant amount of master alloys (Al-5%Ti-1%B) in the melt before DC casting. The underlying mechanisms of grain refinement are (i) controlled heterogeneous nucleation on the TiB_2, $(TiAl)B_2$, and TiC particles and (ii) retardation of dendrite growth by solute elements and creation of the constitutionally supercooled zone, near the solidifying zone during solidification. In order to ensure the fine grain structure of AA6XXX alloys during elevated temperature processing from hot working to annealing, the abnormal growth of grains should be prevented. (0.1–0.5)% addition of alloying elements such as Mn, Cr, Zr, and Sc are very useful to ensure fine grain structure of AA6XXX alloys. The abnormal grain growth is prevented by the presence of Al_3Zr, Al_3Sc, and Al_6Mn dispersoids of size range (5–50 nm). These trace elements, when added, retard recrystallization and abnormal grain growth by raising the recrystallization temperature (TR):

$$\text{Mn \& Cr (TR} \sim 330^\circ\text{C)} < \text{Zr (TR} \sim 410^\circ\text{C)} < \text{Sc (TR} \sim 605^\circ\text{C)}. \tag{4}$$

4.2. Precipitation Strengthening. The AA6XXX alloys are heat-treatable alloys. Precipitation of the alloying elements as coherent GP zones and coherent β'' and hexagonal coherent β' (Mg_2Si), during aging treatment, provides strengthening due to the presence of coherency strain field around the precipitates, which interacts with the moving dislocations [3, 4, 38]. The presence of coherent lath-type hexagonal nonequilibrium Q' precipitates in quaternary alloys which contain Cu can also give rise to coherent strain field. Formation of nonequilibrium coherent hexagonal precipitation gives the highest precipitation strengthening in AA6XXX alloys [3, 38]. The equilibrium fcc $\beta(Mg_2Si)$ or the quaternary Q phases are incoherent and formation of incoherent equilibrium precipitates culminates in slight loss in hardness and yield strength due to the absence of lattice mismatch strain, that is, coherency strain around the equilibrium precipitates in AA6XXX alloys. Overaging of the fcc $\beta(Mg_2Si)$ or hexagonal Q precipitates leads to coarsening of precipitates and results in further loss in yield strength, which should be prevented. Deformation-induced moving dislocation can interact with precipitates in two ways in AA6XXX alloys. Dislocation can either cut the soft coherent precipitates or bypass the hard incoherent dispersoids (the hard equilibrium or over-aged precipitates; this nomenclature is not similar to dispersion hardening phases which are fabricated by powder metallurgy route). The strengthening occurs because of these dislocation and particle interactions. The lattice mismatch strain field or coherency strain field around the dispersoid does not exist which lowers the hardness and yield strength but dispersoids in AA6XXX play the significant role in work hardening restricting the moving dislocations, which finally free from dispersoids leaving one dislocation loop around the dispersoids at higher stress. Precipitation strengthening, caused by soft coherent precipitates in AA6XXX alloys, is due to coherency strain, stacking fault energy interaction, modulus interaction, interfacial energy, and lattice friction stress [56–62]. The coherency strain field resulting from the mismatch between the precipitate and matrix can be the source of high local strength. An increase in local strength due to coherency increases with the volume fraction of coherent precipitate, radius of the precipitate, the mismatch strain field and reduces with the displacement vector of dislocation. The precipitates can have dissimilar stacking fault energy than matrix. Local variation of stacking fault width, in the precipitate and matrix, which the glide dislocation faces when it enters the precipitate from matrix, can be the source of strengthening in AA6XXX alloys, where the increase in the yield strength increases with the local variation in stacking fault, partial dislocation separation force, stacking fault width and precipitate volume fraction. The degree of strengthening due to variation in modulus of rigidity between precipitate and matrix might not be high but that increases with the size and volume fraction of precipitate, local variation in rigidity and decreases with displacement vector of dislocation. Any increment in the precipitate-matrix interface due to cutting of precipitates preferentially for the GP zones, which offer high surface-to-volume ratio, can act as source of high yield strength which increases with the volume fraction and precipitate-matrix surface energy and decreases with the size of precipitates in AA6XXX alloys. The localized increase in the yield strength due to presence of precipitate with higher yield strength than the matrix can also cause the increase in the lattice friction stress, that is, Peierls stress for the movement of dislocation. This strengthening effect increases with the increase in the volume fraction and size of precipitates and reduces with the displacement vector of dislocation. For incoherent [16] dispersoids, the yield stress is determined by the shear stress needed for bowing out dislocation line between two precipitates with certain interparticle distance. The interparticle distance plays prominent role in strengthening of incoherent precipitates in AA6XXX alloys.

4.3. Work Hardening. The shear strength exhibited by AA6XXX alloys continuously increase with shear strain. This phenomenon of increase in the shear strength due to previous plastic deformation is known as work hardening [3, 4]. The multiplication of dislocations raises the dislocation density during plastic deformation because of dislocation emission from steps and ledges of grain boundary, dislocation regeneration by Frank-Read mechanism when the moving dislocation is pinned between the precipitates or dispersoid or solutes, condensation of vacancies and by multiple cross slip mechanisms in AA6XXX alloys. The plastic deformation induces the movement of dislocations. During the plastic-deformation-induced movement of high population of dislocations, dislocations interact with jogs and sessile dislocations and with the other barriers present in alloys such as substructure cell wall, grain boundary, precipitate, dispersoid, and solute atmosphere [63–68]. The restriction in dislocations movement at the barriers produces the dislocation pile-up on the slip plane which generates high back stress against the applied stress. Work hardening arising from the dislocation pile-up is appreciable over larger distance and known as long-range strength, which is insensitive to temperature [3]. Hence, AA6XXX alloys can show considerable strengthening even at elevated temperature. In fcc AA6XXX alloys, two types of sessile dislocations are produced, namely, Frank partial dislocation and sessile dislocations which produce the Lomer-Cottrell barrier. The edge dislocations can be produced due to condensation of vacancies in AA6XXX alloys with the displacement vector $(a_0/3)[111]$, which is normal to central stacking fault plane, where the glide of that (Frank Partial) dislocation is restricted. Frank partial dislocation cannot move or glide due to the application stress and acts as sessile barrier at ordinary temperature unless elevated temperature driven diffusion controlled climb of that dislocation occurs. At ambient temperature, the sessile barrier produced by Frank partial dislocations can provide barrier against the movement of other dislocations. Another crucial dislocation reaction which can give rise to formation of the sessile barrier is known as Lomer-Cottrel reaction in AA6XXX alloys, where the leading partials on intersecting $\{111\}$ slip planes react with each other and produce $(a_0/6)[011]$ pure edge dislocation in the (100) plane and the displacement vector of the resulting dislocation does not fall in either of its planes of its stacking fault. This barrier can be through either at elevated temperature and/or at high stress. Formation of jogs in edge dislocation can also contribute to work hardening restricting the movement of edge dislocations. The dislocation-dislocation cutting process on the slip plane ranges only (6–11) interatomic spacing, raising the stress field locally which is not extended over wide range in AA6XXX alloys, hence known as short-range strength, which can be sensitive to temperature and does not prevail at elevated temperature. The metal working process produces the cellular substructure composed of cell wall and cell interior as a function of strain, strain rate, and temperature in many alloys. The shape and size of the cell structure depend also on the composition of the alloys. In aluminium, the cellular substructure, which is produced by metal working is quite stable and does not vary in size to a large extent with stain, strain rate [3]. However, possibility of cell refinement with variable strain rate is not invalid in AA6XXX alloys [32]. Cell walls act as the barrier against the mobile dislocations. The mutual interaction of immobile cell wall and cell interior dislocations with the mobile dislocation is the source of significant work hardening [66], which depends on the volume fraction of cell wall, composed of tight-packed dislocations, and cell interior which contains comparatively less dislocations than the cell wall. Cellular substructure formation during metalworking in AA6XXX alloys might successfully musk the effect the grainboundary on work hardening [66] due the fact that the majority of mobile dislocations meet the cell wall barrier first prior to meeting the grain boundary.

5. Properties of AA6XXX Alloys

The properties of AA6XXX alloys vary with composition and processing such as hot working, cold working, annealing, and aging processes [2, 5, 7, 69, 70]. Yield strength, ultimate tensile strength, elongation, shear strength, fracture strength, and hardness values for the annealed state and T6 state of some widely used AA6XXX alloys are tabulated in Table 10 and Table 11 [7]. The annealed alloys offer yield strength (40–80 MPa) which is about half of the Ultimate Tensile Strength (85–150 MPa), while significant increase in Yield Strength and ultimate tensile strength can be observed after T6 treatment of the same alloys. T6-conditioned alloys exhibit Yield strength ranging from (210–360) MPa and ultimate tensile strength (240–385) MPa. The margin between Yield Strength and Ultimate Tensile Strength is also reduced after T6 conditioning of these alloys. A significant increase in the shear strength and fracture strength after T6 condition is observed for these alloys. However the ductility denoted by % elongation is reduced after T6 conditioning (10–15%) compared to the % elongation of annealed state (18–35%). The T6 conditioning of these alloys is associated with about 3 times increase in the Brinell hardness value (25–43 HB) compared to hardness of annealed state (73–120 HB). Mg, Si, and Cu, alloying additions, play important role in defining the mechanical properties of AA6XXX alloys. Alloying elements such as Mn and Cr increase the strength (10–15)%. Segregation of excess Si as platelets along the grain boundary raises the intergranular brittleness in fully hardened alloys. Addition of Mn counteracts this brittleness. Iron correction by Mn can increase toughness. Alloying elements like B, Ti, and Zr interact with the structure during processing and also contribute to hardening of these alloys. Young's Modulus of these alloys is about 65–70 GPa and shear strength can be 65–70% of that of ultimate tensile strength. The ordinary and elevated temperature compressive strengths are similar

Table 10: Properties of AA6XXX alloys in annealed condition [7]. Y.S: yield strength, UTS: ultimate tensile strength, El: elongation, SS: shear strength, FS: fracture strength, hardness.

Designation	Y.S (MPa)	U.T.S (MPa)	El. (%)	S.S (MPa)	F.S (MPa)	H.V (HB)
AA6053	55	110	35%	75	55	26
AA6061	55	124	25%	83	62	30
AA6063	48	88	25%	68	55	25
AA6066	83	152	18%	96	68	43
AA6070	70	145	20%	94	60	35
AA6463	50	88	23%	70	55	25

Table 11: Properties of AA6XXX alloys in T6 condition [7]. Y.S: yield strength, UTS: ultimate tensile strength, El: elongation, SS: shear strength, FS: fracture strength, hardness.

Designation	Y.S (MPa)	U.T.S (MPa)	El. (%)	S.S (MPa)	F.S (MPa)	H.V (HB)
AA6053	220	255	13	160	88	80
AA6061	276	310	12	207	94	94
AA6063	214	235	11	148	65	73
AA6066	360	388	12	234	110	120
AA6070	352	378	10	234	94	120
AA6463	214	241	12	152	68	74

as tensile strengths. Notch sensitivity of these alloys is not high. The low-temperature strength exhibited by these at the temperature −200°C can be about 75% higher than room temperature strength. The fracture toughness and ductility are not altered much at low temperature. Deformation at elevated temperature shows low strength but enhanced ductility. Creep resistance is quite good at room temperature and strength is just comparable with yield strength, but with the increasing temperature, the creep resistance decreases proportionally. The corrosion resistance of these alloys is fairly good but alloys become susceptible to intergranular corrosion and stress corrosion when Cu content (>0.4%) and Si content are increased. Addition of Fe without Fe corrector increases the tendency of pitting corrosion. The hot shortness problem is severe in AA6XXX alloys along Al-Mg_2Si line but decreases with excess either Mg or Si content. Hot workability of AA6XXX alloys depends on the earlier homogenization process. The two-step homogenization with second step temperature as high as 580°C can provide large Mg_2Si dispersoids, which offer enhanced hot workability than the single-step homogeneization. The later produces fine dispersoids, which interact with the moving dislocation and cause hot working difficult [2].

6. Use of AA6XXX Alloys

In the last decades, great efforts have been made by the automotive giants of the world to become highly advanced and innovative to increase the fuel efficiency, reduce green house gas emissions by proper designing light weight cars and vehicles based on the intensive use of the latest developed AA6XXX aluminium alloys. In recent years, the growing attention on fuel economy and environmental factors has significantly increased the use of aluminium alloys for automotive purposes. According to the Government regulations in many countries (like USA, JAPAN and EUROPE), the cars and the vehicles have to be 15–20% more fuel efficient by 2000–2012 [71].

In view of this, the endeavor was to replace the heavier materials by lighter aluminium alloys which can reduce the weight of the vehicles and can make them fuel efficient. The data shows about 7% savings in fuel economy can be possible by reduction of weight of the car by 10% and that is also associated with significant reduction in the green house gases [72]. The average use of the aluminium per car in Western Europe lies between 100 and 400 kg [72]. Recently developed steels, magnesium and fiber-reinforced plastics are the competent candidate materials for this purpose; however, aluminium alloys are still standing tall among the competitors due to the associated advantages like mass production capability, good availability at a reasonable price, and good recyclability [73]. The data analysis from USA shows that 32% of the total aluminium used was in the transportation sectors [71].

The position of aluminium alloys for the transportation sectors all over the world was strengthened due to several novel concepts in light-weight solutions and applications which might need further developments in the key technology areas like flexible space frames, chassis components like axles, wheels, and body in white applications. The advancement in AA6XXX aluminium alloy use was due to its easy functional integration, tailor-made properties, and reliability which its competitors are unable to render. The availability of diversified semis such as castings, sheets, and extrusions

Table 12: Use of AA6XXX alloy Rolled (R) and Extruded (F).

Designation	Car and light weight vehicle	Aircraft, aerospace	Marine structure
AA6061	F	F	F
AA6063	F	F	F
AA6013	—	R	—
AA6010	R	—	—
AA6016	R	—	
AA6013	—	R	—
AA6065	F	—	—
AA6111	R	—	—
AA6005	F	—	—

which can be produced in mass production and can be integrated easily with their tailor-made properties has put the AA6XXX aluminium alloy development and use as strategic part of evolution of transportation industry. Extrusions of AA6XXX alloys are widely used (Table 12) in transportation sectors for aircraft, automobile, and marine structures. Underlying innovative technology and proper light-weight design with integrated functions allow these alloys to be used in flexible car concepts for the parts which required complicated shape forming such as space frames, chassis parts, bumper, crash elements, and air bags. AA6XXX alloys are generally used where quenching occurs during processing. The final strength and formability are determined by subsequent age-hardening techniques. Particularly for space frames, the properties required, such as good extrudability, excellent corrosion resistance, and weldability, are satisfied by AA6XXX alloys which can be formed at T4 condition while strength can be increased by subsequent T5 or T6 conditioning [74, 75].

Since the inception of car and light-weight vehicle manufacturing and aluminium alloy fabrication, the rolled aluminium alloys sheets are used for the body in white application. The potential alloys for sheet application are non-heat-treatable AA5XXX and heat-treatable AA6XXX alloys (Table 12) due to possibility of mass production, availability at reasonable price, excellent formability which can be tailored by proper alloy design and controlling the processing. The surface appearance and age-hardening response are also crucial criteria which enhance the use of AA6XXX alloys than the non-heat-treatable AA5XXX alloys. In the class of AA6XXX alloys, the sheet of AA6010, AA6016, and AA6111 alloys are generally used for sheet body of cars while recently the AA6181A alloy has been enlisted due its recycling aspects. These alloys show very high ductility in T4 condition and high strength after age hardening during paint bake cycle. The high strength and good formability of AA6111 alloys for up to 1.0 mm thick outer panel gauge is used in USA while in Europe the 6016 alloy is preferred for up to 1.2 mm thick panel gauge, which shows superior formability and higher corrosion resistance but less bake-hardened strength than AA6111 alloy.

References

[1] W. Hufnagel, *Key to Aluminium Alloys*, Aluminium Publication, Dusseldorf, Germany, 1999.

[2] T. R. Ramachandran, "Advances in Aluminium Processing and Its Automotive Application," Workshop Lecture Notes, pp. 28–32, Indian Institute of Metals, Pune Chapter, 2006.

[3] G. Dieter, *Mechanical Metallurgy, SI Metric Edition*, McGraw-Hill, London, UK, 1988.

[4] W. D. Callister, *Fundamentals of Materials Science and Engineering*, John Wiley & Sons, Hoboken, NJ, USA, 2001.

[5] J. Hirsch, B. Skrotzki, and G. Gottstein, *Aluminium Alloys, Their Physical and Mechanical Properties*, Wiley-VCH, Weinheim, Germany, 2008.

[6] 2010, http://aluminium.matter.org.uk/aluselect/06_composition_browse.asp.

[7] D. Kopeliovich, Wrought aluminum-magnesium-silicon alloys (6xxx), 2010, http://www.substech.com/dokuwiki/doku.php?id=wrought_aluminum-magnesium-silicon_alloys_6xxx.

[8] 2010, http://www.aluminum.org.

[9] G. Gottstein, *Physical Foundations of Materials Science*, Springer, Berlin, Germany, 2004.

[10] F. J. Humphreys and M. Hatherly, *Recrystallization and Related Annealing Phenomena*, Elsevier, Oxford, UK, 2004.

[11] E. Paul, J. T. Black, and R. A. Kohser, *Materials and Processes in Manufacturing*, John Wiley & Sons, Hoboken, NJ, USA, 2003.

[12] R. E. Sanders Jr., "Technology innovation in aluminum products," *The Journal of Minerals*, vol. 53, no. 2, pp. 21–25, 2001.

[13] T. R. Ramachandran, "Advances in Aluminium Processing and Its Automotive Application," Workshop Lecture Notes, pp. 41–44, Indian Institute of Metals, Pune Chapter, 2006.

[14] A. L. Dons, "The Alstruc homogenization model for industrial aluminum alloys," *Journal of Light Metals*, vol. 1, no. 2, pp. 133–148, 2001.

[15] O. Engler and J. Hirsch, "Texture control by thermomechanical processing of AA6xxx Al–Mg–Si sheet alloys for automotive applications—a review," *Materials Science and Engineering A*, vol. 336, no. 1-2, pp. 250–262, 2002.

[16] N. C. W. Kuijpers, W. H. Kool, P. T. G. Koenis, K. E. Nilsen, I. Todd, and S. van der Zwaag, "Assessment of different techniques for quantification of α-Al(FeMn)Si and β-AlFeSi intermetallics in AA 6xxx alloys," *Materials Characterization*, vol. 49, no. 5, pp. 409–420, 2002.

[17] N. C. W. Kuijpers, J. Tirel, D. N. Hanlon, and S. van der Zwaag, "Quantification of the evolution of the 3D intermetallic structure in a 6005A aluminium alloy during a homogenisation treatment," *Materials Characterization*, vol. 48, no. 5, pp. 379–392, 2002.

[18] S. N. Samaras and G. N. Haidemenopoulos, "Modelling of microsegregation and homogenization of 6061 extrudable Al-alloy," *Journal of Materials Processing Technology*, vol. 194, no. 1–3, pp. 63–73, 2007.

[19] R. Vissers, M. A. van Huis, J. Jansen, H. W. Zandbergen, C. D. Marioara, and S. J. Andersen, "The crystal structure of the β' phase in Al–Mg–Si alloys," *Acta Materialia*, vol. 55, no. 11, pp. 3815–3823, 2007.

[20] N. C. W. Kuijpers, F. J. Vermolen, C. Vuik, P. T. G. Koenis, K. E. Nilsen, and S. van der Zwaag, "The dependence of the β-AlFeSi to α-Al(FeMn)Si transformation kinetics in Al–Mg–Si alloys on the alloying elements," *Materials Science and Engineering A*, vol. 394, no. 1-2, pp. 9–19, 2005.

[21] Z. Ahmad, "The properties and application of scandium-reinforced aluminum," *JOM*, vol. 55, no. 2, pp. 35–39, 2003.

[22] D. Kuhlmann-Wilsdorf, "Theory of plastic deformation," *Materials Science and Engineering A*, vol. 113, pp. 1–42, 1989.
[23] M. Schikorra, L. Donati, L. Tomesani, and A. E. Tekkaya, "Microstructure analysis of aluminum extrusion: prediction of microstructure on AA6060 alloy," *Journal of Materials Processing Technology*, vol. 201, no. 1–3, pp. 156–162, 2008.
[24] R. Lapovok, I. Timokhina, P. W. J. McKenzie, and R. O'Donnell, "Processing and properties of ultrafine-grain aluminium alloy 6111 sheet," *Journal of Materials Processing Technology*, vol. 200, no. 1–3, pp. 441–450, 2008.
[25] P. Mukhopadhyay, M. Loeck, and G. Gottstein, "A cellular operator model for the simulation of static recrystallization," *Acta Materialia*, vol. 55, no. 2, pp. 551–564, 2007.
[26] P. Mukhopadhyay and I. Samajdar, "Sources of recrystallized grains and their contributions in recrystallization of an AA3104 aluminium alloy," *Transactions of the Indian Institute of Metals*, vol. 61, no. 4, pp. 329–339, 2008.
[27] P. Mukhopadhyay, S. Biswas, and H. A. Chokshi, "Deformation characterization of superplastic AA7475 alloy," *Transactions of the Indian Institute of Metals*, vol. 62, no. 2, pp. 149–152, 2009.
[28] P. Mukhopadhyay, "Modelling concurrent static recovery of aluminium alloy during recrystallization in cellular operator model," *Aluminium*, vol. 82, article 91, 2006.
[29] M. Goerdeler, M. Crumbach, P. Mukhopadhyay, G. Gottstein, L. Neumann, and R. Kopp, "Modelling the evolution of texture, microstructure and mechanical properties during hot rolling, cold rolling and annealing of VIR [*] AA5182," *Aluminium*, vol. 80, article 666, 2004.
[30] P. Mukhopadhyay, "Simulation of recrystallization texture of AA5182 and AA3104 with experimental validation," *Aluminium*, vol. 78, article 912, 2002.
[31] L. P. Troeger and E. A. Starke Jr., "Particle-stimulated nucleation of recrystallization for grain-size control and superplasticity in an Al–Mg–Si–Cu alloy," *Materials Science and Engineering A*, vol. 293, no. 1, pp. 19–29, 2000.
[32] W.-S. Lee, J.-C. Shyu, and S.-T. Chiou, "Effect of strain rate on impact response and dislocation substructure of 6061-T6 aluminum alloy," *Scripta Materialia*, vol. 42, no. 1, pp. 51–56, 2000.
[33] Abstract of Japanese Patent Publication no. 05 263203, Patent Abstract of Japan, 1993.
[34] Abstract of Japanese Patent Publication no. 63 109146, Patent Abstracts of Japan, 1988.
[35] *Alloy and Temper Designation Systems for Aluminum and Aluminum Alloys*, vol. 2, Metals Handbook, 10th edition, 1990.
[36] S . H. Avner, *Introduction to Physical Metallurgy*, TATA McGraw Hill, New Delhi, India, 2001.
[37] S. Esmaeili and D. J. Lloyd, "Modeling of precipitation hardening in pre-aged AlMgSi(Cu) alloys," *Acta Materialia*, vol. 53, no. 20, pp. 5257–5271, 2005.
[38] D. J. Chakrabarti and D. E. Laughlin, "Phase relations and precipitation in Al–Mg–Si alloys with Cu additions," *Progress in Materials Science*, vol. 49, no. 3-4, pp. 389–410, 2004.
[39] S. J. Andersen, H. W. Zandbergen, J. Jansen, C. Træholt, U. Tundal, and O. Reiso, "The crystal structure of the β'' phase in Al–Mg–Si alloys," *Acta Materialia*, vol. 46, no. 9, pp. 3283–3298, 1998.
[40] B. Chalmers, *Proceedings of the Royal Society A*, vol. 193, p. 89, 1948.
[41] R. Clark and B. Chalmers, "Mechanical deformation of aluminium bicrystals," *Acta Metallurgica*, vol. 2, no. 1, pp. 80–86, 1954.
[42] G. J. Taylor, "Plastic strain in metals," *Japan Institute of Metals*, vol. 62, p. 307, 1938.
[43] J. F. W. Bishop and R. Hill, "XLVI. A theory of the plastic distortion of a polycrystalline aggregate under combined stresses," *Philosophical Magazine*, vol. 42, pp. 414–427, 1951.
[44] J. F. W. Bishop and R. Hill, " CXXVIII. A theoretical derivation of the plastic properties of a polycrystalline face-centred meta," *Philosophical Magazine*, vol. 42, pp. 1298–1307, 1941.
[45] U. F. Kocks, "Polyslip in polycrystals," *Acta Metallurgica*, vol. 6, no. 2, pp. 85–94, 1958.
[46] U. F. Kocks, "The relation between polycrystal deformation and single-crystal deformation," *Metallurgical and Materials Transactions*, vol. 1, no. 5, pp. 1121–1143, 1970.
[47] E. O. Hall, "The deformation and ageing of mild steel: III discussion of results," *Proceedings of the Physical Society B*, vol. 64, no. 9, pp. 747–753, 1951.
[48] N. J. Petch, "The cleavage strength of polycrystals," *Journal of the Iron and Steel Institute London*, vol. 173, pp. 25–28, 1953.
[49] N. Hansen and B. Ralph, "The strain and grain size dependence of the flow stress of copper," *Acta Metallurgica*, vol. 30, no. 2, pp. 411–417, 1982.
[50] P. K. Sharma and T. R. Ramachandran, *An Update on Grain Refinement of Aluminium Alloys*, Workshop Lecture Notes, Indian Institute of Metals, Pune Chapter, 2006.
[51] A. Cibula, "The mechanism of grain refinement of sand castings in aluminum alloys," *The Journal of the Institute of Metals*, vol. 786, pp. 321–360, 1950.
[52] M. Easton and D. St John, "Grain refinement of aluminum alloys: part I. The nucleant and solute paradigms—a review of the literature," *Metallurgical and Materials Transactions A*, vol. 30, no. 6, pp. 1613–1623, 1999.
[53] M. Easton and D. St John, "Grain refinement of aluminum alloys: part II. Confirmation of, and a mechanism for, the solute paradigm," *Metallurgical and Materials Transactions A*, vol. 30, no. 6, pp. 1625–1633, 1999.
[54] M. Easton and D. St John, "Grain refinement of aluminum alloys: part I. The nucleant and solute paradigms—a review of the literature," *Metallurgical and Materials Transactions A*, vol. 30, no. 6, pp. 1613–1623, 1999.
[55] T. R. Ramachandran, *Trace Elements in Aluminium and Aluminium Alloys*, Workshop Lecture Notes, Indian Institute of Metals, Pune Chapter, 2006.
[56] J. W. Martin, *Precipitation Hardening*, Pergamon Press, New York, NY, USA, 1968.
[57] E. Orowan, *Symposium on Internal Stresses*, Institute of Metals, London, UK, 1947.
[58] A. Kelly, "Precipitation hardening," *Progress in Materials Science*, vol. 10, no. 3, pp. 151–391, 1963.
[59] H. Gleiter and E. Hornbogen, "Precipitation hardening by coherent particles," *Materials Science and Engineering*, vol. 2, no. 6, pp. 285–302, 1967.
[60] N. F. Mott and F. R. N. Nabarro, "An attempt to estimate the degree of precipitation hardening with a simple model," *Proceedings of the Physics Society London*, vol. 52, article 86, 1940.
[61] V. Gerold and H. Haberkorn, "On the critical resolved shear stress of solid solutions containing coherent precipitates," *Physica Status Solidi*, vol. 16, no. 2, pp. 675–684, 1966.
[62] P. B. Hirsch and A. Kelly, "Stacking-fault strengthening," *Philosophical Magazine*, vol. 12, no. 119, pp. 881–900, 1965.
[63] L. M. Brown and R. K. Hamm, in *Strengthening Methods in Crystals*, A. Kelly and R. B. Nicholson, Eds., p. 10, Elsevier, London, UK, 1971.

[64] D. McLean, *Mechanical Properties of Metals*, John Wiley & Sons, New York, NY, USA, 1962.
[65] M. A. Meyers and K. K. Chawla, *Mechanical Metallurgy*, Prentice Hall, Englewood Cliffs, NJ, USA, 1984.
[66] F. Roters, D. Raabe, and G. Gottstein, "Work hardening in heterogeneous alloys," *Acta Materialia*, vol. 48, no. 17, pp. 4181–4188, 2000.
[67] H. Conrad, "Thermally activated deformation of metals," *Journal of Metals*, vol. 145, no. 3638, pp. 582–588, 1964.
[68] A. Seeger, *Dislocations and Mechanical Properties of Crystals*, John Wiley & Sons, New York, NY, USA, 1957.
[69] R. P. Garrett, J. Lin, and T. A. Dean, "An investigation of the effects of solution heat treatment on mechanical properties for AA 6xxx alloys: experimentation and modelling," *International Journal of Plasticity*, vol. 21, no. 8, pp. 1640–1657, 2005.
[70] L. P. Troeger and E. A. Starke Jr., "Microstructural and mechanical characterization of a superplastic 6xxx aluminum alloy," *Materials Science and Engineering A*, vol. 277, no. 1-2, pp. 102–113, 2000.
[71] T. R. Ramachandran, P. K. Sharma, and K. Balasubramanian, *Aluminium for Automotive Applications*, Workshop Lecture Notes, Indian Institute of Metals, Pune Chapter, 2006.
[72] *Proceeding of Automotive Committee*, The Japan Aluminium Association, 2002.
[73] J. Hirsch, *Automotive Trends in Aluminium, The European Perspective*, Workshop Lecture Notes, Indian Institute of Metals, Pune Chapter, 2006.
[74] T. Sheppard, *Extrusion of Aluminium Alloys*, Kluwer Academic, Dodrecht, The Netherlands, 1999.
[75] G. B. Burger, A. K. Gupta, P. W. Jeffrey, and D. J. Lloyd, "Microstructural control of aluminum sheet used in automotive applications," *Materials Characterization*, vol. 35, no. 1, pp. 23–39, 1995.

11

Densification Behaviour in the Fabrication of Al-Fe Metal Matrix Composite Using Powder Metallurgy Route

Siddhartha Tiwari,[1,2] Priyanka Rajput,[1] and Sanjay Srivastava[1]

[1] *Department of Materials Science and Metallurgical Engineering, University Institute of Engineering & Technology, CSJM University, Kanpur 208024, India*
[2] *Department of Metallurgical and Materials Engineering, Indian Institute of Technology Kharagpur, Kharagpur 721302, India*

Correspondence should be addressed to Siddhartha Tiwari, iitkgp.siddhartha@gmail.com

Academic Editors: Y. S. Chang and S. C. Wang

Densification behaviour of the powder particles in two- and/or multicomponent system is unpredictable. The present work deals with the densification behaviour of Al-Fe powder particles during compaction and sintering in order to fabricate the Al-Fe metal matrix composites by powder metallurgy route. Green compacts of Al-6.23 wt.% Fe powder particles were fabricated under varying compaction pressures, and these fabricated green compacts were sintered over a series of temperatures (430°C–590°C). The sintered products have been characterised with the help of X-ray diffraction (XRD) and scanning electron microscope attached with energy dispersive spectroscopy (EDS). It has been found that the green density of the compacts increases with increase in compaction pressure. The rate of change in green density is found to be lower at higher compaction pressures. Sintered density increases with increasing sintering temperature up to 550°C whereas a drop in sintered density is perceived at 590°C. This decrease in sintered density is considered to occur due to swelling which has been explained on the basis of the Kirkendall effect. The XRD and EDS analyses of sintered products indicated the presence of Al and Fe particles with the trace amount of intermetallics.

1. Introduction

Aluminium matrix composites (AMCs) are one of the most demanding engineering materials in the category of metal matrix composites (MMCs) due to combination of their light weight and excellent mechanical and tribological properties. These composites have been widely used for structural, nonstructural, and functional applications in automotive and aerospace industries [1–3]. Powder metallurgy (P/M) route is one of the most widely used methods for producing AMCs due to its low processing costs as well as the ease and the accuracy to obtain near net shaped components of complex geometry. The most essential steps of P/M technique are mixing of powders, compaction, and sintering [4–6]. Compaction of loose powders under externally applied pressure and sintering of green compacts produced in earlier stage are the main processing parameters, and these parameters affect the density of the final product. Densification behaviour under compaction depends mainly on the applied pressure and yield stress of the powder particles. One of the important processing steps in P/M route is sintering, which is used to produce density-controlled materials and components from metal and/or ceramic powders by applying thermal energy. It can be divided into two categories: solid-state sintering and liquid-state sintering, below and above the melting temperature respectively [7].

It is easy to predict the densification behaviour of single-component powder particles where the hardness and/or yield stress of powder particles is approximately the same. But, in two or multicomponent systems, where the hardness of each of the powder particles is different, it is difficult to apply any generalised theory or model [8]. Bouvard [8] briefly described the densification behaviour of hard and soft powders under applied pressure in order to fabricate metal ceramic composites and concluded that the mechanism of densification varies with the amount of load applied relative to the hardness of the powder particles. A number of models have been proposed to understand compaction of loose powder and densification during sintering, but most of the models are based on single-component system [9–12]. Interestingly, proposed models and theories of sintering construct fundamental ideas to understand sintering behaviour of one, two, and/or multiphase system, but in practice sintering behaviour of two or multicomponent system is unpredictable, as it may be accompanied with shrinkage, swelling, or no volume change [13–15]. For example, Kown and savitskii [15] summarised the experimental results related to the densification behaviour of two-component systems during solid-state sintering such as W-Fe, Cr-Ni, Cr-Pd, Co-Mo, Cu-Ag, Ni-Mo, Co-Fe, Fe-Co, and Fe-Cu and discussed the sintering mechanisms. It is noticed that in all the above binary component systems, the densification characteristics during sintering, like swelling, compaction, and/or no volume change, are strongly dependent on the concentration of the elements and exhibit different densification behaviour in different systems. Therefore in P/M process, to achieve optimum density of a particulate composite of two-component system, it is necessary to construct preliminarily investigations on the densification behaviour with the help of experimental results. In a comprehensive study, it is found that most of the earlier investigations, related to Al-Fe binary phase system, are focused either on liquid-phase sintering and/or iron-rich composition (more than 20 wt.% Fe) [2, 16]. Therefore in the present investigation, an attempt has been made to examine the compaction and sintering behaviour of Al-rich Al-Fe binary phase system.

This experimental study deals with the densification behaviour of mixture of Al-6.23 wt.% Fe powders under compaction and sintering to produce Al-Fe metal matrix composite by P/M route. The major objectives of present investigation are to study (a) the effect of compaction pressure on the green density of compacts, prepared under 133–399 MPa compaction pressure and (b) the effect of sintering temperature on the densification of each green compacts in the temperature range of 430°C–590°C. The microstructures of sintered products have been characterised using scanning electron microscope (SEM) and energy dispersive spectroscopy (EDS). X-ray diffraction (XRD) analysis has been carried out to examine the existing phases after sintering. The obtained results are briefly explained to understand the mechanisms involved during the densification under compaction and sintering. The swelling behaviour at higher sintering temperature is also discussed.

2. Experimental

Commercially atomized aluminium powder with 99.7% purity and electrolytic iron powder with 99.5% purity of 250–300 mesh size (supplied by Loba Chemie Private Ltd., India) have been selected for the present investigation. The appropriate amount (Al-6.23 wt.% Fe) of selected powders was properly mixed for approximately 150–180 minutes, using agate mortar and pestle. The mixture of powders (approximately 1 gm) was then subjected to compaction at pressure range of 133–399 MPa to prepare the cylindrical pellets (green compacts) of 13 mm diameter and 3 mm height. A single acting hydraulic press was used for compaction, and zinc stearate was sprayed on the inner walls of die for lubrication. Weighing was done using electronic pan balance. The prepared green compacts were then dried for 30 minutes at temperature 200°C for complete removal of moisture, entrapped gases, and other impurities. Furthermore, sintering of all the green compacts was carried out using a ceramic tubular furnace in argon atmosphere at a heating rate of 5°C/min. To investigate the variation in sintered density and change in phases, these prepared pellets were sintered for 120 minutes at temperatures 430°C, 470°C, 510°C, 550°C, and 590°C followed by furnace cooling. Theoretical density and porosity were calculated using the following equations:

$$\text{Theoretical density} = \frac{(\text{Density of Fe} \times \%\text{Fe}) + (\text{Density of Al} \times \%\text{Al})}{100},$$

$$\text{Porosity} = \left(\frac{\text{Theoretical density} - \text{Experimental density}}{\text{Theoretical density}}\right) \times 100 \quad (1)$$

The green and sintered densities of pellets were measured by weight/volume ratio as well as the Archimedes principle (water displacement technique) to ensure the appropriateness of values of densities. The microstructure of the prepared composite was examined with the help of a scanning electron microscope (SEM, model: JEOL, JSM-5800, JEOL Ltd., Japan). The elemental analysis of the specimens was performed using an energy dispersive X-ray (EDX) microanalyser equipped with scanning electron microscope (SEM). Metallographic samples were prepared by standard procedure, that is, first ground up to 400 grade emery papers and then polished up to 0.25 μm diamond finish using Meta Serve 250 Grinder-Polisher made by BUEHLER group. The polished samples were then etched by Keller's reagent. X-ray diffraction (XRD) analyses have been performed for the identification of the phases with PHILIPS PW1792 diffractometer, operated with Cu Kα radiation. The obtained peaks, corresponding to the phases, were identified by matching with the JCPDS standards using X'Pert High Score Plus software.

3. Results and Discussion

3.1. Effect of Compaction Pressure: Variation in Green Density. The study deals with the densification behaviour of Al-rich Al-Fe metal matrix composite during sintering. Al-Fe green compacts of well-mixed Al and Fe powders (6.23 wt.% Fe) were prepared. Theoretical density of these green compacts was calculated to be 3.02 g/cm^3. The compaction pressure within the range of 133–399 MPa was applied for preparing the green compacts. Figure 1(a) shows the variation in green density and Figure 1(b) illustrates the relative change in theoretical density (compaction) as well as relative change in

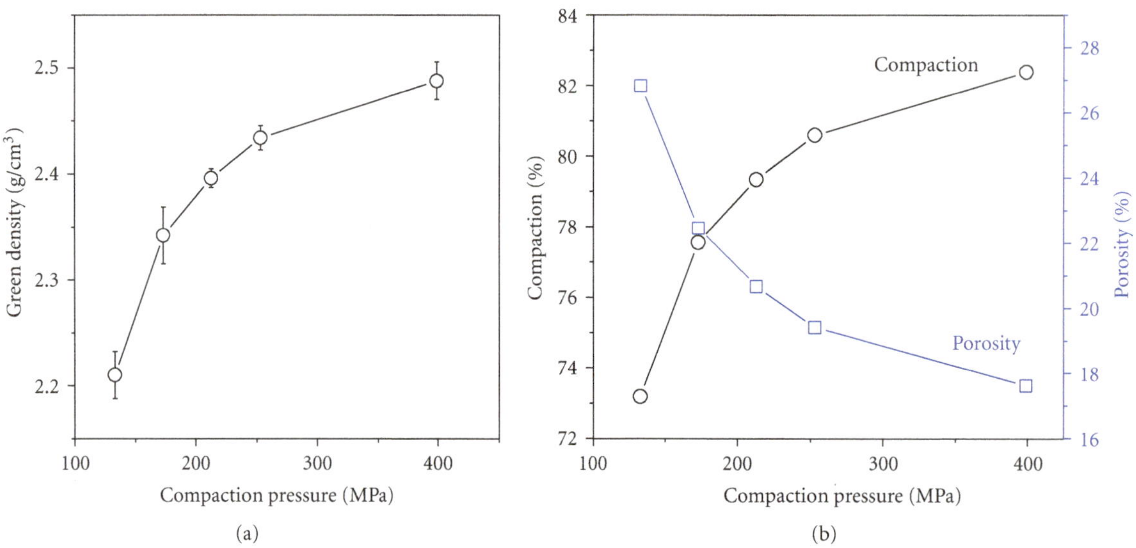

FIGURE 1: Variation in (a) green density and (b) compaction (theoretical density) with respect to compaction pressure.

porosity with respect to compaction pressure. The maximum green density of the prepared green compact is found ~2.5 g/cm^3 at 399 MPa pressure which is ~82% of the theoretical density. The percentage change in theoretical density of the green compacts is termed as compaction. It is found that the green density of the compacts increases with increasing compaction pressure, but the rate of increase in green density decreases with the increase in compaction pressure. A similar kind of trend is reported earlier also [8, 17]. The most basic and efficient processes that occur during compaction under pressure are the bulk movement or the rearrangement of particles and the deformation of particles. Bulk movement of particles depends upon the particle to particle contact and free space into which the particle may move. This movement is limited by the frictional forces developed between neighbouring particles and between particles and die surfaces. Therefore, the relative ease of such motion increases with decrease in apparent density and particle size. In the case of die compaction, there is some lateral motion due to restraining action of blocking particles and the availability of free spaces. The bulk movement of the particles is thus restrained by the powder characteristics which increase the frictional forces. The movement of particles within the powder mass takes place at relatively low pressures and accounts for the early densification of the material. The densification during compaction due to plastic deformation of individual particles is also an important mechanism. In this densification process the effectiveness of external pressure is highest at low pressures and becomes progressively less effective with increasing pressure [8, 17, 18].

A relative decrease in porosity from ~27% to ~17% can also be seen in Figure 1(b) with increase in compaction pressures and green density. Generally, in case of pure aluminium, around 90% of theoretical density can be achieved in the selected pressure range, but due to the presence of the iron particles in the Al-Fe powder mixture, only ~82% of theoretical density is achieved. As aluminium particles are soft and have lower yield strength with regards to the applied pressures, the soft aluminium particles play the major role in the densification. On the other hand the iron particles are relatively hard and hinder the densification [8, 17, 19].

3.2. Effect of Sintering Temperature: Variation in Sintered Density. The densification behaviour during sintering of selected Al-Fe green compacts is illustrated in Figure 2. Figure 2(a) represents variation in sintered density with respect to green density at each sintering temperature. It is observed that the rate of change of sintered density is higher for the compacts of lower green density whereas its magnitude is lower for the compacts of higher green density. The reason behind this kind of behaviour is associated with the driving force required for sintering. In the earlier stages of sintering, the driving force is produced by the surface energy, which is associated with the internal surface area of the particles. For sintering, a decreasing green density signifies an increase in the amount of internal surface area and, consequently, a greater driving force. Therefore, rate of change in sintered density resulting from Figure 2(a) increases with decrease in green density [7].

In Figure 2(b), it has been found that the sintered density increases with increase in sintering temperature up to 550°C (which is 75% of the melting temperature of the matrix); however, a decrease in sintered density is observed above this temperature for all the green compacts. Since sintered density is associated with the dimensional changes of green compacts, therefore it can be stated that the volume shrinkage is taking place up to 550°C whereas a swelling of compact is taking place at 590°C. It is difficult to generalise any theory to describe this kind of sintering behaviour of two- or multicomponent system. In general, sintering involves initial bonding among particles, neck growth, pore channel closure,

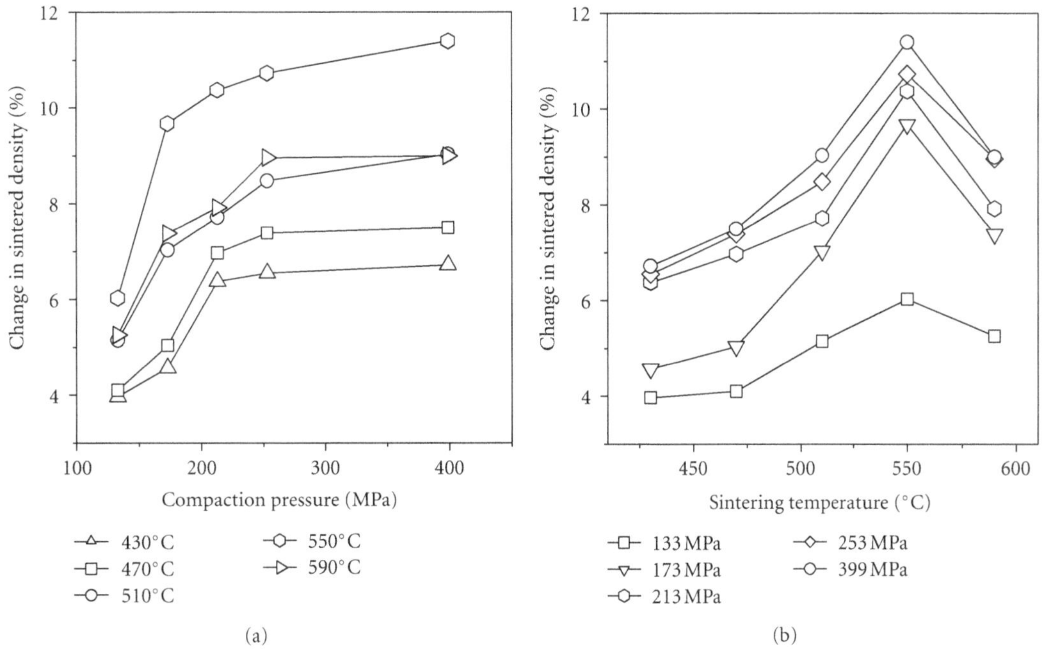

FIGURE 2: Variation in % change in sintered density with respect to (a) compaction pressure and (b) sintering temperature.

pore rounding, and densification or pore shrinkage. The alliance of these stages is basically accomplished by lowering of free energy of the system which is associated with the diffusion of particles and vacancy [8, 13–15]. Various models have been proposed for the neck formation and growth of spherical particles due to vacancy and grain boundary diffusion [20, 21].

In actual practice, particles associated with the sintering mechanism are irregular in size and morphology. The basic mechanism which involves transportation of the matter during sintering is proposed by diffusion process for both single- and two-component systems and can be well described with the help of the Kirkendall effect [15, 22]. Here, up to a certain temperature range, volume shrinkage takes place, and after that temperature, volume expansion is observed which can also be pronounced as swelling. The Kirkendall effect is nothing but diffusion of vacancies and/or atoms due to concentration and/or temperature gradient [23]. In the present situation, this expected mechanism is taking place between aluminium–aluminium particles and aluminium–iron particles. In both cases, grain boundaries and exterior surfaces are probable positions for both the creation and elimination of vacancies [15, 24]. The experimental facts best concur the concept that is the most important mechanism for the creation of vacancies is dislocation climb and that dislocation climb and void formation account for most of the absorbed vacancies, positive climb being associated with the removal of vacancies and negative climb with the creation of vacancies. In aluminium, vacancy diffusion is the most favourable process, because in face centred cubic (FCC) system, the least amount of thermal energy is required to activate the vacancy movement [25]. Let us assume a situation when two aluminium particles are in contact with each other (no grain boundary between the particles) and a small neck of high curvature forms. This high curvature is the main cause for the vacancy accumulation near the neck region and results into a concentration gradient of vacancies, which provides a driving force for diffusion. Vacancies moving away from the neck translate material moving towards the neck, and neck radius rapidly increases as material is deposited there, which results in neck growth and system shrinkage [20]. If the particles are dissimilar in size, then differences in curvature along the particle surface produce stresses that produce a difference in chemical potential. These chemical potential differences are translated to vacancy concentration differences. In most of the cases where single-component system is considered, volume shrinkage takes place. It is also well known that sintering is extremely dependent on temperature. An increase in the sintering temperature greatly increases the rate and magnitude of any changes occurring. At higher temperature, translation of vacancies increases which helps in increase of density and/or volume shrinkage [15, 20, 21].

In body-centered cubic (BCC) metals like α-iron, diffusion by ring mechanism requires less thermal energy than vacancy diffusion; the discovery of the Kirkendall effect in BCC metals indicates that these metals also diffuse by a vacancy mechanism [23]. Now the question arises, how a diffusion mechanism takes place between two particles of different materials where substitutional solid solution has been produced as a result of formation of intermetallic compound, which also may be a cause of swelling [16, 21]. This type of mechanism is well described by Kwon and savitskii [15] with the help of the Kirkendall effect for many binary metal powder mixtures. In this proposed mechanism, concentration stresses arise between two different powder

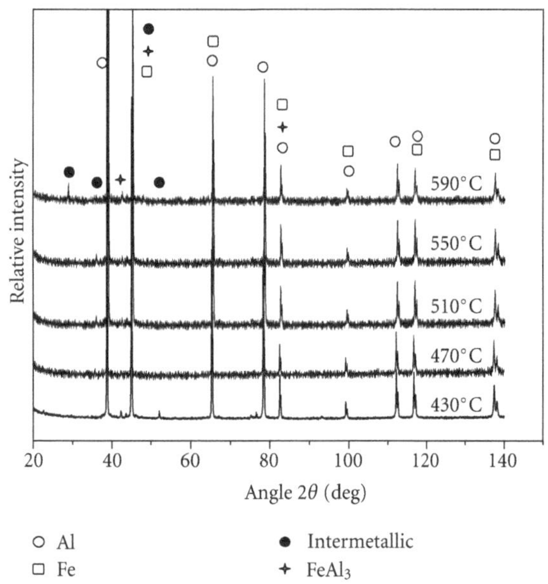

FIGURE 3: The XRD results of Al-6.23 wt.% Fe compacted at 399 MPa and sintered at various temperatures for 120 minutes.

components mixed in diffusion zone because of the differences in type of crystal lattice, lattice parameters, elastic modules, and so on. Also, the presence of grain boundaries between the particles acts as a sink for vacancies [15, 21]. Diffusion process is motivated by driving forces which come up due to temperature gradient, concentration gradient, mobility of atoms, and so forth. The Kirkendall experiment demonstrates the rate at which the diffusion of two types of atoms of binary solution is not the same; therefore mobility of atoms is one of the important parameters [23]. In Al-Fe powder mixture, aluminium has lower melting point than iron; therefore when temperature is raised up to 590°C, the mobility of aluminium particles increases drastically as this temperature is close to the melting temperature of aluminium. High diffusivity of aluminium [26] causes more vacancy creation which results in void formation as well as substitutional solid solution of Al-Fe particles.

3.3. X-Ray Diffraction and Microstructural Analyses. XRD analysis has been carried out in order to find out the phases present in a sintered product. It should be noted that as illustrated in Section 3.2 (Figure 2(b)), green compacts prepared under the compaction pressure of 399 MPa produced maximum change in sintered density in the temperature range of 430°C to 590°C. Therefore, only the above specimens have been considered for further studies like phase, elemental, and microstructural analyses. The diffraction patterns of the selected samples are shown in Figure 3.

The most prominent and relatively high intense peaks of Al and α-Fe were identified, as matched with the JCPDS file no. 00-001-1180 for Al and 00-006-0696 for Fe. Relatively high intensities of Al and α-Fe corresponding to their utmost amount in sintered products. On the other hand, a number of tiny peaks correspond to the intermediate phases like $FeAl_3$ (JCPDS file no. 00-001-1265) have been examined. Interestingly, the existence of some new peaks has been found at 590°C which does not have any correspondence to either Al or Fe and neither with aluminium oxide. On rigorous analysis, the closest match found to the referred peaks was of $Al_{13}Fe_4$ (JCPDS file no. 00-050-0797), but any strong evidence of existence of $Al_{13}Fe_4$ phase is still unavailable. Although these peaks were absent at 430°C and 470°C, a deficient amount has been noticed at 510°C and 550°C. Consequently, appearance of this intermediate phase at 590°C could be a decisive cause of decrease in sintered density (swelling).

A brief study of earlier investigations related to Al-Fe binary system describes detailed ideas regarding the formation of intermetallic compounds, produced by a variety of techniques like mechanical alloying and solid- and liquid-phase sintering. For example, Kang and Hu [16] reported the swelling behaviour of reactive sintering of Fe-29 at.% Al mixtures due to the formation of an intermediate phase having chemical formula Fe_2Al_5 at the temperature range of 500°C–750°C. One of the authors of the present investigation has already reported the formation of Al_3Fe intermetallics in the Al-Fe composites which has been prepared by dispersion of iron in aluminium melt by impeller mixing and bottom pouring chill casting technique [2]. The iron content for the above investigation varies from 1.67 to 11.2 wt%. This report suggests that the formation of intermetallics can also take place at low concentration of iron in aluminium–iron binary mixtures. Olszówka-Myalska et al. [27] characterised intermetallic compounds in a aluminium–iron metal matrix composite in the composition range of 3–6 vol.% Fe at hot pressing temperature of 630°C for the time duration of 5–30 minutes. But they have not mentioned the chemical formula of the characterised intermetallic compound. In a study of structural evaluation of mechanically alloyed Al-Fe powders, the authors reported a mixed amorphous phase and solid solution of Fe in Al, having a composition of Al-10 at.% Fe. It has also been recommended that the formation of $FeAl_3$ intermetallic can take place after heat treatment of milled powders within a range of 4 to 10 at.% Fe [28]. As discussed in Section 3.2, in solid-state sintering, formation of intermetallic compound takes place due to interdiffusion of atoms. This interdiffusion strongly depends on diffusivity of each element. Although iron has extremely low solubility in aluminium, the diffusivity of iron in aluminium was found to be maximum at temperature range 550°C–630°C in an investigation carried out from 350°C to 630°C [26].

Moreover, in the diffusion couple of aluminium and iron, aluminium can move faster than iron because of its low melting temperature. On the basis of earlier reports and the present investigation, it can be emphasized that the formation of intermetallic compounds or solid solution can take place in aluminium-rich composition of aluminium-iron composites or alloys below the melting point of aluminium. Here, it is expected that the formation of a new phase is taking place at 590°C in a considerable amount and a remarkable intense peak of intermetallic most likely $Al_{13}Fe_4$ has been marked. Hence, the existence of this kind of solid

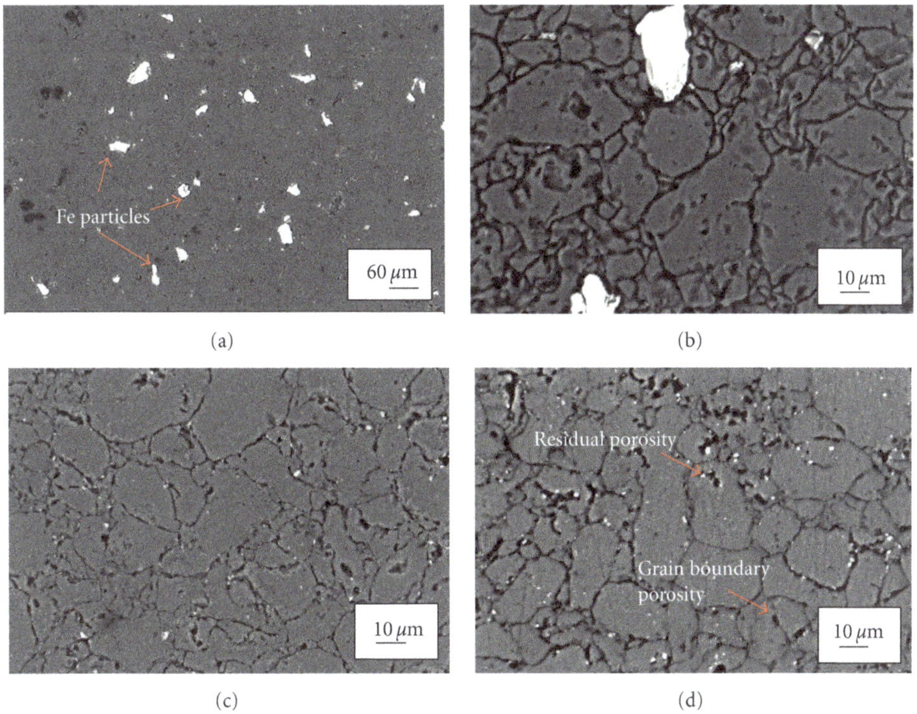

FIGURE 4: Typical SEM microstructure of Al-Fe composite shows (a) distribution of Fe particles over Al matrix sintered at 550°C, residual and grain boundary porosities (b) sintered at 510°C, (c) sintered at 550°C, and (d) 590°C.

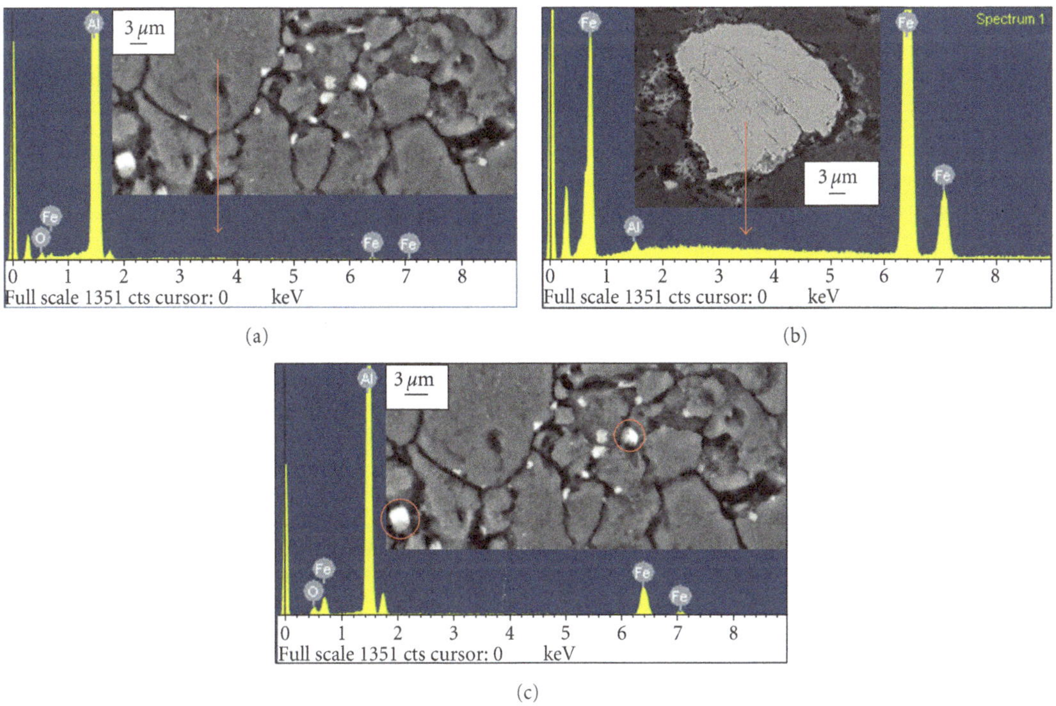

FIGURE 5: EDS spot analysis of sintered Al-Fe metal matrix composite: (a) Al matrix, (b) Fe particle consists small amount of Al, and (c) small particles (intermetallics) exhibit considerable amount of Al and Fe.

solution can also account as a reason for decreased sintered density at this temperature.

Figure 4 represents the SEM images of the Al-Fe sintered product. The distribution of Fe in Al matrix sintered at 550°C can be easily seen in Figure 4(a) at relatively lower magnification. Figures 4(b)–4(c) show the matrix of sintered product, sintered at 510°C, 550°C, and 590°C, respectively. It is also noticed that very tiny iron particles are distributed throughout the aluminium matrix along the grain boundary and throughout the grains. Microstructures also represent reduction of porosity along the grain boundaries and residual porosity within grains. The concentration of the residual and grain boundary porosity is noticed more in the microstructure of specimen sintered at 590°C.

The EDS profile corresponding to the microstructures is also given as an evidence of elements present in the sintered product. Figure 5 illustrates the points at which EDS is performed, and the intensity of peaks is corresponding to the elements present at that point. The existence of elements in all possible phases such as presence of Al in Fe particle and Fe in Al matrix is shown in Figures 5(a) and 5(b), respectively. The EDS result of tiny particles shows existence of both Al and Fe in the particles in Figure 5(c).

4. Conclusions

In order to achieve maximum sintered density in the fabrication of Al-rich Al-Fe (Al-6.23 wt.% Fe) metal matrix composites, optimum sintering temperature is found to be 550°C (which is 75% of the melting temperature of the matrix). Sintered density increases with increase in sintering temperature up to 550°C and decreases with further increase in temperature. A drop in sintered density, that is, swelling, is found at 590°C for all the green compacts. This swelling behaviour is considered to occur due to the formation of excess vacancies and Al-Fe solid solution of lower density. These phenomena can be well described with the help of the Kirkendall effect. Sintered density is found to increase with a higher rate for the green compacts of lower green density as compared to that of higher green density. Maximum green density of the compacts, that is, ~2.5 g/cm^3, is found at 399 MPa which is ~82% of the theoretical density. As expected, the green density of the compacts increases with increase in compaction pressure whereas the rate of increase in green density decreases with the increasing compaction pressure.

Acknowledgments

The authors are thankful to Dr. Rakesh Bhatnagar (Faculty, UIET, CSJM University, Kanpur) and DMSRDE DRDO, Kanpur, India, for the technical discussion and support throughout the work.

References

[1] M. K. Surappa, "Aluminium matrix composites challenges and opportunities," *Sadhana*, vol. 28, p. 319, 2003.

[2] S. Mohan and S. Srivastava, "Surface behaviour of asCast Al-Fe intermetallic composites," *Tribology Letters*, vol. 22, no. 1, pp. 45–51, 2006.

[3] A. Heinz, A. Haszler, C. Keidel, S. Moldenhauer, R. Benedictus, and W. S. Miller, "Recent development in aluminium alloys for aerospace applications," *Materials Science and Engineering*, vol. 280, no. 1, pp. 102–107, 2000.

[4] J. M. Torralbaa, C. E. da Costab, and F. Velasco, "P/M aluminum matrix composites: an overview," *Journal of Materials Processing Technology*, vol. 133, pp. 203–206, 2003.

[5] J. R. Pickens, "Aluminium powder metallurgy technology for high-strength applications," *Journal of Materials Science*, vol. 16, no. 6, pp. 1437–1457, 1981.

[6] Y. B. Liu, S. C. Lim, L. Lu, and M. O. Lai, "Recent development in the fabrication of metal matrix-particulate composites using powder metallurgy techniques," *Journal of Materials Science*, vol. 29, no. 8, pp. 1999–2007, 1994.

[7] S.-J. L. Kang, *Sintering: Densification, Grain growth and Microstructure*, Jorden Hill, Oxford, UK, 1st edition, 2005.

[8] D. Bouvard, "Densification behaviour of mixtures of hard and soft powders under pressure," *Powder Technology*, vol. 111, no. 3, pp. 231–239, 2000.

[9] M. Braginsky, V. Tikare, and E. Olevsky, "Numerical simulation of solid state sintering," *International Journal of Solids and Structures*, vol. 42, no. 2, pp. 621–636, 2005.

[10] J. Svoboda and H. Riedel, "New solutions describing the formation of interparticle necks in solid-state sintering," *Acta Metallurgica et Materialia*, vol. 43, no. 1, pp. 1–10, 1995.

[11] C. L. Martin and D. Bouvard, "Study of the cold compaction of composite powders by the discrete element method," *Acta Materialia*, vol. 51, no. 2, pp. 373–386, 2003.

[12] T. Kraft and H. Riedel, "Numerical simulation of solid state sintering; model and application," *Journal of the European Ceramic Society*, vol. 24, no. 2, pp. 345–361, 2004.

[13] G. S. Upadhyaya, "Future directions in sintering research," *Science of Sintering*, vol. 43, no. 1, pp. 3–8, 2011.

[14] G. S. Upadhyaya, "Some issues in sintering science and technology," *Materials Chemistry and Physics*, vol. 67, no. 1–3, pp. 1–5, 2001.

[15] Y.-S. Kwon and A. Savitskii, "Solid-state sintering of metal powder mixtures," *Journal of Materials Synthesis and Processing*, vol. 9, pp. 299–317, 2001.

[16] H.-Z. Kang and C. -T. Hu, "Swelling behavior in reactive sintering of Fe-Al mixtures," *Materials Chemistry and Physics*, vol. 88, no. 2-3, pp. 264–272, 2004.

[17] N. Showaiter and M. Youseffi, "Compaction, sintering and mechanical properties of elemental 6061 Al powder with and without sintering aids," *Materials & Design*, vol. 29, no. 4, pp. 752–762, 2008.

[18] M. N. Rahaman, *Ceramic Processing and Sintering*, CRC Press, New York, NY, USA, 2nd edition, 2005.

[19] H. Masuda, K. Higashitani, and H. Yoshida, *Powder Technology Handbook*, Taylor and Francis group, New york, NY, USA, 3rd edition, 2006.

[20] H. Djohari, J. I. Martínez-Herrera, and J. J. Derby, "Transport mechanisms and densification during sintering: I. Viscous flow versus vacancy diffusion," *Chemical Engineering Science*, vol. 64, no. 17, pp. 3799–3809, 2009.

[21] H. Djohari and J. J. Derby, "Transport mechanisms and densification during sintering: II. Grain boundaries," *Chemical Engineering Science*, vol. 64, no. 17, pp. 3810–3816, 2009.

[22] P. F. Stableinjr and G. C. Kuczynskit, "Sintering in multicomponent metallic systems," *Acta Metallurgica*, vol. 11, no. 12, pp. 1327–1337, 1963.

[23] R. E. Reed-Hill and R. Abbaschian, *Physical Metallurgy Principles*, PWS Pub. Co., Boston, UK, 3rd edition, 1994.
[24] R. W. Balluffi and L. L. Seigle, "Growth of voids in metals during diffusion and creep," *Acta Metallurgica*, vol. 5, no. 8, pp. 449–454, 1957.
[25] R. S. Barnes, "The climb of edge dislocations in face-centred cubic crystals," *Acta Metallurgica*, vol. 2, no. 3, pp. 380–385, 1954.
[26] K. I. Hirano, R. P. Agarwala, and M. Cohen, "Diffusion of iron, nickel and cobalt in aluminum," *Acta Metallurgica*, vol. 10, no. 9, pp. 857–863, 1962.
[27] A. Olszówka-Myalska, J. Szala, and J. Cwajna, "Characterization of iron aluminides formed in situ in an aluminium matrix composite," *Materials Characterization*, vol. 56, no. 4-5, pp. 379–383, 2006.
[28] D. K. Mukhopadhyay, C. Suryanarayana, and F. H. Froes, "Structural evolution in mechanically alloyed Al-Fe powders," *Metallurgical and Materials Transactions A*, vol. 26, no. 8, pp. 1939–1946, 1995.

12

A Kinetic Study of the Emerging of Grains and Block of Grains from the Inner Volume to the Free Surface of a Cd-Zn Alloy Superplastically Deformed

J. A. Montemayor-Aldrete,[1] R. Mendez-Olvera,[1] P. Ugalde-Velez,[2] G. Torres-Villaseñor,[3] and M. del Castillo-Mussot[1]

[1] *Departamento de Estado Solido, Instituto de Física, Universidad Nacional Autónoma de México, Apartado Postal 20-364, 01000 Mexico City, DF, Mexico*
[2] *Departamento de Materiales (CBI), Universidad Autónoma Metropolitana-Azcapotzalco, Avenida San Pablo 180, Col. Reynosa, 02200 Mexico City, DF, Mexico*
[3] *Instituto de Investigación en Materiales, Universidad Nacional Autónoma de México, Apartado Postal 70-360, 04510 Mexico City, DF, Mexico*

Correspondence should be addressed to J. A. Montemayor-Aldrete, jamonal@fisica.unam.mx

Academic Editors: R. Hebert and P. Lukac

A local measurement technique for the study of the kinetic processes of emerging of grains or blocks of grains from the inner volume to the free surface of superplastic materials during deformation is presented and used for the case of the Cd-Zn eutectic alloy deformed at room temperature. This technique could be used to evaluate the approximate time of fracture due to fissure or cavitation growth in samples under superplastic deformation. In principle, this technique will be useful for the development of physical procedures, which allows retarding the process of formation of low mismatch angle, $\overline{\theta}$, between neighboring grains, process which gives place to blocks of grains which dynamically behave as units under the shear stress action. For materials with nanocrystalline structures, such processes are expected to be higher than those of the case of microcrystalline materials.

1. Introduction

Superplasticity is the ability of polycrystalline material to exhibit, in a generally isotropic way, very high tensile elongations prior to failure [1]. Such phenomena occur mainly by grain-boundary sliding (GBS), between individual grains, with size lower than 10 micrometers, which glide relative to each other with little or no change in shape during long plastic deformation [2, 3]. Also, there exists some experimental evidence for cooperative grain-boundary sliding during superplastic deformation [4–7]. Most of the early studies dealt with microduplex metallic alloys which make the grain growth very difficult, and the unabated scientific interest on superplasticity has from long ago broadened the scope to include intermetallic compounds and nanocrystalline materials [8]. This interest has arisen partly from a scientific viewpoint and partly from the increasing awareness that superplastic materials can be utilized in forming complex shapes in simple and inexpensive forming operations. There are two main advantages in utilizing superplastic materials for metal forming operations. First, large strains can be achieved without necking. Second, the stresses required for superplastic deformation are generally low [9]. Recently, a model which was validated against experimental observations concerning metals, alloys and ceramics of micrometer and submicrometer grain sizes, nanostructured materials, and intermetallics has been proposed to account for optimal superplasticity; their central assumption is that the rate controlling deformation process on GBS is confined to high-angle grain/interphase boundary regions that are essential for grain-boundary sliding developing to a mesoscopic scale (defined to be of the order of a grain diameter or more) and for superplastic flow setting in [10]. There are numerous reports of cavity nucleation and growth at grain boundaries during superplastic flow in several different alloys, and some experimental data in Zn-Al alloy shows that the cavities are initially reasonably spherical, but they become elongated when the sample pulls out to achieve the maximum superplastic elongation [11]; there are also experimental data which states that superplastic deformation is usually limited by the development of voids or cavities which are seen to form at triple point grain boundaries during the process of GBS [12]. And also there exists direct correlation between the number of cavities nucleated per unit volume and the cumulative average GBS displacement which provides experimental evidence that GBS is the driving force for creep cavity nucleation [13].

It is well known [14, 15] that superplastic flow of a material is often described by a constitutive equation of the form

$$\dot{\epsilon} = \frac{ADGb}{kT}\left(\frac{b}{d}\right)^{p}\left(\frac{\sigma}{G}\right)^{n}, \tag{1}$$

where $\dot{\epsilon}$ is the deformation rate, D is the coefficient of grain-boundary diffusion, G is the shear modulus, b is the modulus of Burgers vector, k is the Boltzmann constant, T is the test temperature in Kelvin degrees, d is the average grain size involved in the grain-boundary sliding processes, p is an exponent depending on the grain size (commonly equal to 2), σ is the flow stress, n is a parameter inverse to the coefficient of strain rate sensitivity of the flow stress, and A is a constant.

The main purpose of this paper is twofold: (1) to presents a new scanning electron microscopy technique which is applied for the case of the Cd-17.4% Zn alloy to measure the kinetically evolution of the emerging of grains or blocks of grains from the inner volume to the free surface of superplastic material during deformation and (2) to develop a simple thermodynamical model about the time evolution of block size of sliding grains during superplastic deformation in order to give some explanation about our experimental data.

2. Model

It is well known that there are some general thermodynamical properties of grain boundaries in polycrystalline solids which are relevant to the superplasticity phenomenon: (a) the grain-boundary surface energy, $\gamma(\theta)$, is an approximately linear increasing function of the mismatch angle between neighboring grains to an angle of approximately 20 degrees and then is essentially constant for all larger angles [16]. It is generally recognized that in a completely recrystallized solid, the driving force for grain growth lies in the surface energy of the grain boundaries [17]. The information on the previous section suggests that in a sample without the previous deformation, superplastic deformation starts with grain-boundary sliding of individual grains, and later according to the second thermodynamic law during the superplastic deformation such grains it tends to grow (however such phenomena, it is almost controlled by different experimental techniques) and also grows a tendency to diminish the free energy per unit volume, $G_{Pcs}(\overline{\theta})$, by diminishing the average mismatch angle, $\overline{\theta}$, between neighboring grains. This last phenomenon explains the tendency to grow the effective size on the gliding blocks formed by several grains, tendency described by the time evolution of the Gibbs energy per unit volume of the solid:

$$G_{Pcs}\left(\overline{\theta}\right) = \left(4N\pi r^{2}\right)\gamma\left(\overline{\theta}\right) - TS, \tag{2}$$

where N is the number of grains per unit volume, r is the mean grain radius, and S is the configurationally entropy per unit volume due to the grains. In other words, if during a superplastic deformation test the number of grains remains constant, the only possibility to fulfill the second thermodynamic law for a system deforming superplastically far from thermodynamic equilibrium is

$$\frac{dG_{Pcs}\left(\overline{\theta}\right)}{dt} = \left(4N\pi r^{2}\right)\frac{d\gamma\left(\overline{\theta}\right)}{dt} \leq 0. \tag{3}$$

Such condition implies that as far as the deformation test proceeds, the possibility of growing on the block size of sliding grains grows which is given by the time evolution of $\gamma(\theta)$,

$$\frac{d\gamma\left(\overline{\theta}\right)}{dt} \leq 0. \tag{4}$$

Many physical phenomena are described by the first-order differential equations whose solution is an exponential decay which are common in solid state physics, medicine, biology and biophysics, geophysics, optics, engineering, chemistry, and electrochemistry [18]. In particular for polycrystalline materials, it is known that the general equations for the annealing of vacancies in metals containing impurities has shown that the decay curve is exponential [19], and also there is an exponential decay in nonlinear thermoelasticity [20], and in dislocation creep during a total unloading test, there is an exponential decay [21]; in all these cases, the system evolution occurs in a monotonic way and with an exponential decay of the free energy to its equilibrium value. Based on the previous data, it is possible to consider that during superplastic deformation the Gibbs free energy per unit volume $G_{Pcs}(\overline{\theta})$ decays exponentially with time, in other words,

$$\frac{dG_{Pcs}\left(\overline{\theta}\right)}{dt} = -\alpha e^{-\beta t} \leq 0, \tag{5}$$

where α is a positive constant and $\beta \geq 0$ is the decay constant. By using (3), it is straightforward that

$$\frac{d\gamma\left(\overline{\theta}\right)}{dt} = -\left(\frac{\alpha}{4N\pi r^{2}}\right)e^{-\beta t}. \tag{6}$$

In such scheme, it is possible to suppose that the local or global cooperative grain-boundary sliding occurs by the gliding of blocks of grains which glide as an physical entity and has been formed by many neighboring grains with low grain boundaries energy states (low values of $\gamma(\overline{\theta})$). As the surface energy from grain boundaries provides the driving force for grain agglomeration which gives place to blocks of grains, then the simplest way to relate the change on the grain-boundary surface energy, $\gamma(\theta)$, and the rate of grain agglomeration is the following:

$$\left(\frac{dr_{\text{Block}}}{dt}\right)\left(\frac{d\gamma(\overline{\theta})}{dt}\right) = C, \quad (7)$$

where C is assumed to be a constant and (dr_{Block}/dt) is the rate of grain agglomeration which gives place to blocks of grains. Then,

$$\left(\frac{dr_{\text{Block}}}{dt}\right) = C\left(\frac{4N\pi r^2}{\alpha}\right)e^{+(\beta t)}. \quad (8)$$

With $\beta^{-1} = \tau$ and with τ as the exponential time constant, or grow time constant, it is clear that such thermodynamical process eventually gives place to the fracture of the sample under superplastic deformation.

2.1. Experimental Procedure. The material used for this experiment was a Cd-17.4% Zn alloy. The alloy was prepared by melting cadmium and zinc in air furnace at 823°K and, then, poured into a copper ingot mold, and it was allowed to cool at room temperature. In order to avoid crack propagation during rolling due to the presence of dendrites, an annealing at 453°K was applied. The alloying element contents were measured by chemical method [22]. The material was laminated at room temperature at 4.6 m/s with three rolling reduction steps 0.25 cm each, the plate obtained by lamination resulting in a grain size of 4.9 μm. After that, flat standard specimens for tensile tests parallel to the rolling direction were prepared with the following dimensions: 0.5 cm thick, 0.20 cm wide, and 1.6 cm of gage length. The samples were mechanical polished until reaching 0.5 micrometers diamond polishing paste. The tension tests, with an average constant crosshead velocity of $1.5 * 10^{-3}$ cm/s, were conducted at room temperature into a tension plate of a scanning electron microscope (SEM) model Jeol T20 which allows amplifications from 35 to 10,000.

The new technique is a variation of one previously proposed [23–25] which provides a mesoscopic coordinate system inscribed on the surface of the tension test specimen. Such technique allows to establish in a repeatable way any angle relative to any axis of any coordinate system, or distances between grains, or to measure local or global true deformation in parallel or perpendicular directions relative to the tension axis. Here, such technique was adapted to measure grain area and perimeters of specific grains for a case where plastic deformation is so high that it is not possible to inscribe durable Vickers indentations on the surface of the deforming sample and, then, to take some particular defect of the sample as the origin of coordinated system is required.

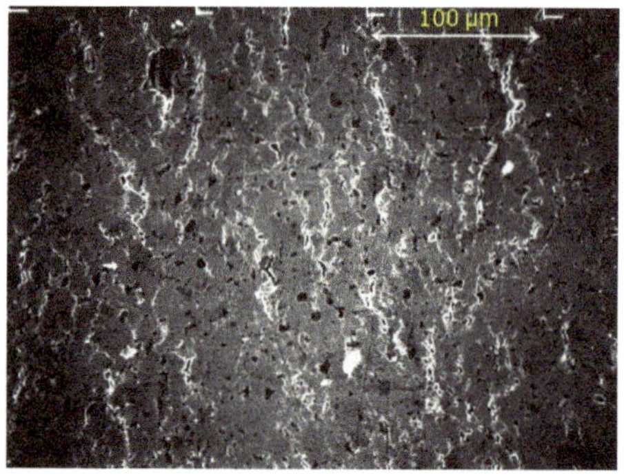

FIGURE 1: Scanning electron microscope micrograph for Cd-Zn eutectic alloy after 0.75 mm of elongation at room temperature, depicting GBS, which appears like bright lines.

3. Experimental Data

The local superficial deformation on different zones of the tension test sample was studied by using a coordinated reference system. During certain periods of time, the sample was deformed at the previously mentioned crosshead velocity; in each occasion after reaching some predetermined deformation test was interrupted to proceed to take photos of the superficial microstructure in different regions. The same procedure was repeated several times until fracture of the sample under tension occurs. The dispersion of experimental data was high, and only some data were presented to show the essential features of the corresponding phenomena. Among other parameters, the corresponding areas of emerging of grains or blocks of grains from the inner volume to the external surface of the sample, and their perimeters were measured for different times and deformations.

Figure 1 is a scanning electron microscope micrograph for Cd-Zn eutectic alloy after 0.75 mm of elongation at room temperature, depicting GBS, which appears like bright lines and corresponds to a typical zone in the surface of the deforming sample. During the experiments of sample deformation, two types of emerging populations of grains or blocks of grains from the inner volume to the external surface of the sample were found: emerging individual equiaxial grains and blocks of grains which arising from cavitation with form of cracks or fissures, as could be observed in Figure 2. These populations have different properties in their temporal evolution. In Figure 3, the rate of change of area of one emerging equiaxial grain is presented; it is clear that once the grains have emerged their changes in area are due to the interaction with neighboring grains during the plastic deformation of the sample. In Figure 4, the emergence of a block of grains appears arising from a fissure. The same zone appears in Figure 5 as a temporal sequence in a collage, where the kinetically process of the emerging of a block of grains to free surface could be appreciated; the quantitative data for such block appears in Figure 2. In Figure 6, some perimeter measurements are presented.

3.1. Analysis of Experimental Data. In order to obtain the scanning electron microscope micrograph for Cd-Zn

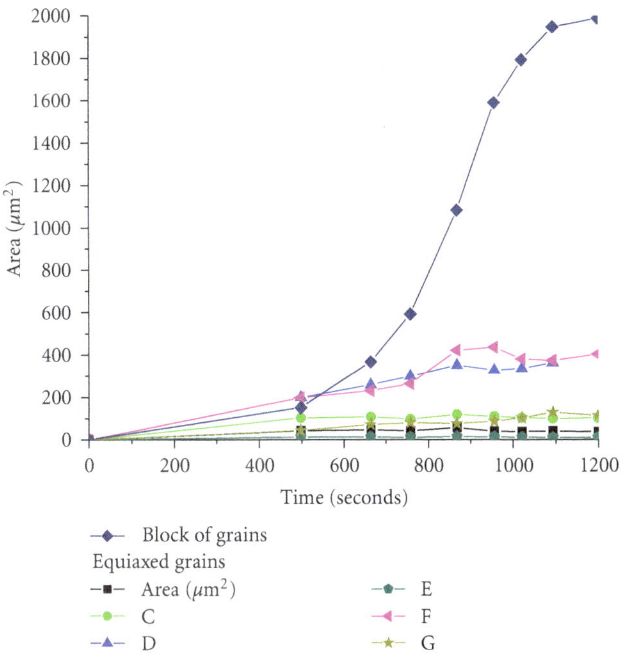

FIGURE 2: The area of grains from the inner volume emerging on the external surface of the sample during plastic deformation on a Cd-Zn eutectic alloy as a function of time.

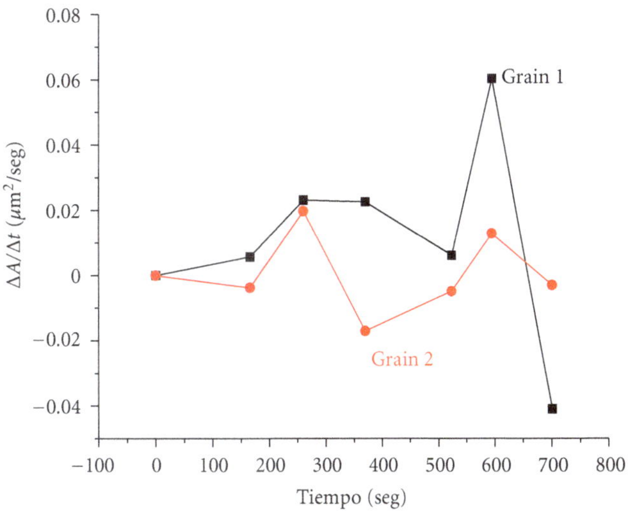

FIGURE 3: The rate of change of area of one emerging equiaxial grain emerging to the external surface from the inner volume as a function of time is presented for a Cd-Zn eutectic alloy deformed at room temperature.

eutectic alloy, it was necessary to stop the deformation for several minutes; this procedure may have led to possible changes in the microstructure observed, due to relaxation of internal stresses at the free surface. From Figure 3, it is immediate that during the plastic deformation the area of emerging equiaxed grains is changing with time by the interaction with their neighbors (above, below, and lateral neighboring grains). The positive change in area corresponds

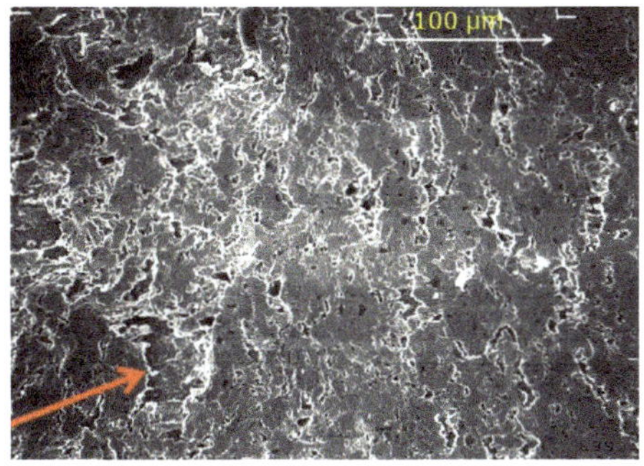

FIGURE 4: The emergence of a block of grains which travels from the inner volume and produces a fissure at the external surface.

to an emerging stage of some particular grains, and the negative area changes to a partial sinking process of the grain. Their shapes keep changing in the short term, but finally tend to reach a state of thermodynamic equilibrium (in particular mechanical equilibrium); please see Grain 2 in the figure. The materials emerging from the inside of cracks or fissures are block of grains formed by contiguous or neighboring grains which have common low-energy grain boundaries, that is, their common grain boundaries are strongly linked and slide like a unit, as can be seen from Figure 2, in which it is possible to see that the area of a totally emerged block of grains is about 26 times greater than the area of an average equiaxed grain. For our experimental data, showed in Figure 2, about material arising from the inside of cracks or fissures, it is clear that during the first 950 seconds an exponential growth on the area of the arising block of grain occurs, with a grow time constant of about 210.4 s. Therefore, with our technique, it is possible to determinate some critical time (in our case $\tau_{\text{Crit}} = 950\,\text{s}$) for which the cracks or fissures can develop in two directions: (1) decreasing its speed, due to the total emerging of blocks of grains towards the surface or (2) continuing to grow exponentially as a crack that gives rise to fracture of the tension test sample under the action of the applied stress, because the block of grains partially under the free surface is not able to glide fast enough as required to fulfill the corresponding fissure. As shown in Figure 4, which corresponds to a block of grains that appears emerging from a fissure such fissure corresponds also to one of the areas depicted in Figure 2, for the block of grains where it is possible to see that the area of this block of grains as measured at the free surface of the sample under superplastic deformation tends to an equilibrium value. The corresponding kinetically processes indicated in Figure 5.

It is clear that the technique proposed and used here for Cd-Zn eutectic alloy could be applied to the study of the superplasticity phenomena as a complementary technique to the previously existing. In particular, such technique will be useful for the study of superplasticity in materials with nanocrystalline structures where, due to thermodynamical reasons, the kinetically evolution of the emerging of grains or blocks of grains from the inner volume to the free surface of superplastic material during deformation is expected to be higher than for the case of microcrystalline materials.

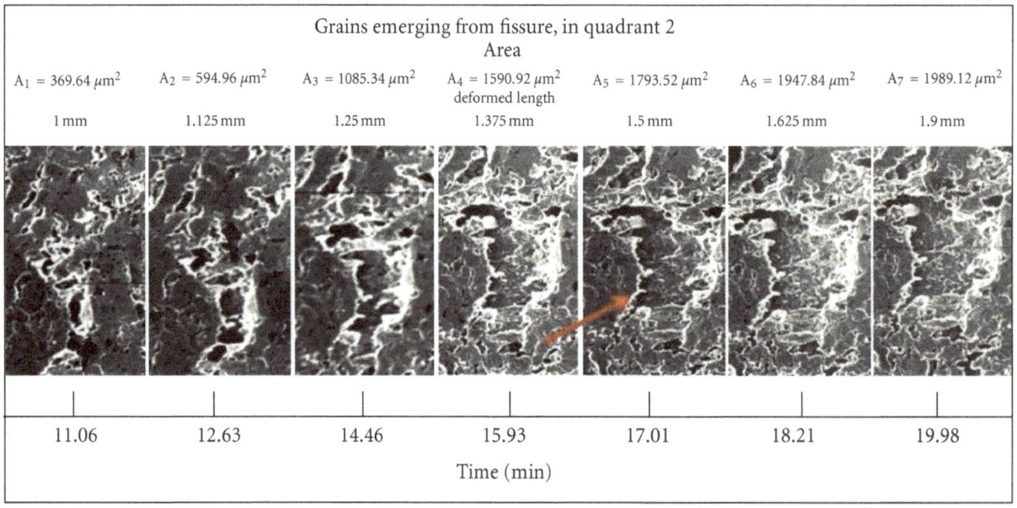

FIGURE 5: The same zone that appears in Figure 4 for the alloy Cd-Zn 17.4% deformed at room temperature is presented here as a temporal sequence in a collage. The temporal emerging process of a block of grains to the external surface of the deforming sample could be appreciated; the plot of the quantitative data of area appears in Figure 2.

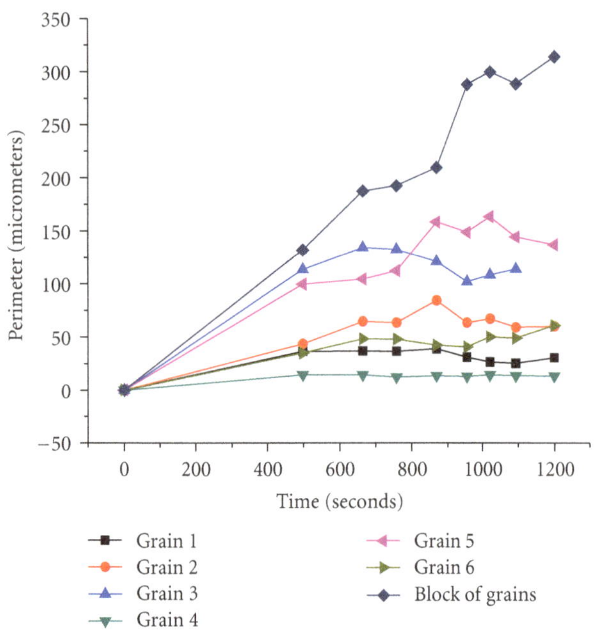

FIGURE 6: Some perimeter measurements for the alloy Cd-17.4% Zn deformed at room temperature is presented.

4. Discussion

Our experimental results is to support the basis of a theory recently published [10], as well as in a more quantitative way the model here presented (based on the second law of thermodynamics) for the diminishing on the free energy per unit volume, $G_{Pcs}(\overline{\theta})$, by diminishing the average mismatch angle, $\overline{\theta}$, between neighboring grains. And also, our results are in agreement with some experimental data for superplastic deformation in a Zn-Al alloy which shows that cavities are initially reasonably spherical, but they become elongated when the sample pulls out to achieve the maximum superplastic elongation [11]; however, in our case, such cavities are mainly fulfilled with the grains or block of grains that merge to external surface from inferior layers. And obviously, for the case of superplastic deformation of the Cd-Zn eutectic alloy deformed at room temperature, we provide direct experimental evidence that GBS is the driving force for creep cavity nucleation, as deduced previously by an experimental correlation [13]. The technique presented here will be useful for the development of physical procedures, which allows to retarding the process of formation of low mismatch angle, $\overline{\theta}$, between neighboring grains; process which gives place to blocks of grains which dynamically behaves as units under the shear stress action. For materials with nanocrystalline structures, such processes are expected to be higher than for the case of microcrystalline materials.

5. Conclusions

(1) A new local measurement technique for the study of the kinetic processes of emerging of grains from the inner volume at the free surface of superplastic materials during deformation has been presented and used for the case of the Cd-Zn eutectic alloy deformed at room temperature.

(2) This technique can be used to evaluate the approximate time to fracture by fissure growth in samples under superplastic deformation.

(3) The technique presented here will be useful for the development of physical procedures, which allows to retarding the process of block by grains agglutination in materials with nanocrystalline structures, where by thermodynamical reasons, the process of formation of low mismatch angle, $\overline{\theta}$, between neighboring grains, which gives place to blocks of grains, is

expected to be higher than for the case of microcrystalline materials.

(4) In principle, the present technique may be used to study the local surface forces which are acting over individual surface grains during the superplastic deformation.

References

[1] T. G. Langdon and J. Wadsworth, "Summary and topics of ICSAM-91," in *Proceedings of the International Conference on Superplasticity in Advanced Materials*, S. Hori, M. Tokizane, and N. Furishiro, Eds., pp. 847–852, The Japan Society for Research on Superplasticity, Osaka, Japan, 1991.

[2] M. F. Ashby and R. A. Verrall, "Diffusion-accommodated flow and superplasticity," *Acta Metallurgica*, vol. 21, no. 2, pp. 149–163, 1973.

[3] M. G. Zelin, M. R. Dunlap, R. Rosen, and A. K. Mukherjee, "The direct observation of cooperative grain-boundary sliding and migration during superplastic deformation of lead-tin eutectic in shear," *Journal of Applied Physics*, vol. 74, no. 8, pp. 4972–4982, 1993.

[4] M. G. Zelin and M. V. Alexsandrova, "Cooperative grain-boundary sliding during superplastic deformation," in *Proceedings of the International Conference on Superplasticity in Advanced Materials*, S. Hori, M. Tokizane, and N. Furishiro, Eds., pp. 95–101, The Japan Society for Research on Superplasticity, Osaka, Japan, 1991.

[5] V. V. Astanin, O. A. Kaibyshev, and S. N. Faizova, "Cooperative grain boundary sliding under superplastic flow," *Scripta Metallurgica et Materiala*, vol. 25, no. 12, pp. 2663–2668, 1991.

[6] H. S. . Yang, M. G. Zelin, R. Z. Valiev, and A. K. Mukherjee, "Strain induced morphological changes of α_2 and β phases in Ti_3Al alloys during superplastic deformation," *Scripta Metallurgica et Materialia*, vol. 26, no. 11, pp. 1707–1712, 1992.

[7] M. G. Zelin and A. K. Mukherjee, "Study of cooperative grain boundary sliding by using macroscopic marker lines," *Metallurgical and Materials Transactions A*, vol. 26, no. 3, pp. 747–750, 1995.

[8] S. Krishnamurthy and A. H. Chokshi, "High temperature deformation and densification in alumina-yttria Composites," in *Proceedings of the International Conference on Superplasticity in Advanced Materials (ICSAM '97)*, A. H. Chokshi, Ed., vol. 243–245 of *Materials Science Forum*, pp. 369–374, Trans Tech Publications, 1997.

[9] F. A. Mohamed, "Micrograin superplasticity: characteristics and utilization," *Materials*, vol. 4, no. 7, pp. 1194–1223, 2011.

[10] K. A. Padmanabhan, "Grain boundary sliding controlled flow and its relevance to superplasticity in metals, alloys, ceramics and intermetallics and strain-rate dependent flow in nanostructured materials," *Journal of Materials Science*, vol. 44, no. 9, pp. 2226–2238, 2009.

[11] M. Kawasaki and T. G. Langdon, "The development of internal cavitation in a superplastic zinc-aluminum alloy processed by ECAP," *Journal of Materials Science*, vol. 43, no. 23-24, pp. 7360–7365, 2008.

[12] V. N. Perevezentsev, V. V. Rybin, and V. N. Chuvil'deev, "The theory of structural superplasticity-IV. Cavitation during superplastic deformation," *Acta Metallurgica Et Materialia*, vol. 40, no. 5, pp. 915–924, 1992.

[13] C. R. Blanchard, R. A. Page, and S. Spooner, "The measurement of compressive creep deformation and damage mechanisms in a single-phase alumina part II correlation of creep cavitation and grain boundary sliding," *Journal of Materials Science*, vol. 33, no. 20, pp. 5049–5058, 1998.

[14] T. G. Langdon, "A unified approach to grain boundary sliding in creep and superplasticity," *Acta Metallurgica et Materialia*, vol. 42, no. 7, pp. 2437–2443, 1994.

[15] R. Z. Valiev, R. K. Islamgaliev, and N. F. Yunusova, "Superplasticity of nanostructured metallic materials obtained by methods of severe plastic deformation," *Metal Science and Heat Treatment*, vol. 48, no. 1-2, pp. 47–53, 2006.

[16] R. E. Reed-Hill, *Physical Metallurgy Principles*, Editorial D. Van Nostrand Co., Princeton, NJ, USA, 1968.

[17] R. E. Reed-Hill, *Physical Metallurgy Principles*, Editorial D. Van Nostrand Co., Princeton, NJ, USA, 1968.

[18] A. A. Istratov and O. F. Vyvenko, "Exponential analysis in physical phenomena," *Review of Scientific Instruments*, vol. 70, no. 2, pp. 1233–1257, 1999.

[19] A. C. Damask and G. J. Dienes, "Theory of vacancy annealing in impure metals," *Physical Review*, vol. 120, no. 1, pp. 99–104, 1960.

[20] J. E. Muñoz Rivera and R. Kamei Barreto, "Existence and exponential decay in nonlinear thermoelasticity," *Nonlinear Analysis: Theory, Methods and Applications*, vol. 31, no. 1-2, pp. 149–162, 1998.

[21] J. A. Montemayor-Aldrete and M. López de Haro, "A statistical mechanical approach to dislocation creep and the total unloading test on Al-11wt Zn alloy," in *Advances in Science, Technology and Applications of Zn-Al Alloys*, G. Torres Villaseñor, Y. H. Zhu, and C. Piña, Eds., pp. 115–1120, U.N.A.M., México, Mexico, 1994.

[22] J. E. Moreno and G. Torres-Villaseñor, "In situ observations of the superplastic deformation of Cd-Zn eutectic alloy," in *Advanced Materials' 93, III/B: Composites, Grain Boundaries and Nanophase Materials*, M. Sakai et al., Ed., vol. 16 of *Transactions of the Materials Research Society of Japan*, pp. 1023–1026, 1994.

[23] J. D. Muñoz-Andrade, A. Mendoza-Allende, G. Torres-Villaseñor, and J. A. Montemayor-Aldrete, "Mapping of the granular flow during superplastic deformation of microsample of Zn-20.2%Al-1.8%Cu alloy at room temperature," *Journal of Materials Science*, vol. 36, no. 3, pp. 795–799, 2001.

[24] J. A. Montemayor-Aldrete, J. D. Muñoz-Andrade, G. Torres-Villaseñor, and A. Mendoza Allende, "Superplasticity: fundamental problems to be solved," *Recent Research Developments in Metallurgical & Materials Sciences*, vol. 5, pp. 11–13, 2001.

[25] J. D. Muñoz-Andrade, A. Mendoza-Allende, G. Torres-Villaseñor, and J. A. Montemayor-Aldrete, "Cooperative grain boundary sliding at room temperature of a Zn-20.2%Al-1.8%Cu superplastic alloy," *Materials Science Forum*, vol. 357-359, pp. 551–558, 2001.

13

Defect Studies in bcc and fcc Iron

A. Ghorai and Arjun Das

Department of Physics, Chandernagore College, Chandernagore 712136, India

Correspondence should be addressed to A. Ghorai, amitavaghorai@rediffmail.com

Academic Editor: A. Dlouhy

Variation of vacancy formation energy (E_F^{1v}) with r_c of Ashcroft's empty core model potential (AECMP) model for different exchange and correlation functions (ECFs) show almost independent nature but slight variation with ECF for both bcc α iron and fcc γ iron.

1. Introduction

Recently in some papers, defect structures in α-Fe were discussed using Monte Carlo (MC) technique [1, 2], *ab initio* density functional (PP) model [3–5], and molecular dynamics (MD) simulation [6]. Iron exists in different phases, namely α-Fe (room temperature to 768°C, bcc structure, lattice constant $a = 5.48$ AU, and ferromagnetic), β-Fe (768°C to 910°C, bcc structure, $a = 5.48$ AU, and nonmagnetic), γ-Fe (910°C to 1400°C, fcc structure, $a = 6.73$ AU, and paramagnetic), and δ-Fe (1400°C to 1535°C, bcc structure again). Korhonen et al. [7] predicted that the stability of a self-interstitial in bcc-Fe depends on the range of potential but not on the type, while Osetsky et al. [8] predicted significant larger values of vacancy formation energy than the experimental ones in cases of bcc V and Cr using LDA.

Söderlind et al. [9] incorporated the results of above two groups of researchers using a full-potential, linear muffin-tin-orbital (FP-LMTO) method in conjunction with both the local-density approximation (LDA) and the generalized-gradient approximation (GGA) in bcc metals. A complementary *ab initio* pseudopotential (PP) method has also been used. They predicted FP-LMTO-LDA and PP-LDA formation energies are nearly identical within or close to experimental error bars for all bcc metals except Cr, and the overall agreement with experiment is better for the 4*d* and 5*d* metals than the 3*d* metals. GGA and LDA formation energies are very similar for the 4*d* and 5*d* metals but for the 3*d* metals, and especially Fe, GGA performs better. The dominant structural effects are an approximate 5% inward relaxation of the first near-neighbor shell for group V metals and a corresponding 1% inward relaxation for group VI metals, with the exception of Mo, for which the second-shell atoms also relax inward by about 1%. Thus it will be interesting to use in this paper the one parameter (r_c) Ashcroft's empty core model potential (called here after AECMP) [10] to study the variation of monovacancy formation energy (E_F^{1v}) in iron with the help of nine different types of exchange and correlation functions (called ECF) [11–20].

2. Formulations

The structure-dependent energy of a crystal depends on ion-ion, ion-electron, and electron-electron interactions and is also dependent on the modified lattice wave numbers. The modifications in the lattice wave numbers from their perfect lattice value are necessary to maintain the lattice volume and the number of lattice ions constant. The ion-ion interaction is determined from electrostatic energy and the last two interactions are included in the band structure energy, which is calculated using the second-order perturbation theory incorporating pseudopotential model form [16]. When a vacancy is created the Brillouin zone volume has to be scaled up in order to keep the lattice volume constant and so the lattice wave numbers are modified. Finally,

Table 1: Different forms of ECF $f(q)$ and corresponding fitted value of parameter r_c of AECMP in atomic unit (AU).

	Different forms of $f(q)$	bcc α-Fe	fcc γ-Fe
King and Cutler [11, 12]	$\dfrac{q^2}{2(q^2+2k_F^2)}$	1.1017	1.0028
Sham [13]	$\dfrac{q^2}{2(q^2+k_F^2+2k_F/\pi a_0)}$	1.1176	1.0204
Geldert and Vosko [14]	$\dfrac{q^2}{2[q^2+2k_F^2/(1+(0.026m'/m)(3\Omega/4\pi z a_0^3)^{1/3})]}$	1.1069	1.0083
Kleinman [15]	$\dfrac{1}{4}\left[\dfrac{q^2}{k_F^2+2k_F/\pi a_0}+\dfrac{q^2}{q^2+k_F^2+2k_F/\pi a_0}\right]$	1.1536	1.0496
Harrison [16]	$\dfrac{q^2}{2(q^2+k_F^2+k_F^2/3)}$	1.1432	1.0478
Vashishta and Singwi [17]	$A(1-e^{-Bq^2/k_F^2})$	1.1539	1.0501
Taylor [18]	$\dfrac{q^2}{4k_F^2}\left(1+\dfrac{0.1534}{\pi k_F}\right)$	1.1687	1.0566
Hubbard [19]	$\dfrac{q^2}{2(q^2+k_F^2)}$	1.1723	1.0789
Mahanti and Das [20]	$\dfrac{q^2}{k_F^2[1+(2/(1+(0.026m'/m)(3\Omega/4\pi z a_0^3)^{1/3}))^2]}+\dfrac{q^2}{q^2+k_F^2[1+(2/(1+(0.026m'/m)(3\Omega/4\pi z a_0^3)^{1/3}))^2]}$	1.2334	1.1282

one gets the expressions for vacancy formation energy E_F^{1v} considering relaxation energy associated with these defect formations as

$$E_F^{1v} = \sum_{q_0}{}' \frac{q_0}{3}\frac{\partial U(q_0)}{\partial q_0} + \frac{\Omega}{2\pi^2}\int_0^{\infty} U(q)q^2 dq, \quad (1)$$

where

$$U(q) = \underset{\eta\to\infty}{Lt}\frac{2\pi e^2 z^2}{\Omega q^2}e^{-q^2/4\eta} + [w(q)]^2\varepsilon(q)\chi(q), \quad (2)$$

$$w(q) = -\frac{4\pi z e^2 \cos q r_c}{\Omega q^2}, \quad (3)$$

$$\varepsilon(q) = 1 - \frac{8\pi e^2}{\Omega q^2}[1-f(q)]\chi(q), \quad (4)$$

$$\chi(q) = -\frac{3z}{8E_F}\left[1+\frac{4k_F^2-q^2}{4k_F q}\ln\frac{|2k_F+q|}{|2k_F-q|}\right]. \quad (5)$$

Here q_0 and q are the lattice and quasi-continuous wave numbers, respectively, Ω is the atomic volume, e the electronic charge, z the valency, η the convergence factor, k_F the Fermi wavenumber, E_F the Fermi energy, $w(q)$ the AECMP with parameter r_c, $\varepsilon(q)$ the dielectric function or screening factor, $\chi(q)$ the perturbation characteristics, and $f(q)$ the ECF whose nine different forms have been shown in Table 1.

3. Discussions

The calculation of (1) needs integration over quasi-continuous wavenumbers by quadrature technique and the discrete sum over lattice wave numbers. The input and output parameters for this purpose for fcc iron (γ-Fe) and bcc iron (α-Fe) are shown in Table 2. In the first step the variation of E_F^{1v} with parameter r_c of AECMP is plotted for nine different ECF from 0 to 5 AU as shown in Figure 1 for fcc iron (γ-Fe) and Figure 2 for bcc iron (α-Fe). The two graphs have positive peaks and they look almost similar due to $\cos^2 q r_c$ term of AECMP for all exchange and correlations but there is a slight variation. It is observed that experimental value of E_F^{1v} lies near the nodal point corresponding to the condition $E_F^{1v} \to 0$ rather than that to the maxima. Fitted value of r_c has been chosen corresponding to the condition $a_0 \le r_c < 2\pi/k_F$, where a_0 is the Bohr radius.

It is observed that the experimental value of E_F^{1v}, obtained from positron annihilation technique, lies within the range of the theoretical value of it obtained from the empirical relation [23–25]:

$$T_m(K) = 1200E_F^{1v} = 660Q_0 = 360E_{\text{coh}}. \quad (6)$$

Here T_m is the melting temperature, Q_0 the activation energy, and E_{coh} the cohesive energy of the metal. From the experimental value of E_F^{1v} we note that the fitted value of r_c is within the first peak value of 2 AU and we note from Table 2 that r_c values lie close to Bohr radius (a_0 = 1 AU). ECF's of Kleinman [15], Harrison [16], Vashishta and Singwi [17],

Table 2: Input and output parameters for fcc and bcc iron (1 eV = 13.605 Rydberg and 1 atomic unit (AU) = 5.29177 × 10^{-2} nm).

	fcc γ-Fe	bcc α-Fe
Valency z	2	2
Lattice constant a AU	6.727	5.424
Volume AU^3	76.118	79.765
k_F AU^{-1}	0.920	0.905
E_F Ryd.	0.846	0.820
Melting temperature $T_m(K)$	1536	1811
Cohesive energy E_{coh} eV	4.28	4.28
Activation energy for self-diffusion Q_0 eV	2.943	2.658
Vashishta-Singwi parameters		
A	0.929633	0.938470
B	0.309178	0.312208
E_F^{1v}(theo) in eV[1]	1.284–1.619	1.284–1.509
E_F^{1v}(expt) in eV[2]	1.40	1.60
Range of r_c fitted to E_F^{1v}(expt) in AU	1.0028–1.1282	1.1017–1.2234
Mean r_c in AU	1.0492	1.1501

[1] Using (6) [21]; [2] [22].

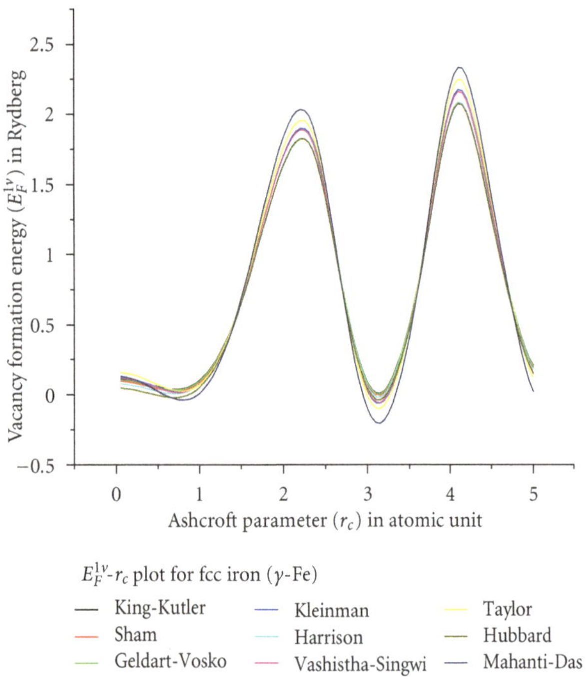

Figure 1: E_F^{1v}-r_c plot for fcc γ-iron (γ-Fe).

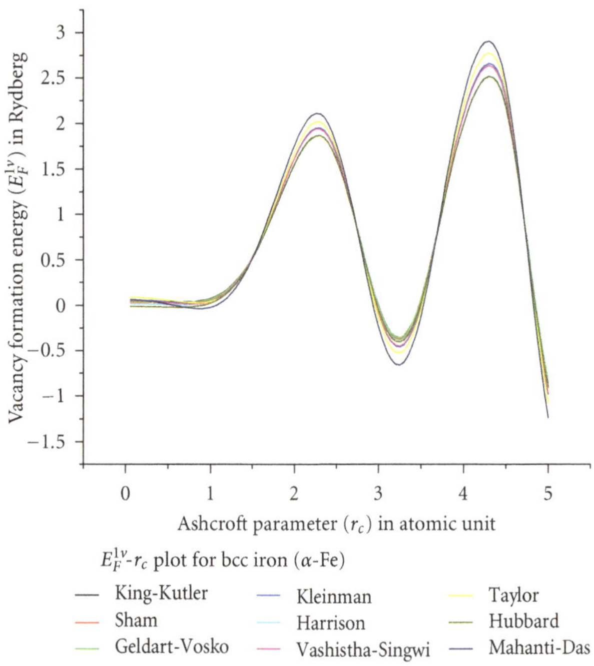

Figure 2: E_F^{1v}-r_c plot for bcc α-iron (α-Fe).

and Taylor [18] give reasonably close values of r_c while others give the range over of r_c.

In conclusion, it should be noted that the inherent simplicity of AECMP makes it difficult to have a universal r_c parameter for all types of atomic property calculations and we have to use different ECFs of which Taylor, Harrison, Kleinmann, Vashishta and Singwi type of ECF give better results in this case.

References

[1] J. Rottler, D. J. Srolovitz, and R. Car, "Point defect dynamics in bcc metals," *Physical Review B*, vol. 71, no. 6, Article ID 064109, 12 pages, 2005.

[2] S. M. J. Gordon, S. D. Kenny, and R. Smith, "Diffusion dynamics of defects in Fe and Fe-P systems," *Physical Review B*, vol. 72, no. 21, Article ID 214104, 10 pages, 2005.

[3] C. Domain and C. S. Becquart, "Diffusion of phosphorus in α-Fe: An *ab initio* study," *Physical Review B*, vol. 71, no. 21, 13 pages, 2005.

[4] C. C. Fu and F. Willaime, "*Ab initio* study of helium in α-Fe: dissolution, migration, and clustering with vacancies," *Physical Review B*, vol. 72, no. 6, Article ID 064117, 6 pages, 2005.
[5] P. Olsson, C. Domain, and J. Wallenius, "*Ab initio* study of Cr interactions with point defects in bcc Fe," *Physical Review B*, vol. 75, no. 1, Article ID 014110, 12 pages, 2007.
[6] C. Bos, J. Sietsma, and B. J. Thijsse, "Molecular dynamics simulation of interface dynamics during the fcc-bcc transformation of a martensitic nature," *Physical Review B*, vol. 73, no. 10, Article ID 104117, 7 pages, 2006.
[7] T. Korhonen, M. J. Puska, and R. M. Nieminen, "Vacancy-formation energies for fcc and bcc transition metals," *Physical Review B*, vol. 51, no. 15, pp. 1926–1927, 1995.
[8] Yu. N. Osetsky, M. Victoria, A. Serra, S. I. Golubov, and V. Priego, "Computer simulation of vacancy and interstitial clusters in bcc and fcc metals," in *Proceedings of the International Workshop on Defect Production, Accumulation and Materials Performance in an Irradiation Environment*, vol. 251, pp. 34–48, Journal of Nuclear Materials, 1997.
[9] P. Söderlind, L. H. Yang, J. A. Moriarty, and J. M. Wills, "First-principles formation energies of monovacancies in bcc transition metals," *Physical Review B*, vol. 61, no. 4, pp. 2579–2586, 2000.
[10] N. W. Ashcroft, "Electron-ion pseudopotentials in metals," *Physics Letters*, vol. 23, no. 1, pp. 48–50, 1966.
[11] W. F. King III and P. H. Cutler, "Lattice dynamics of beryllium from a first-principles nonlocal pseudopotential approach," *Physical Review B*, vol. 2, no. 6, pp. 1733–1742, 1970.
[12] W. F. King III and P. H. Cutler, "Lattice dynamics of magnesium from a first-principles nonlocal pseudopotential approach," *Physical Review B*, vol. 3, no. 8, pp. 2485–2496, 1971.
[13] L. J. Sham, "A Calculation of the phonon frequencies in sodium," *Proceedings of the Royal Society A*, vol. 283, no. 1392, pp. 33–49, 1965.
[14] D. J. W. Geldert and S. H. Vosko, "The screening function of an interacting electron gas," *Canadian Journal of Physics*, vol. 44, pp. 2137–2171, 1965.
[15] L. Kleinman, "Exchange and the dielectric screening function," *Physical Review*, vol. 172, no. 2, pp. 383–390, 1968.
[16] W. A. Harrison, *Pseudopotentials in the Theory of Metals*, W. A. Benjamin, New York, NY, USA, 1966.
[17] P. Vashishta and K. S. Singwi, "Electron correlations at metallic densities. v," *Physical Review B*, vol. 6, no. 3, pp. 875–887, 1972.
[18] R. Taylor, "A simple, useful analytical form of the static electron gas dielectric function," *Journal of Physics F*, vol. 8, no. 8, pp. 1699–1702, 1978.
[19] H. Hubbard, "The description of collective motions in terms of many-body perturbation Theory. II. The correlation energy of a free-electron gas," *Proceedings of the Royal Society A*, vol. 243, no. 1234, pp. 336–352, 1958.
[20] S. D. Mahanti and T. P. Das, "Theory of knight shifts and relaxation times in alkali metals-role of exchange core polarization and exchange-enhancement effects," *Physical Review B*, vol. 3, no. 5, pp. 1599–1610, 1971.
[21] C. Kittel, *Introduction to Solid State Physics*, Wiley Eastern, New Delhi, India, 5th edition, 1979.
[22] H. Matter, J. Winter, and W. Triftshäuser, "Phase transformations and vacancy formation energies of transition metals by positron annihilation," *Applied Physics A*, vol. 20, no. 2, pp. 135–140, 1979.
[23] A. Ghorai, "Dependence of mono-vacancy formation energy on the parameter of Ashcroft's potential," *Defect and Diffusion Forum*, vol. 278, pp. 25–32, 2008.
[24] A. Ghorai, "A study of the variation of monovacancy formation energy with the parameter of Ashcroft's potential and different exchange and correlation functions for some group-IIa and group-VIII FCC metals in the active divalent state," *Defect and Diffusion Forum*, vol. 293, pp. 11–18, 2009.
[25] A. Ghorai, "Exploration of parameters of Ashcroft's potential using monovacancy formation energy and different exchange and correlation functions for some trivalent FCC metals," *Defect and Diffusion Forum*, vol. 294, pp. 113–118, 2009.

14

Structural and Electronic Properties of Pure Ta, TaNO, and TaZrNO with Ab Initio Calculations

T. Chihi,[1] M. Fatmi,[2,3] M. Guemmaz,[4] and J. C. Parlebas[5]

[1] *Laboratory for Elaboration of New Materials and Characterization (LENMC), University of Ferhat Abbas, Setif 19000, Algeria*
[2] *Research Unit on Emerging Materials (RUEM), University of Ferhat Abbas, Setif 19000, Algeria*
[3] *Laboratory of Physics and Mechanics of Metallic Materials (LP3M), University of Ferhat Abbas, Setif 19000, Algeria*
[4] *DAC Laboratory, Faculty of Sciences, University of Ferhat Abbas, Setif 19000, Algeria*
[5] *IPCMS, UMR 7504 CNRS-UDS, 23 Rue de Loess, 67034 Strasbourg Cedex 2, France*

Correspondence should be addressed to M. Fatmi, fatmimessaoud@yahoo.fr

Academic Editors: A. Chrysanthou, Y. Hao, M.-C. Wang, and Y.-d. Wang

This paper presents the results of self-consistent first-principle calculations for the crystal structure and electronic structure of pure tantalum, TaNO, and TaZrNO within density functional theory (DFT) for the sake of comparison and shows the influence of allowing elements on the interatomic distance and the Fermi level. The large total densities of states (TDOS) value for TaZrNO implies the highest electronic conductivity. The difference in values is due to the Zr metallic atoms presence in TaZrNO compound. There is a strong interaction between Ta and (N, O) (Ta−N = 0.39, Ta−O = 0.21) in TaON compound, and Zr presence increases this interaction (Ta−N = 1.74, Ta−O = 0.30) in TaZrON compound.

1. Introduction

The elemental tantalum Ta crystallizes in three crystalline phases, bcc-Ta (α-phase), f.c.c-Ta, and a new phase which is now generally referred to as β-tantalum. The discoverers of the tetragonal tantalum β-Ta (a metastable phase), in 1965 are Read and Altman [1]. It has been attracting much interest in most applications because of its high resistivity (170–210 $\mu\Omega$ cm) [2–5]. It is preferred for fabricating capacitors and resistors. The chemical stability and robust mechanical properties of Ta make it a particularly desirable material. Numerous crystal structures have been reported for β-Ta. A tetragonal unit cell Ta was proposed by Read and Altman [1] Das [6] proposed a bcc-based superlattice structure, while Burbank [7] proposed a hexagonal hcp structure and β-uranium model that was also proposed by Arakcheeva et al. [8, 9] on the basis of X-ray diffraction (XRD) study on single crystals of β-Ta produced through electrolytic crystallization, and in the end, the anomalous f.c.c-Ta structure was observed in very thin films of tantalum [10, 11].

On the other hand, nitride formation is common to most transition elements. Many compositional and structural forms exist, with many transition elements forming several different nitride phases. In many of these compounds, nitrogen atoms occupy interstitial lattice sites because they are smaller than the metal atoms. For this reason, they are often referred to as interstitial compounds. Transition metal nitrides are refractory metals that possess technologically useful properties including superconductivity and ultrahigh hardness, and they combine various physical and chemical properties, such as high melting points (around 3000°C). They also possess electronic and magnetic properties that make them useful as electronic and magnetic components and as superconductors [12].

Although monometallic nitrides have been the object of considerable studies [13–15], bimetallic transition metal nitrides have attracted only limited attention. Van Dover et al. [16] investigated the Ternary Transition-Metal Nitride Y-Nb-N and Gd-Cr-N Systems by reactive sputtering, providing evidence for a new superconducting (Nb,Y) N solid solution.

Similarly, the literature on oxynitrides has been scarce. Oxynitrides of transition metals are a new exciting class of materials [17] that possess interesting refractory behaviour, higher elastic modulus, and hardness. They also offer great potential for their optical properties and recently have received much attention because of their potential use as pigment materials [18]. It has been known that oxygen atoms can substitute nitrogen atoms in monometallic nitrides due to the similarity in their radius. In many of the oxynitrides compounds, the N and O atoms are found in interstitial lattice positions in between the metal atoms. For this reason, their phases can exist over broad composition ranges with appreciable vacancy concentrations (both metal and nonmetal) and their physical properties are quite sensitive to composition. Yashima et al. [19] investigate neutron diffraction for confirmation of anion ordering and synchrotron powder diffraction for high-precision analysis of the crystal structure and electron density of an active TaON photocatalyst sample under visible-light excitation. Yashima et al. [20] employ the density functional theory (DFT) for theoretical calculations of the electron density distribution and partial density of states of TaON compound.

Our primary aim was therefore to present the results of a theoretical investigation of the structural and electronics of metastable β-Ta, bcc-Ta, and f.c.c-Ta. The bcc-Ta ((f.c.c-Ta)) structures are cubic; the space group is Im-3 m (no.229) (Fm-3 m (no.225)) with two (four) formula units per unit cell, the metastable β-Ta (β-uranium, Distorted A15, and Hexagonal) structures are (tetragonal $P4_2/mnm$ (no.136), distorted Pm(-3)m (no.223), and hexagonal $P6_3/mmc$ (no.194)). The bimetal (monometal) transition metal oxynitride TaZrNO (TaNO) structures are hexagonal (cubic), and the space group is $P\bar{6}m2$ (no.187) ($F\bar{4}3m$ (no.216)) with one (four) formula unit(s) per unit cell. Until now, there has been no report on the electronic properties of pure tantalum, TaNO, and TaZrNO.

This paper presents the results of self-consistent first-principles calculations for the crystal structure and electronic structure of pure tantalum, TaNO, and TaZrNO within DFT for the sake of comparison and shows the influence of allowing elements on the interatomic distance and the Fermi level.

The paper is organized as follows. The computational method is described in Section 2. In Section 3, the results are presented and compared with available experimental and theoretical data. Conclusion is given in Section 4.

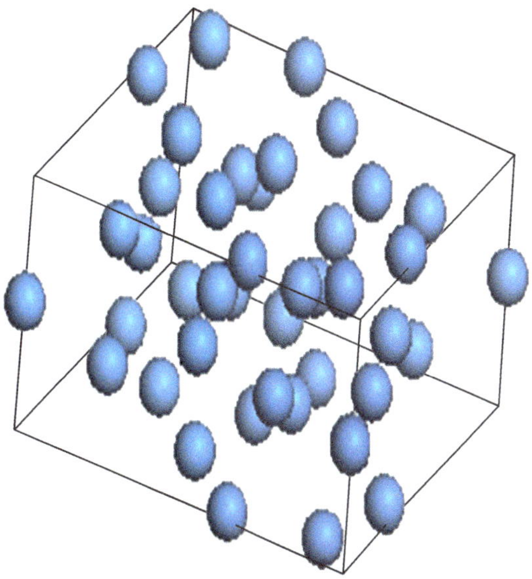

FIGURE 1: (Ta) β-uranium structure.

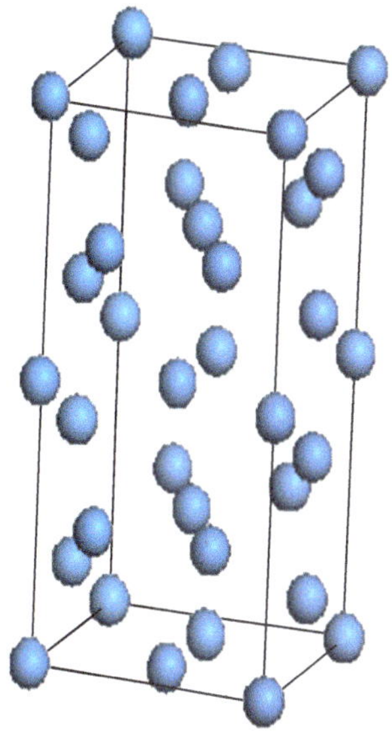

FIGURE 2: (Ta) Distorted A15 β-structure.

2. Computational Method

All calculations were performed by using the CASTEP (Cambridge Serial Total Energy Package) simulation program [21] that solves the Schrodinger-like Kohn-Sham equations according to the formalism of the density functional theory (DFT) [22, 23]. We used the Generalized Gradient Approximation (GGA) and a Perdew-Burke-Ernzerhof (PBE) scheme [24] for handling the electronic exchange-correlation potential energy. Also, the pseudopotentials constructed using the ab initio norm conserving scheme describe the valence electron interaction with the atomic core, in which the Ta ($4f^{14}$ $5d^3$ $6s^2$), Zr ($4d^2$ $5s^2$), N ($2s^2$ $2p^3$), and O ($2s^2$ $2p^4$) orbitals are treated as valence electrons. Using for all structures high cut-off energy (280 eV) even at the price of spending long computational time is the condition to obtain accurate results. Brillouin zone (BZ) sampling is carried out using a $6 \times 6 \times 6$ Monkhorst-Pack mesh set [25]. For Ta (distorted A15) and hexagonal (type Cd or Zn) structures, the BZ sampling is carried out using a $10 \times 10 \times 6$ Monkhorst-Pack mesh and a cut-off energy of (280 eV). Atomic positions are relaxed and optimized within a density mixing scheme, based on a Conjugate Gradient (CG) method for eigenvalues minimization. Actually, the equilibrium lattice parameter is determined from a structural optimization, using the Broyden-Fletcher-Goldfarb-Shenno (BFGS) minimization technique. This technique provides a fast way of finding the lowest energy structure, with the

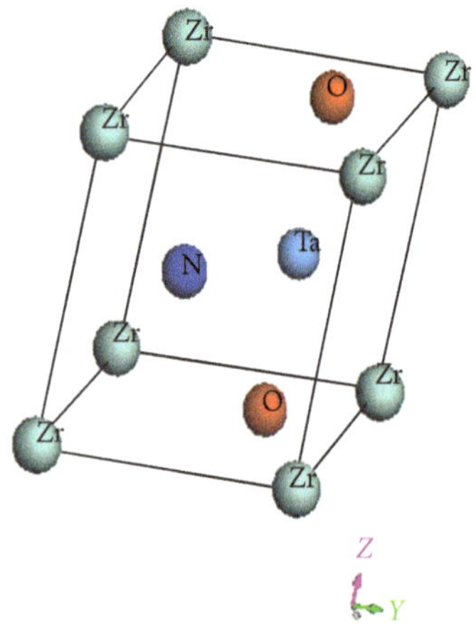

FIGURE 3: TaZrNO structure.

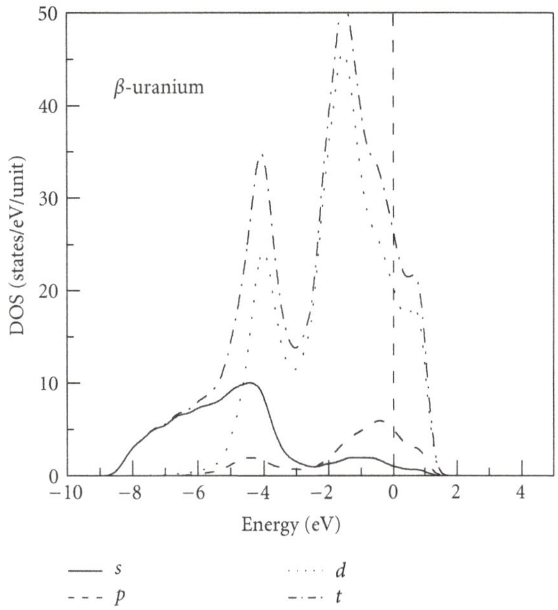

FIGURE 4: Total and partial densities of states for β-Ta (β-uranium).

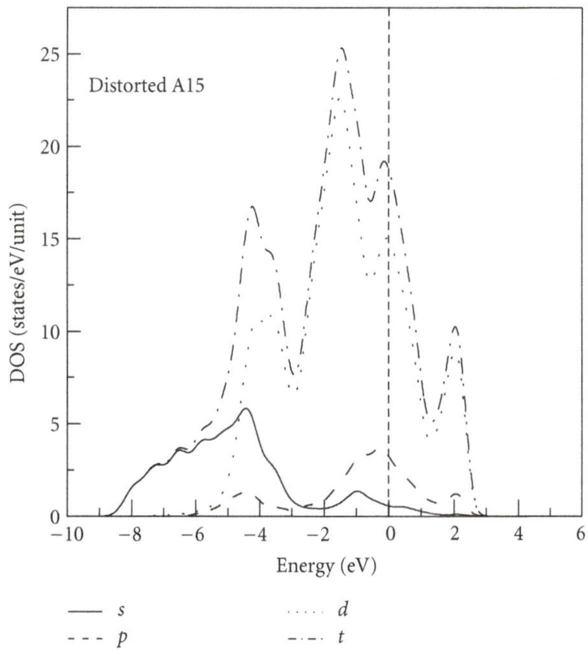

FIGURE 5: Total and partial densities of states for β-Ta (distorted A15).

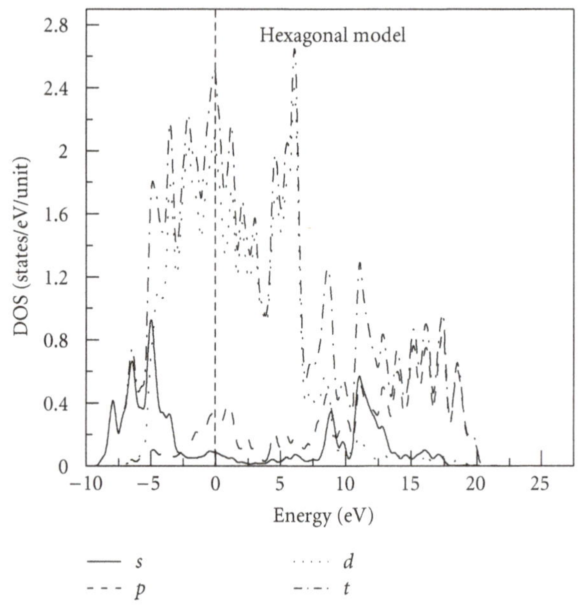

FIGURE 6: Total and partial densities of states for β-Ta (hexagonal model).

following thresholds for converged structures: (i) energy change per atom less than 2×10^{-5} eV, (ii) residual force less than 0.05 eV/Å, (iii) atom displacement during geometry optimization less than 0.002 Å, and (iv) maximum stress within 0.1 GPa. The crystal structures of Ta: β-uranium, Ta: Distorted A15 β-structure, and TaZrNO structure are given in (Figures 1, 2, and 3).

3. Results and Discussion

3.1. Structural Properties. Results for the lattice parameter *a*, *c* and *c*/*a* are reported (Table 1) and compared with experimental and theoretical calculations. Our calculated GGA values for *a* for pure tantalum are in good agreement with the experimental data [1, 3, 7]. We have obtained the same value for *a* of TaNO as experimental [26] one. For TaZrNO, our computed lattice constant (*a*) is bigger than the experimental one, within 7.4% [27]. Also, for both structures, the calculated lattice parameter ratio *c*/*a* (0.523, 1.978, and 1.792 for0020β-uranium, distorted A15, and hexagonal, resp.) is in reasonable accord with the previously considered data (0.531, 1.860, and 1.890, [1, 7, 28–31] and 1.098 for TaZrNO that is also in reasonable accord with of 1.064) [27]. In order to attempt an understanding the interatomic distances of the various compounds, we have found excellent agreement between our calculated

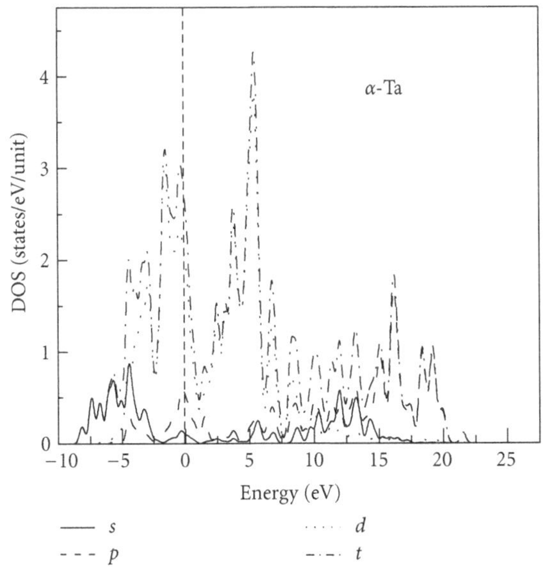

FIGURE 7: Total and partial densities of states for α-Ta.

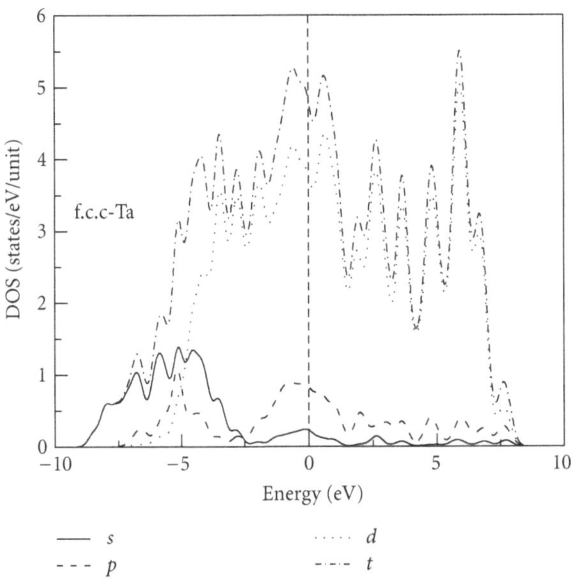

FIGURE 8: Total and partial densities of states for f.c.c-Ta.

and available experiment interatomic distances for different phases of pure tantalum. The results of distances are reported in Table 1.

3.2. Electronic Structures. Density of states and electronic band structure often provide sufficient information for a thorough characterization of the electronic properties of the material. Total density of states (DOS) of all β-Ta (metastable phase), α-Ta, and f.c.c-Ta structures were calculated in order to understand differences in the chemical bonding between them. The DOS for β-Ta structure, shown in Figures 4 and 5, is concentrated in two peaks. The low energy peak, between −9 and −3 eV (the zero of energy is taken at the Fermi energy E_f), is comprised mostly of *s* and *d* states. The other broad peak, lying at and above E_f, has *d* character with small admixture of *p* characters and is responsible for the N(E_f) of 26.024 and 18.268 states/eVÅ^3 for β-uranium and distorted A15 models, respectively. For the hexagonal model, the DOS is mainly due to the *d* states, with small *s* characters at around −5 and 10 eV (Figure 6). Figures 7 and 8 shows respectively the density of states for α-Ta and fcc-Ta that are mainly dominated by *d* states and little *s* contributions especially around −5 eV. Both structures are metallic because of the finite DOS at the Fermi level (E_f). The DOS values at E_f of all structures are shown in Table 2. Let us mention a large DOS at the Fermi level; this implies a high electronic conductivity and various applications to electronic conductors. Figures 9 and 10 show the total density of states (TDOS) and partial density of states (PDOS) for TaNO and TaZrNO compounds. The TDOS of TaNO (TaZrNO) can be mainly distributed in the following three energy regions in the valence bonding region.

(1) In the region one from approximately −26 eV to −23 eV (−25 eV to −22 eV). In this region, the DOS mainly consists of *s* orbitals of O, together with a small portion of *s* orbitals of Ta (*s* orbitals of O, together with a small portion of *s* orbitals of Ta and Zr). These states represent the electronic core structures;

(2) In region two, it is from −18.0 eV to −16.0 eV (−19 eV to −17 eV). In this region, the DOS is contributed mainly from the hybridized *s* orbitals of N, *d* orbitals of Ta, together with a small portion of *s* orbitals of Ta for TaNO, and the same behavior is observed for TaZrNO compounds;

(3) In region three; it is from −14 eV to 0 eV (−12 eV to 0 eV). This is due to the *p*-O, *p*-N, and *d*-Ta orbitals (*p*-O, *p*-N, *d*-Ta, and *d*-Zr with a small portion of *s*-Ta and *s*-Zr contributions). Above the Fermi energy level, the DOS is predominated by *d*-Ta and *p*-N orbitals with a small portion of *p*-O contribution (*d*-Ta, *d*-Zr, *p*-N, and *p*-O orbitals with a small portion of *p*-Ta and *p*-Zr contributions), corresponding to the electronic anti-bonding states. The two TDOS have some similarities, however, in Figures 9 and 10.

TDOS at the Fermi level is 1.493 for TaNO.

TDOS at the Fermi level is 2.010 for TaZrNO, and the large value for TaZrNO implies the highest electronic conductivity.

The difference in values is due to the Zr metallic atoms presence in TaZrNO compound.

The pure tantalum has the highest (E_f), and generally speaking, the smaller the (E_f) is, the unstable the compound will be.

3.2.1. Bond Orders between Atoms. Bond order is the overlap population of electrons between atoms, and this is a measure of the strength of the covalent bond between atoms. If the overlap population is positive (+), a bonding-type interaction is operating between atom, whereas if it is negative (−), an antibonding-type interaction is dominant between

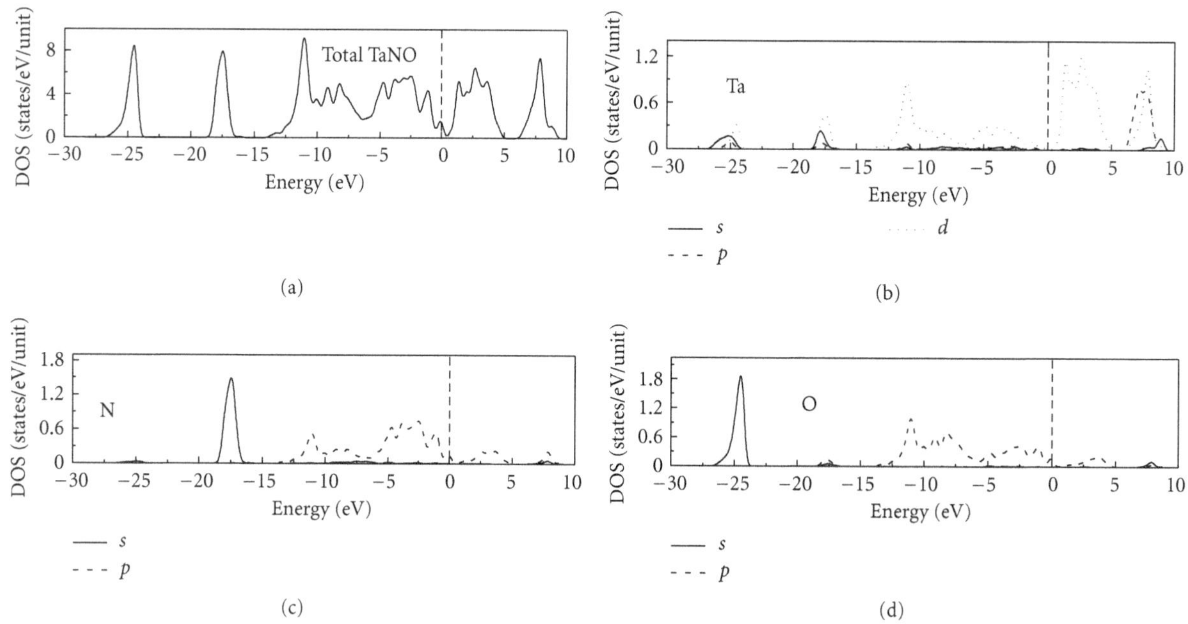

Figure 9: Total and partial densities of states for TaNO.

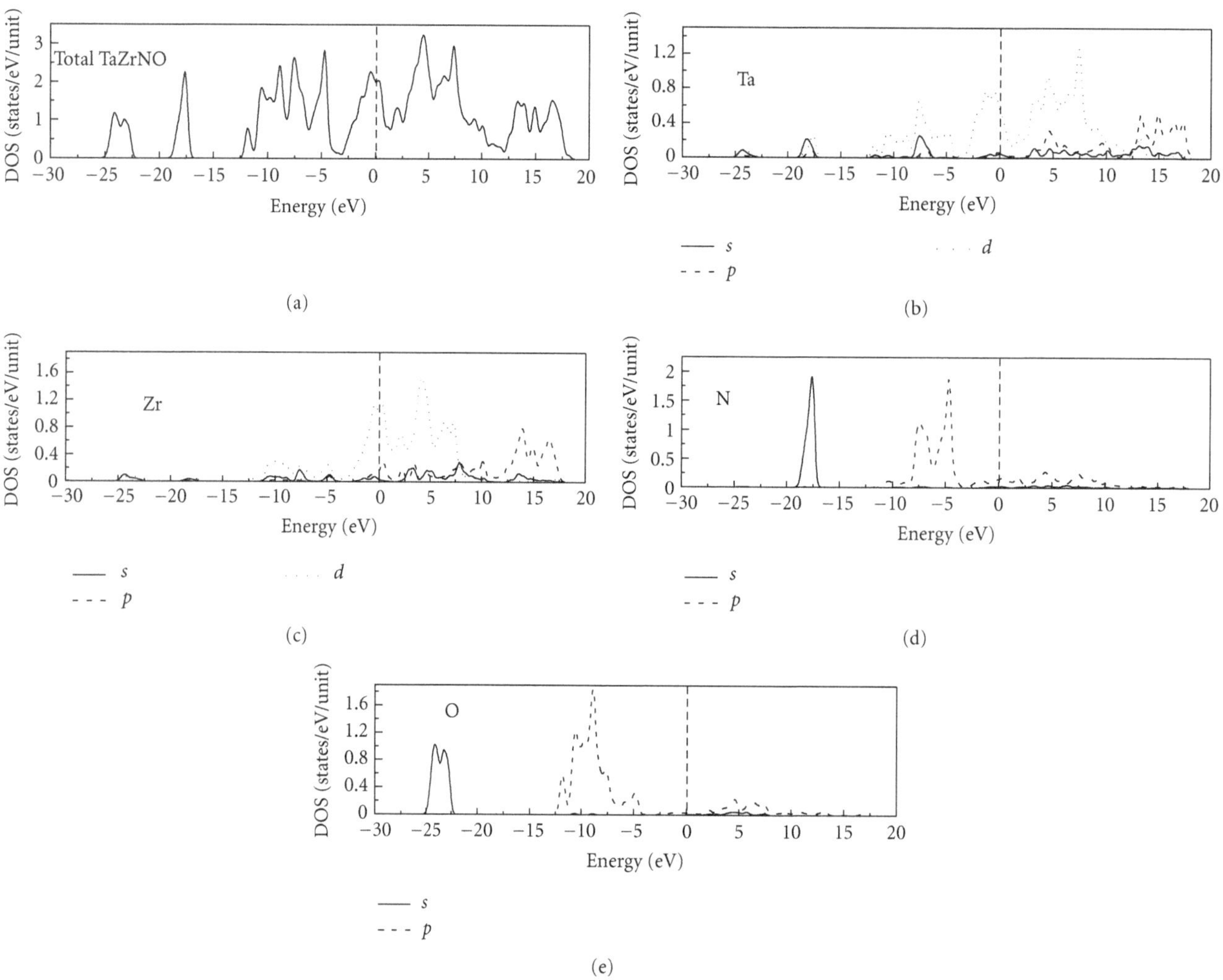

Figure 10: Total and partial densities of states for TaZrNO.

Table 1: Lattice constant, shortest Ta–Ta, density, and atomic/unit cell of the all metastable phase.

Materials	Lattice constant a_0 (Å)	c/a	Shortest Ta-Ta (Å)	Density (g/cm^{-3})	Atom/unit cell	Reference
β-uranium	9.9345	0.523	2.608	17.582	30	[28, 29]
	10.194	0.531	2.650	16.330	30	
Distorted A15	5.1551	1.978	2.550	17.736	16	[1, 30, 31]
	5.3400	1.860	1.890	16.960	16	
Hexagonal	2.804	1.792	2.805	17.605	2	[7]
	2.830	1.890	2.830	16.230	2	
α–Ta	3.217	—	2.786	18.050	2	[3]
	3.31	—	2.860	16.550	2	
f.c.c-Ta	4.108	—	2.905	17.329	4	
	4.390	—	3.100	14.21	4	[11–31]
TaNO	4.827	—	3.031	17.798	12	[26]
	4.827	—	—	—	—	
TaZrNO	3.373	1.098	3.707	13.733	4	[27]
	3.645	1.064	—	—	—	

Table 2: Fermi level E_f of the all metastable phase.

	β-uranium	Distorted A15	Hexagonal	α-Ta	f.c.c-Ta	TaNO	TaZrNO
E_f	26.272	18.268	2.418	2.579	4.704	1.493	2.010

atoms. It is apparent that the bonding-type interactions are operating between the metal M (M = Ta, Zr) (4d and 5d) and the nonmetal (N, O) 3d electrons. Thus, there is a strong interaction between Ta and (N, O) (Ta − −N = 0.39, Ta − −O = 0.21) in TaON compound, and Zr presence increases this interaction (Ta − −N = 1.74, Ta − −O = 0.30) in TaZrON compound.

4. Summary and Conclusion

Using the first principles based on the DFT, we studied the total (partial) density of states TDOS (PDOS) of pure tantalum, TaNO, and TaZrNO. The large value for TaZrNO implies the highest electronic conductivity. The difference in values is due to the Zr metallic atoms presence in TaZrNO compound. A bonding-type interaction is operating between atoms, and thus there is a strong interaction between Ta and (N, O) (Ta − −N = 0.39, Ta − −O = 0.21) in TaON compound, and Zr presence increases this interaction (Ta − −N = 1.74, Ta − −O = 0.30) in TaZrON compound.

References

[1] M. H. Read and C. Altman, "A new structure in tantalum thin films," *Applied Physics Letters*, vol. 7, no. 3, article 51, 2 pages, 1965.

[2] N. Schwartz, W. A. Reed, P. Polash, and M. H. Read, "Temperature coefficient of resistance of beta-tantalum films and mixtures with b.c.c.-tantalum," *Thin Solid Films*, vol. 14, no. 2, pp. 333–346, 1972.

[3] P. N. Baker, "Preparation and properties of tantalum thin films," *Thin Solid Films*, vol. 14, no. 1, pp. 3–25, 1972.

[4] A. Schauer and M. Roschy, "R.F. sputtered β-tantalum and b.c.c. tantalum films," *Thin Solid Films*, vol. 12, no. 2, pp. 313–317, 1972.

[5] Y. Chouan and D. Collobert, "Plasma influence in tantalum sputtering," *Journal of Applied Physics*, vol. 48, no. 6, pp. 2274–2279, 1977.

[6] G. Das, "A new structure of sputtered tantalum," *Thin Solid Films*, vol. 12, no. 2, pp. 305–311, 1972.

[7] R. D. Burbank, "X-ray study of β-tantalum," *Journal of Applied Crystallography*, vol. 6, no. 3, pp. 217–224, 1973.

[8] A. Arakcheeva, G. Chapuis, and V. Grinevitch, "The self-hosting structure of β-Ta," *Acta Crystallographica Section B*, vol. 58, no. 1, pp. 1–7, 2002.

[9] A. Arakcheeva, G. Chapuis, H. Birkedal, P. Pattison, and V. Grinevitch, "The commensurate composite σ-structure of β-tantalum," *Acta Crystallographica Section B*, vol. 59, no. 3, pp. 324–336, 2003.

[10] R. B. Marcus and S. Quingley, "Formation of f.c.c., b.c.c. and β-tantalum films by evaporation," *Thin Solid Films*, vol. 2, no. 5-6, pp. 467–477, 1968.

[11] P. N. Dendigh and R. B. Marcus, "Structure of very thin tantalum and molybdenum films," *Journal of Applied Physics*, vol. 37, no. 12, article 4325, 6 pages, 1966.

[12] T. H. Geballe, B. T. Matthias, J. P. Remeika et al., *Physics (Long Island City, N.Y.)*, vol. 2, p. 293, 1966.

[13] D. A. Papaconstantopoulos, W. E. Pickett, B. M. Klein, and L. L. Boyer, "Electronic properties of transition-metal nitrides: the group-V and group-VI nitrides VN, NbN, TaN, CrN, MoN, and WN," *Physical Review B*, vol. 31, no. 2, pp. 752–761, 1985.

[14] C. Stampfl, W. Mannstadt, R. Asahi, and A. J. Freeman, "Electronic structure and physical properties of early transition metal mononitrides: density-functional theory LDA, GGA, and screened-exchange LDA FLAPW calculations," *Physical Review B*, vol. 63, no. 15, Article ID 155106, 11 pages, 2001.

[15] P. Ojha, M. Aynyas, and S. P. Sanyal, "Pressure-induced structural phase transformation and elastic properties of transition metal mononitrides," *Journal of Physics and Chemistry of Solids*, vol. 68, no. 2, pp. 148–152, 2007.

[16] R. B. van Dover, B. Hessen, D. Werder, C. H. Chen, and R. J. Felder, "Investigation of ternary transition-metal nitride systems by reactive cosputtering," *Chemistry of Materials*, vol. 5, no. 1, pp. 32–35, 1993.

[17] T. Das, "Oxynitride glasses—an overview," *Bulletin of Material Science*, vol. 23, no. 6, pp. 499–5047, 2000.

[18] E. Guenther and M. Jansen, "Optical properties of $Ta_{(3-x)}Zr_{(x)}N_{(5-x)}O_{(x)}$ semiconductor pigments," *Materials Research Bulletin*, vol. 36, no. 7-8, pp. 1399–1405, 2001.

[19] M. Yashima, Y. Lee, and K. Domen, "Crystal structure of tantalum oxynitride TaON, a visible light responsive photocatalyst," *Experimental Reports*, vol. 15, no. 395, 2008.

[20] M. Yashima, Y. Lee, and K. Domen, "Crystal structure and electron density of tantalum oxynitride, a visible light responsive photocatalyst," *Chemistry of Materials*, vol. 19, no. 3, pp. 588–593, 2007.

[21] M. D. Segall, P. J. D. Lindan, M. J. Probert et al., "First-principles simulation: ideas, illustrations and the CASTEP code," *Journal of Physics Condensed Matter*, vol. 14, no. 11, pp. 2717–2744, 2002.

[22] P. Hohenberg and W. Kohn, "Inhomogeneous electron gas," *Physical Review*, vol. 136, no. 3, pp. B864–B871, 1964.

[23] W. Kohn and L. J. Sham, "Self-consistent equations including exchange and correlation effects," *Physical Review*, vol. 140, no. 4, pp. A1133–A1138, 1965.

[24] J. P. Perdew, S. Burke, and M. Ernzerhof, "Generalized gradient approximation made simple [Phys. Rev. Lett. 77, 3865 (1996)]," *Physical Review Letters*, vol. 78, no. 7, article 1396, 1997.

[25] H. J. Monkhorst and J. D. Pack, "Special points for Brillouin-zone integrations," *Physical Review B*, vol. 13, no. 12, pp. 5188–5192, 1976.

[26] H. Wolff, H. Schilling, M. Lerch, and R. Dronskowski, "A density-functional and molecular-dynamics study on the physical properties of yttrium-doped tantalum oxynitride," *Journal of Solid State Chemistry*, vol. 179, no. 8, pp. 2265–2270, 2006.

[27] N. Schonberg, "The structure of the metallic quaternary phase ZrTaNO," *Acta Chemica Scandinavica*, vol. 8, pp. 627–629, 1954.

[28] P. T. Moseley and C. J. Seabrook, "The crystal structure of β-tantalum," *Acta Cryst B*, vol. 29, no. 5, pp. 1170–1171, 1973.

[29] A. C. Lawson and C. R. Olsen, "Structure of β-uranium," *Acta Cryst B*, vol. 44, no. 2, pp. 89–96, 1988.

[30] P. Catania, R. A. Roy, and J. J. Cuomo, "Phase formation and microstructure changes in tantalum thin films induced by bias sputtering," *Journal of Applied Physics*, vol. 74, no. 2, pp. 1008–1014, 1993.

[31] D. Mills, *Journal of the CanadianCeramic Society*, vol. 35, p. 48, 1966.

15

Nickel-Containing Alloys for Medical Application Obtained by Methods of Mechanochemistry and Powder Metallurgy

D. D. Radev

Institute of General and Inorganic Chemistry, BAS, "Acad. G. Bonchev" Street, Bl.11, Sofia 1113, Bulgaria

Correspondence should be addressed to D. D. Radev, ddradev@gmail.com

Academic Editors: T. Czujko, Y. Hao, and P. Markovsky

The methods of mechanochemistry, in combination with cold pressing and pressureless sintering, were used to obtain the most popular nickel-based and nickel-containing alloys used in dentistry and implantology. It was shown that the intense mechanical treatment of Ni, Ti, and Cr powders used as reagents, and the application of the above-mentioned simple powder metallurgical technique for densification allows obtaining NiCr and NiTi alloys with controlled structural properties. The nickel-based dental alloys obtained by mechanically activated sintering possess excellent mechanical, technological, and aesthetic properties. These alloys are suitable as dental restorative materials and for production of porcelain veneered constructions like crowns and bridges using the so-called metal-to-ceramic dental technique. It was shown that the method of mechanically assisted synthesis allows obtaining nanosized NiTi alloy at significantly lower temperature in comparison with the traditional high-temperature alloying. It was also shown that after 40 hours intense mechanical treatment of reagents, a direct synthesis of NiTi alloy proceeds. The product has excellent sinterability which enables to produce bodies with controlled porosity appropriate for application in implantology.

1. Introduction

The range of traditional application of nickel-based alloys covers different areas that require high performance at elevated temperatures. The formation of continuous matrix austenitic γ phase, that usually contains various percentages of Cr, Mo, W, Fe, and Co, the presence of coherently precipitating γ' phases, and carbide or boride grain boundary segregates gives possibilities to form variety of alloys working at extremely high temperatures subjected to a combination of mechanical stresses and thermal shocks. The large number of Ni-based superalloys and the multiplicity of their properties and areas of application are a good example in this respect. Due to the attractive combination of good mechanical properties, chemical resistivity and biocompatibility Ni-based alloys also find a broad application in the fields of dentistry and implantology. Nickel-chromium and nickel-titanium alloys are the most popular representatives in these two medical branches.

1.1. Nickel-Chromium Dental Alloys. Dental alloys could be categorized as noble, containing as a main element gold or palladium and base nickel or cobalt metal alloys [1]. The significant increases of the noble metals price since 1960s stimulate the development of nonprecious nickel-based dental alloys. Nowadays Ni-based alloys used in dentistry have become superior to the gold-based alloys in hardness, elasticity, tensile strength, and particularly in their prices [2]. The ideal dental alloy is biocompatible, and possesses high mechanical strength and corrosion resistance. These alloys also have low prices, appropriate value of the thermal expansion coefficient conformable to that of the veneering ceramic mass, and good technological and aesthetic properties. The current nickel-based dental alloys are multicomponent systems containing in addition to the main metals of Ni, Cr, and Mo (totally about 90–95 wt.%) also elements like Be, Mn, Nb, and Si (5–10 wt% totally) and some additives like Ce, B, C, and N in amounts of part of the percent. Dental alloys are traditionally prepared by consequent melting and thermal alloying of their components in the form of cylindrical ingots weighting 5–10 g. Using centrifugal lost-wax casting of pellets and metal-to-ceramic techniques for porcelain veneering, different dental constructions such as crowns and bridges are produced.

The homogeneity of the multicomponent melt and the exact composition of every pellet are of decisive importance for processes of production and for the quality of the end dental constructions. The traditional method of thermal alloying and casting of nickel-chromium dental alloys is characterized by significant difficulties in obtaining products with constant composition. The considerable differences in the melting and boiling points and densities of the starting components strongly impede to obtain homogeneous melt in the process of thermal alloying and consequently to achieve equal and desired composition and the corresponding set of properties of each dental alloy ingot/pellet. To maintain the high level of similar production processes, expensive melting techniques and strict composition control of every dental pellets batch are required.

1.2. Nickel-Titanium Alloys. The effects of shape memory, two-way shape memory, and superelasticity are among the remarkable properties for equiatomic NiTi or some close in composition NiTi-based alloys. These characteristics are based on the martensitic transformation and its reversion and can be provoked by thermal or mechanical impact. These alloys show an unexpectedly low elastic anisotropy and as a consequence very high ductility determining a cold working up to 60%. The excellent biocompatibility and the high corrosion resistance resulting from the stable TiO_2 layers formation are other remarkable properties of NiTi alloys. The combination of these unique properties makes NiTi the most successful commercial shape-memory alloys [3]. In the last decade these alloys have attracted much interest as functional materials, sensors and actuators, guide wires, medical devices and surgical tools, orthodontic wires, and so forth. Spring-like temperature sensitive stents for cardio-vascular surgery produced from shape-memory NiTi alloy are a good example in that respect. Compressed and supercooled stent is inserted into the artery where under the influence of the human body temperature it recovers its initial shape, thus expanding the blood vessel. The Ni-based shape memory alloys (SMAs) attracted also attention for their potential use as implants because of their appropriate from a medical viewpoint, mechanical properties, and the technological possibility to obtain bodies with a porous microstructure similar to that of the human bone tissue. The traditional method of NiTi synthesis includes serial technological steps such as melting, hot rolling, cold rolling, intermediate annealing, and final cold rolling. During these processes the high-temperature oxidation causes material losses and degradation of the product properties [4]. The described disadvantages make the traditional metallurgical method for synthesis of NiTi time consuming and reduced their economical efficiency.

Here we show how to overcome disadvantages of the traditional metallurgical methods in obtaining Ni-containing alloys with application in dentistry and implantology. The methods of mechanically assisted synthesis and sintering are based on intense milling of components of alloys. In combination with cold pressing and sintering or thermal treatment, high-quality Ni-based dental alloys and NiTi SMAs have been obtained.

2. Experimental

Figure 1 shows SEM of the initial nickel particles used in the synthesis of Ni-based dental alloys. The Ni powder is produced by *Novamet* using carbonyl synthesis method. The particles have regular globular shape and mean particle size of 3–5 μm. As a result of their chemical purity and small particle size, characteristic of the carbonyl synthesis method, these particles have very high sinterability. Chromium particles with mean particle size 150–200 μm were obtained by milling of chromium lumps (99.9% purity) and as a consequence of the mechanical impact have an absolutely flat shape.

Figure 2 shows a mixture of initial Ni-20% Cr particles used as main components in synthesis of multicomponent nickel-based dental alloy. Molybdenum powder with mean particle size of 5.0 μm was obtained by double hydrogen reduction of MoO_3 at 1100°C. Ni-based dental ingots containing Cr and Mo-95.0%, totally and the rest to 100%: Si (*Elkem Silicon Materials*), Mn, and Ce, were obtained by mechanical treatment of reagents, cold pressing of the homogeneous powdery mixture at 15.0 GPa to obtain pellets with desire shape. These green bodies were vacuum sintering at 1295°C for 30 min to achieve high density ingots. Using lost-wax centrifugal casting method, standard bars suitable for mechanical tests were produced. These tests included Vickers hardness (HV_{10}), tensile strength (Rm), elongation limit ($Rp_{0.2}$), and ductile yield (A_5). The thermal expansion coefficient of the Ni-based alloy was determined by dilatometric measurements in the temperature range 20–600°C. Using the same casting method and metal-to-ceramic technique for porcelain veneering, different dental constructions appropriate to evaluate the technological and aesthetic properties of the alloy were produced.

Titanium powder (98.5% purity) obtained by *Fluka* and Ni particles with globular shape and particle size below 10 μm obtained by hydrogen reduction at 600°C of NiO (black) were used as reagents for the syntheses of equiatomic NiTi alloys. Figure 3 shows SEM of the initial Ti powder used in the synthesis of NiTi.

The particles have smooth shape and most of them are interconnected. The size of aggregates and some separated particles exceeds 250 μm. Ni and Ti powders in atomic ratio 1 : 1 were mixed and mechanically treated in a planetary ball mill (*Pulverizette 5/4*, Fritsch GmbH) up to 40 h at a rotation speed of 200 rpm. Stainless steel bowls (80 cm^3) and balls with a diameter of 10 mm and total volume of 25 cm^3 were used. In this way about 30% of the reactor volume was kept free of charge, thus ensuring optimal contact between milling bodies and metal particles, and thus achieving highly efficiency of the mechanical treatment. The milling containers were fitted with semipass flaps allowing mechanical treatment of metal powders under protective Ar atmosphere, thus preventing metal particles from oxidation. Some amount of reagents was removed periodically

FIGURE 1: SEM image of starting Ni particles, ×4500.

FIGURE 2: SEM image of starting NiCr powder mixture (3 : 1), ×150.

FIGURE 3: SEM image of starting Ti particles, ×100.

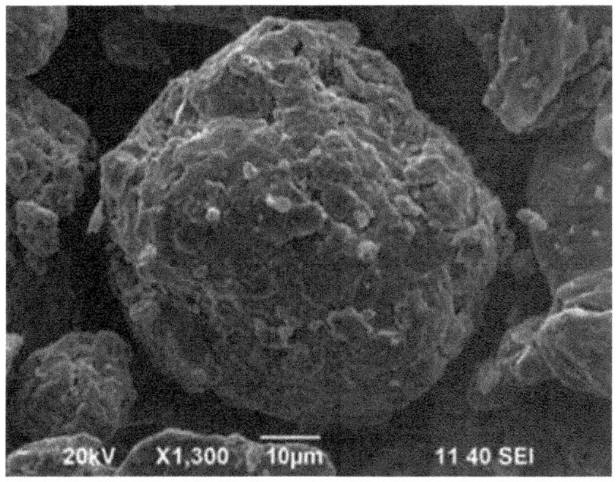

FIGURE 4: SEM image of NiCr after 4 h mechanical treatment, ×1300.

for structural and phase analyses. The mechanically assisted thermal synthesis of NiTi alloy was carried out after 30 h intense mechanical treatment of reagents in a corundum tube furnace in a hydrogen stream at 550°C for 30 min. Sintering of dense NiTi bodies with controlled porosity was performed at 800°C under protective hydrogen atmosphere.

3. Results and Discussion

3.1. Ni-Based Dental Alloys. Figure 4 shows SEM image of Ni-20 wt.% Cr powder mixture after 6 hours intense mechanical treatment in a planetary ball mill. The picture reveals globular aggregates with mean size of about 70–80 μm consisting of submicrometric particles.

This is a typical view of mechanically treated plastic metal particles and demonstrates the morphological and structural changes resulted by the mechanical impact. Homogenization and crashing of particles, appearance of highly reactive oxygen-free surfaces, accumulation different types of structural defects, development of the processes of plastic deformation, and cold welding leading to formation of aggregates are among the effects of mechanically treated solids. Similar morphological pictures were observed after intense mechanical treatment of Ti-Ni and TiB powders in high energetic apparatuses like SPEX and planetary mills [5–7]. The formation of globular aggregates containing components of the alloy is very important from technological point of view for successful implementation of the cold-pressing sintering processes and achievement of dental pellets with maximal density. Sintering of dental alloy pellets with almost theoretical density is decisive for the lost-wax casting process during which the end dental constructions are formed. The eventual presence of pores and cavities in the structure of the pellets leads to oxidation and slag formation during the process of dental construction casting. The practice of the powder metallurgy implies that Cr particles shown in Figure 2 would have poor pressability and sinterability resulting from their flat shape and big sizes. As a consequence, the powder mixture inherited the low sinterability of the Cr particles. The low density of the ingots is a precondition for bad quality of the end dental construction. The homogeneous distribution of Ni and Cr particles after mechanical treatment is demonstrated in Figure 5.

The BSEI image is obtained from the metallographic section of mechanically treated Ni-Cr powder mixture. The content of Ni and Cr exceeds 85 wt% which makes

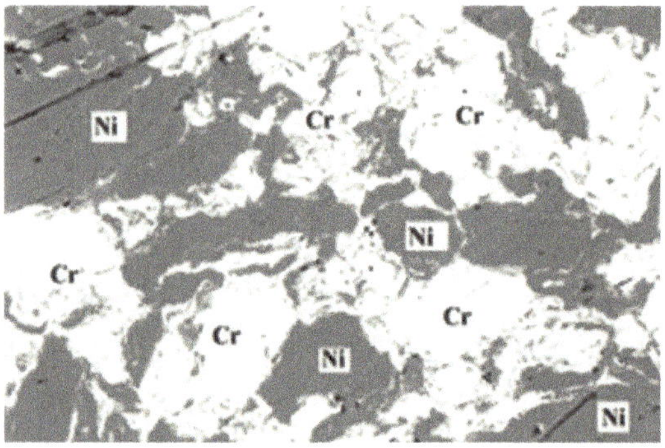

FIGURE 5: BSEI image of NiCr after 4 h mechanical treatment, ×500.

FIGURE 6: Ni-based dental alloy pellets.

the picture of homogeneous distribution of components representative for the alloy. Figure 6 shows images of densified by sintering dental alloy pellets produced on the basis of mechanically treated Ni-Cr powder mixtures.

The picture reveals the technological advantages and flexibility of the production method which includes mechanical treatment of component, densification by the methods of powder metallurgy and obtaining dense pellets having 99% relative density, desired shape and weight. The pellets of the Ni-based alloy obtained in this way have perfect castability demonstrated by casting of a standard tiny lattice model and constant composition of the components [8]. It is very important to notice that the mechanical treatment of the alloy components results in a perfect homogeneity of the composition and morphological changes of metal particles associated with improved pressability and sinterability. The homogeneity of the powder mixture is inherited by ingots, thus ensuring constancy in the properties of the batch.

Figure 7 demonstrates the excellent technological and aesthetic properties of the alloy and the large possibilities to obtain highly complicated porcelain veneered dental constructions. The composition of the nickel-chromium dental alloy is well balanced which means that elements like Si, Mn, and Ce are present in amounts sufficient to improve some properties of the alloy without deteriorating others. For example, it is known that the presence of some amounts of Mn, B, and Ce improves the bonding strength

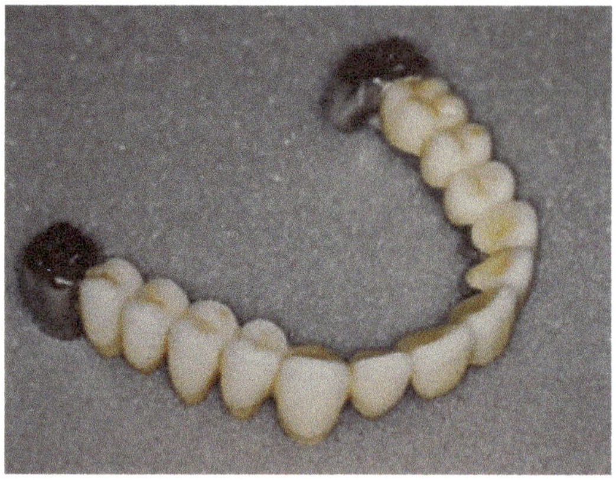

FIGURE 7: Metal-to-ceramic Ni-based dental construction.

between the alloy surface and porcelain veneer. At the same time, excess addition of Mn tends to decrease the corrosion resistance and aesthetic properties of the dental construction, discoloring the porcelain veneer. Furthermore, it is crucial that the B content does not exceed ca. 0.05–0.1 wt.% since it tends to increase the brittleness of the dental constructions. The same could be said for Si, which in amounts exceeding 6 wt.% decreases the mechanical strength of the alloy. These examples clearly indicate the necessity of perfect homogenization of the alloy components, thus providing exact and constant composition of every ingot of the dental alloy. The precisely balanced composition of dental alloys is responsible for the combination of their mechanical, technological, and aesthetic properties. Among these properties the value of the linear coefficient of thermal expansion (CTE) is decisive for the compatibility between the ceramic veneer and metal construction. In order to produce compressive stress in the porcelain during cooling there is a general consensus that the alloy should have higher CTE than the porcelain [9–11]. The value of $13.9 \times 10^{-6} \times C^{-1}$ in the temperature interval 25–600°C was determined by dilatometric tests and is similar to the data for most of the market available Ni-based dental alloys. In ductile materials, at some point, the stress-strain curve deviates from the straight-line relationship, and the Hooke's law no longer applies as the strain increases faster than the stress. From this point on, some permanent deformation in the specimen occurs, and the material is said to react plastically to any further increase in load or stress. That is why the level of the elongation limit ($Rp_{0.2}$) of nickel-chromium dental alloys is a very important mechanical characteristic showing the limit of their application. The standard mechanical tests after casting of ingots show mean value of 482 MPa. This value overcomes the elongation limit of most of the market available nickel-based alloys and is similar to that of some cobalt-based dental alloys which are generally stronger [12–14]. The alloy is characterized by tensile strength (Rm) of 670 MPa, low value of ductile yield (5%), and Vickers hardness (HV_{10}) of 220 units. The well-balanced composition, the chemical purity of the alloy components, and the high density of the dental ingots determine the high level of mechanical and technological properties of the alloy. It should be accentuated

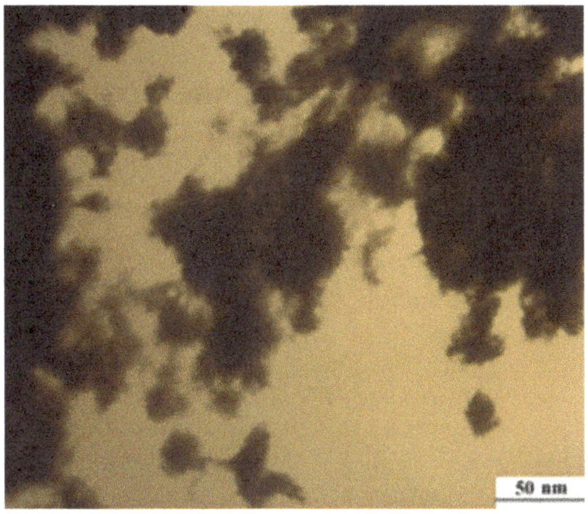

FIGURE 8: TEM of NiTi alloy obtained by mechanically assisted synthesis, ×200 000.

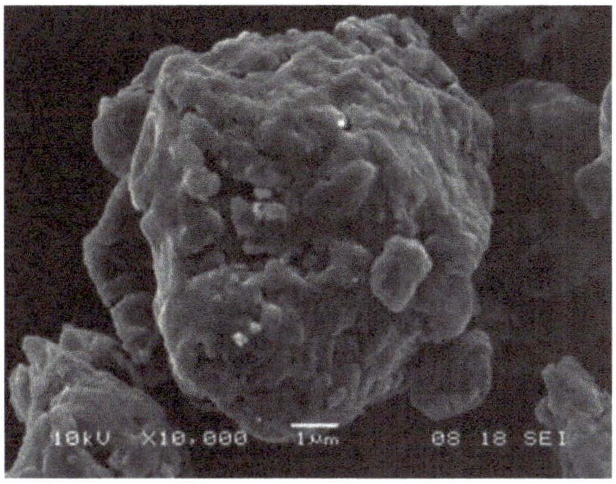

FIGURE 9: SEM image of NiTi alloy obtained by direct mechanical synthesis, ×10 000.

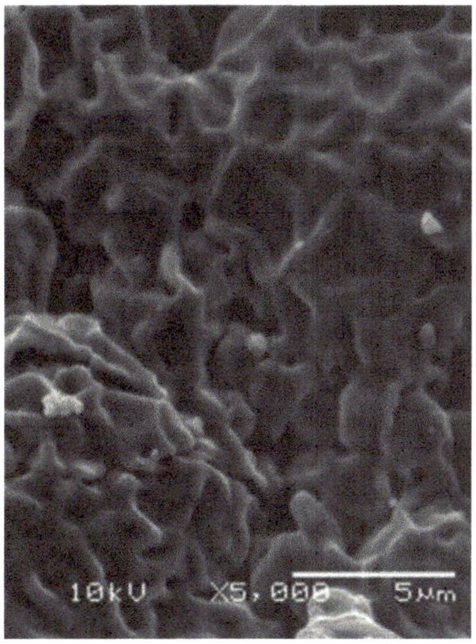

FIGURE 10: SEM of sintered NiTi, ×5000.

that these results were obtained in the absence of Be or other toxic additives to the alloy composition. The role of beryllium which often presents in small amounts (about 2 wt.%) in some base metal dental alloys is to lower the melting temperature and surface tension, thus facilitating castability and improving porcelain-metal bond strength [15–18]. It is believed that acting as a grain refiner, Be improves tensile strength and ductility of the alloy. However, exposure to beryllium vapor or particles is associated with a number of health risks from contact dermatitis to chronic lung disease, known as chronic beryllium disease (CBD). In addition, Be and its compounds in vapor or particulate form have been shown to be carcinogenic [19–21]. The exact composition of nickel-based dental alloys is decisive also for the biocompatibility and for the level of the nickel cations released from contact with the human saliva. Investigations in that area unambiguously show the mutual role of Cr, Mo, and Mn on the formation of stable passive layer protecting from corrosion [22, 23]. Here described new technique for dental alloy production allows obtaining nickel-based ingots with strict composition and constant set of mechanical, technological, and biological properties.

3.2. Mechanochemistry of NiTi Alloys. The mechanochemistry offers variety of synthesis methods like mechanically assisted thermal synthesis, mechanically assisted self-propagating high-temperature synthesis, direct mechanical synthesis, and so forth. Figure 8 shows TEM image of NiTi alloy obtained after 30 h mechanical activation of equiatomic NiTi powder mixture and thermal synthesis at 550°C in a protective atmosphere of pure hydrogen for 30 min.

The data of XRD analysis show presence of the equilibrium TiNi, Ti_2Ni, and $TiNi_3$ phases that limit the region of the system NiTi with respect to its practical application. The data obtained by XRD and SEM investigations show that the product consists of nanosized particles with mean size of 20–30 nm which form an aggregate mass [6, 24]. Nanosized materials possess high sinterability due to their defect structure and large contact surface area resulting in enhanced diffusion mobility. The high activity of such materials allows producing macrosized end bodies at milder conditions, for example, by pressureless sintering or sintering at significantly lower temperatures. Prolonged milling time up to 40 h leads to direct synthesis of NiTi alloy which phase composition is similar to that of the product obtained by mechanically assisted thermal synthesis. Figure 9 shows SEM image of the product which forms globular aggregates typical of the products obtained by mechanical synthesis.

Being built of particles with submicron sizes, these aggregates have complicated structure, possess high sinterability, and are suitable to create dense bodies with controlled porosity. Similar porous structure with uniformly distributed pores being 1-2 μm in size is shown in Figure 10.

The material is densified by sintering at 800°C of NiTi powders obtained by direct mechanical synthesis. The possibility to obtain bodies with controlled porous structure is very important for the application of NiTi alloys in implantology. The presence of open interconnected pores with predetermined dimensions allows penetration of the human bone tissue into NiTi implants. That is a good example for creation of macroproducts with controlled properties due to the use of nanostructured particles [25].

4. Conclusion

The methods of mechanochemistry allow to overcome the difficulties of the traditional thermal alloying methods in synthesis of some Ni-based alloys with medical application. Powder metallurgical route in preparation of Ni-based alloys provides exact composition and random distribution of components in dental ingots and strict control of the nickel-titanium ratio which is decisive for the temperature range of martensite-austenite transformation of the NiTi SMAs. As a result of the intense mechanical treatment, the components of the alloys acquire good press and sinterability which allow, after cold pressing and sintering, producing bodies with controlled porosity. Applying the method of mechanically assisted sintering, high-quality beryllium-free dental alloy ingots, and finally complicated dental constructions could be produced [26]. The method of dental alloy production described here shows how to achieve perfect homogeneity and exact composition of every ingot and probably how to enlarge the area of application of the cheap nickel-containing dental alloys. In the case of NiTi, the mechanical treatment of the reagents leads to a significant lowering of the synthesis temperature or to direct synthesis of nanosized alloy suitable for sintering bodies with controlled porosity.

References

[1] H. F. Morris, M. Manz, W. Stoffer, and D. Weir, "Casting alloys: the materials and clinical effects," *Advances in Dental Research*, vol. 6, pp. 28–31, 1992.

[2] M. Pretti, E. Hilgert, M. Bottino, and R. Avelar, "Evaluation of shear bond strength of the union between two Co-Cr alloys and a dental porcelain," *Journal of Applied Oral Science*, vol. 12, no. 4, pp. 280–284, 2004.

[3] K. Otsuka and T. Kakeshita, "Science and technology of shape-memory alloys: new developments," *MRS Bulletin*, vol. 27, no. 2, pp. 91–100, 2002.

[4] K. Ebato, M. Tsuda, and T. Oomori, US Patent, 5,316,599, 1994.

[5] A. Szajek, M. Makowiecka, E. Jankowska, and M. Jurczyk, "Electrochemical and electronic properties of nanocrystalline $TiNi_{1-x}M_x$ ($M = Mg, Mn, Zr, x = 0, 0.125, 0.25$) ternary alloys," *Journal of Alloys and Compounds*, vol. 403, no. 1-2, pp. 323–328, 2005.

[6] D. D. Radev, "Mechanical synthesis of nanostructured titanium-nickel alloys," *Advanced Powder Technology*, vol. 21, no. 4, pp. 477–482, 2010.

[7] D. D. Radev and D. G. Klissurski, "Mechanochemical synthesis and SHS of diborides of titanium and zirconium," *Journal of Materials Synthesis and Processing*, vol. 9, no. 3, pp. 131–136, 2001.

[8] D. D. Radev and M. I. Marinov, Bg. Patent, 63728, 2002.

[9] P. De Hoff, K. Anusavice, and B. Hojjatie, "Thermal incompatibility analysis of metal-ceramic systems based on flexural displacement data," *Journal of Biomedical Materials Research*, vol. 41, pp. 614–623, 1998.

[10] T. R. Walton and W. J. O'Brien, "Thermal stress failure of porcelain bonded to a palladium-silver alloy," *Journal of Dental Research*, vol. 64, no. 3, pp. 476–480, 1985.

[11] S. C. Lopes, V. O. Pagnano, J. M. D. De Almeida Rollo, M. B. Leal, and O. L. Bezzon, "Correlation between metal-ceramic bond strength and coefficient of linear thermal expansion difference," *Journal of Applied Oral Science*, vol. 17, no. 2, pp. 122–128, 2009.

[12] "US Patent," 6613275, 2003.

[13] "US Patent," 6756012, 2001.

[14] "US Patent Application," 0041820, 2002.

[15] K. F. Leinfelder, "An evaluation of casting alloys used for restorative procedures," *Journal of the American Dental Association*, vol. 128, no. 1, pp. 37–45, 1997.

[16] J. S. Covington, M. A. McBride, W. F. Slagle, and A. L. Disney, "Beryllium localization in base metal dental casting alloys," *Journal of Biomedical Materials Research*, vol. 19, no. 7, pp. 747–750, 1985.

[17] O. L. Bezzon, M. G. de Mattos, R. F. Ribeiro, and J. M. Rollo, "Effect of beryllium on the castability and resistance of ceramometal bonds in nickel-chromium alloys," *The Journal of Prosthetic Dentistry*, vol. 80, no. 5, pp. 570–574, 1998.

[18] H. H. Huang, M. C. Lin, T. H. Lee et al., "Effect of chemical composition of Ni-Cr dental casting alloys on the bonding characterization between porcelain and metal," *Journal of Oral Rehabilitation*, vol. 32, no. 3, pp. 206–212, 2005.

[19] M. Kuschner, "The carcinogenicity of beryllium," *Environmental Health Perspectives*, vol. 40, pp. 101–105, 1981.

[20] I. Fodor, "Histogenesis of beryllium-induced bone tumours," *Acta Morphologica Academiae Scientiarum Hungaricae*, vol. 25, no. 2-3, pp. 99–105, 1977.

[21] W. T. Sanderson, E. M. Ward, and M. R. Petersen, "Re: lung cancer case-control study of beryllium workers," *American Journal of Industrial Medicine*, vol. 39, pp. 133–144, 2001.

[22] J. D. Bumgardner and L. C. Lucas, "Surface analysis of nickel-chromium dental alloys," *Dental Materials*, vol. 9, no. 4, pp. 252–259, 1993.

[23] S. B. Jones, R. L. Taylor, J. S. Colligon, and D. Johnson, "Effect of element concentration on nickel release from dental alloys using a novel ion beam method," *Dental Materials*, vol. 26, no. 3, pp. 249–256, 2010.

[24] R. A. Varin and Ch. Chiu, "Synthesis of nanocrystalline magnesium diboride (MgB_2) metallic superconductor by mechano-chemical reaction and post-annealing," *Journal of Alloys and Compounds*, vol. 407, no. 1-2, pp. 268–273, 2006.

[25] S. Kipp, V. Sepelak, and K. D. Becker, "Chemistry with the hammer: mechanochemistry," *Chemie in Unserer Zeit*, vol. 39, no. 6, pp. 384–392, 2005.

[26] D. D. Radev and M. I. Marinov, "Bg Patent," 002045, 2012.

Modeling to Study the Effect of Environmental Parameters on Corrosion of Mild Steel in Seawater Using Neural Network

Subir Paul

Department of Metallurgical and Material Engineering, Jadavpur University, Kolkata 700032, India

Correspondence should be addressed to Subir Paul, spaulxx@ymail.com

Academic Editors: J. Eckert and J. M. Rodriguez-Ibabe

Prediction of corrosion rate of steel structure in seawater is a challenging task for design and corrosion engineers for existing as well as new structures, due to wide variation of its composition across the global marine environment. The major parameters influencing the rate are salinity, sulphate, dissolved oxygen, pH, and temperature. While the individual effects of these parameters on corrosion are known, the conjoint effect of the parameters together is complex and unpredictable. Endeavors have been made to model the corrosion rate from laboratory experimental data, using Artificial Neural Network to predict corrosion rate at any combinations of the above five parameters and to better understand the effects of these parameters jointly on corrosion behavior. 3D mappings clearly reveal the complex interrelationship between the variables and importance of conjoint effect of the variables rather than single variable on the corrosion rate of steel in seawater.

1. Introduction

Prediction of corrosion rate of steel structure in global marine environment is a challenging task due to wide variation of parameters controlling the rate. The chloride (Cl^-) concentration of seawater varies from about 5.8 gm/Kg to about 24 g/Kg, the sulphate (SO_4^{2-}) from 0.8 gm/Kg to 3.4 gm/Kg and the bicarbonate (HCO_3^-) from 0.01 gm/Kg to 0.2 gm/Kg [1], across the different ocean and sea water. Though the pH of sea water is in the range of neutral to slightly alkaline, local acidity developed due to corrosion products as well as crude petroleum products, lowering the pH to around 4. In general the influence of any of the five parameters, namely, salinity, sulphate, dissolved oxygen, pH, and temperature, on corrosion rate, is independently known. Chloride ion aggravates the degradation of materials in aqueous environment. It helps in breaking passive oxide layer, leading to localized corrosion of crevice and pitting corrosion. PH has a variable effect on corrosion. Corrosion rate is high at lower pH due to acidic corrosion, while, at intermediate pH of 8.5 to 12, it drops down due to formation of passive layer and, at higher pH, the corrosion is severe due to caustic embrittlement. Sulphate ion in general helps the formation of corrosion resistant deposit with Ca and Mg ions, but at lower pH this deposit goes in solution and corrosion rate is enhanced. Both temperature and dissolved oxygen have a strong effect of increasing the corrosion rate due to the fact that the concentration of dissolved oxygen increases the rate of cathodic reaction of water reduction and also the limiting current density of concentration polarization. Temperature increases the diffusion rates of the ions and helps in convection mass transfer, both of which promote depolarization. But conjoint effect of the above-mentioned parameters and the interrelationship among the variables on corrosion rate are complex and not known.

Modeling helps better understanding of corrosion process and the effect of conjoint actions of parameters which are unknown and non-linear functions of the rate. It also helps the construction and corrosion engineers by predicting the life of the structure in a given environment. Complexity of the present problem, involving the effect of the five parameters on single output, necessitates the modeling by artificial neural networks (NNs). It is one of the most powerful modeling techniques based on statistical approach. The method of neural-network simulation is applicable to the corrosion processes as the multifactor systems represented by an aggregate of corrosion-electrochemical parameters (the characteristic potentials and currents are not cymbatical to the parameters of medium, the anionic composition, concentration, temperature, pH value, etc.). This would enable us to reveal the hidden links, which are difficult to analyze.

In the past few years, there has been a constantly increasing interest for NN modeling in different fields of materials science. ANN models have been developed to model different correlations and phenomena in steels [2–5], aluminum alloys [6] and titanium alloys [7, 8]. There has been only limited work on the application of neural network on the effect of the variables influencing marine corrosion process [9, 10].

2. Modeling Methodology

An ANN algorithm [11–14] of studying systems involves a set of uniform modules, named associative neurons. Each of them has a certain number of information inputs, which are equal to the number of independent input parameters of the system, and a certain number of outputs, where the output parameters are formed. The output parameter is calculated by substituting a set of input parameters into a certain function (identical for all of associative neurons of network) taking into account the corresponding set of weight coefficients. The advantage of NN as a tool for the analysis of experimental data is, primarily, its capability of learning, that is, forming the ability to predict the output parameters of a system by any set of input ones. Usually, neural networks are trained using a large number of input with corresponding output data (input/output pairs) so that a particular set of inputs produces, as nearly as possible, a specific set of target outputs, by adjusting the weight associated with each connection (synapse) between neurons. NNs are capable of generalizing experience obtained in the course of learning. As a result, it becomes possible to predict the situations, which are absent in the set of experimental results.

The following matters are important in the design and training of neural networks: (i) architecture of the neural network, (ii) training algorithm, and (iii) transfer function. The term "architecture of the neural network" refers to the number of the layers in the NN and the number of the neurons in each layer. The general structure of input, one hidden and one output layer was used. The numbers of neurons in the input layer and the output layer are determined by the numbers of input and output parameters, respectively. In order to find the optimal architecture, different numbers of neurons in the hidden layer were attempted.

There are many different training algorithms. The details for the NN training algorithms are given in [13]. In order to achieve the best result, different training algorithms were attempted including Batch Gradient Descent, Batch Gradient Descent with Momentum, One-step-secant, Scaled Conjugate Gradient, Resilient Backpropagation, Variable Learning Rate, and Levenberg-Marquardt.

Different training options have been carried out including training on variations of mean square error for better generalization, training against a validation set, and training until the gradient of the error reaches a minimum. Improvement of the generalization has been attempted by means of "regularization" and "early stopping with validation." The transfer function transforms the neuron input value into the output value. For all cases, NN linear transfer function was used in the output layers. In the hidden layers tansigmoid transfer function and log-sigmoid transfer function were employed.

All neural network models were tested by (i) linear regression between NN predictions and new experimental data, (ii) statistical analysis of the error of NN predictions, and (iii) direct comparison of NN predicted and experimental data.

Five input neurons, namely, salinity, sulphate, dissolved oxygen, pH, and temperature, were taken to predict output, the corrosion current. The number of sets of laboratory experimental data (87) was divided as 60% for training, 20% for testing, and the rest 20% for validation. Very good performances of the models are achieved. The neural network models are used for further simulations and predictions at various combinations of input.

3. Experimental Procedure

Series of laboratory experiments were performed to produce the data of corrosion rate for training, testing, and validation of neural network. Rectangular samples of $1 \times 2\,\text{cm}^2$ were cut from the mild steel bar. They were ground carefully in order to make the edges blunt. The samples were then further ground by belt drive and polished up to 3/0 emery paper. The samples were washed with acetone and were observed under a low-magnification microscope to inspect for any surface defect of pit or scratch. Fresh solutions were prepared before each experiment, using KCl, $CaCl_2$, $NaHCO_3$, and $MgSO_4$ salts in double-distilled water with variation of five experimental variables: salinity, sulphate, dissolved oxygen, pH, and temperature. Polarization experiments were carried out as per ASTM standard methods [15–19] using Gamry Potentiostat. Standard Corrosion Cell has been used to perform the electrochemical potentiostatic polarization tests using Gamry Potentiostat. Corrosion rates were determined by Tafel's extrapolation and linear polarization methods, with repetitions of experiments under same conditions for at least thrice till the consistency of the results are obtained.

4. Results and Discussions

The architecture of the neural networks (NNs) is multiplayer perceptron with (1) the layer of neurons receptors (the input layer) realizing a receipt of information from the outside, (2) the layer of associative neurons (hidden), whose functions were chosen, and (3) the layer of output neurons, which form the response of the network to the external stimulus. An ANN of such a type is termed perceptron with one hidden layer of neurons. Most of applied works are associated with the use of such networks, because they are best studied and as a rule, one hidden layer is sufficient to solve the

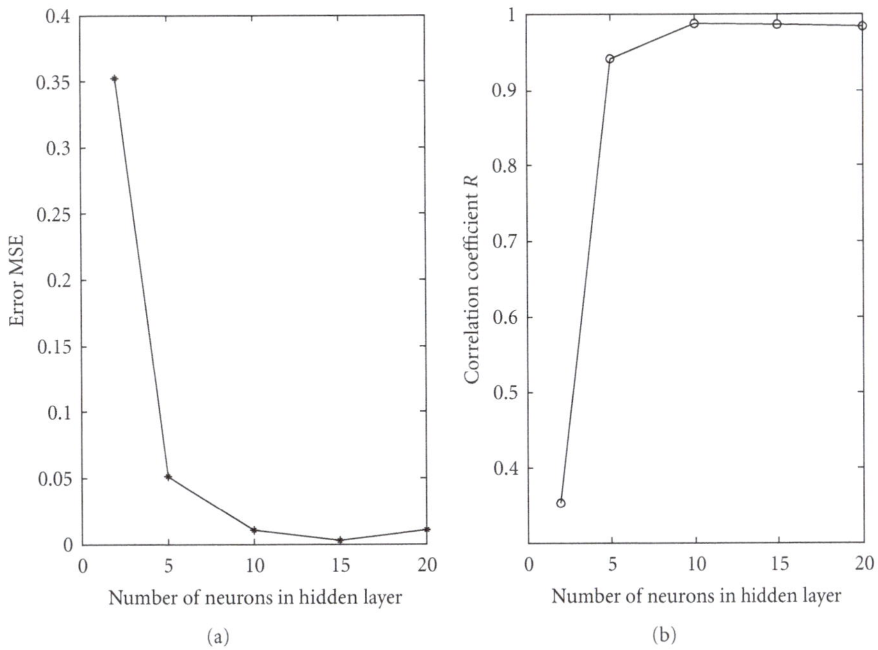

FIGURE 1: Effect of neuron size on performance of neural network.

majority of problems [1, 5, 7]. The input layer consists of five input variables of chloride, sulphate, dissolved oxygen, pH, and temperature of the seawater. The output layer is of single output of corrosion current density. Feedforward backpropagation algorithm was used for network training with transfer functions of tan-sigmoid (tansig) and purelin for hidden and output layers, respectively. The network was trained with Batch Gradient Descent (traingd). The selection of above transfer functions and training function are based on the fact that these combinations of functions with the present experimental data give minimum error. To choose an optimum number of neurons in the hidden layer, different numbers of neurons were selected and the errors MSE were compared (Figure 1). It is found that an optimum number of 15 neurons in the hidden layer gives the minimum error. Using the above architect of the NNs, it was trained. The performance of training, testing and validation has been illustrated in Figure 2. It is seen after about 50 epoch, validation and test lines almost merge with each other. The training curve meets the goal of threshold error of 0.002 after 375 epochs. The result indicates a good performance of training. The outputs of corrosion current obtained from the model were compared with those from experimental determined data to validate the model, and a regression line of near 45° was obtained (Figure 3) with correlation coefficient R of 0.98968. This signifies the network has been well trained and the prediction of corrosion rate by the model can very well match with that of actual experimental determination.

Every model has some limitations and uncertainty. In neural network, a huge number. of experimental data is required; the more the data, the better the simulation and prediction. Out of these data, about 60% are used for

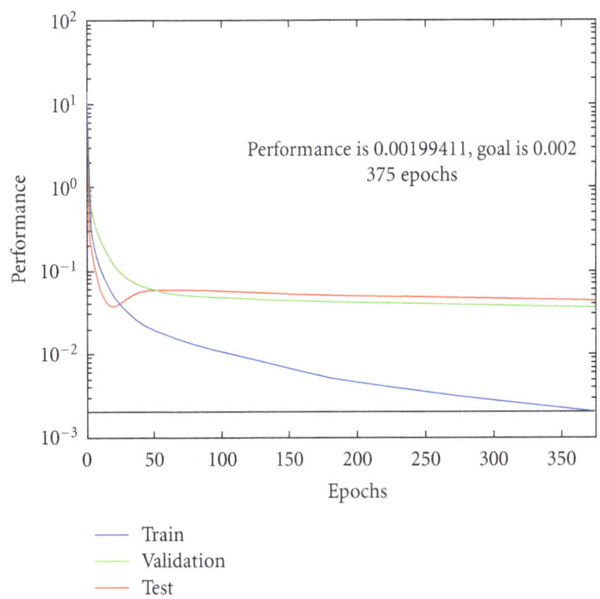

FIGURE 2: Training of neural network with corrosion rate data of mild steel using traingd as training function, Tansig and purlin as transfer functions.

studying the trend (training) and developing the initial model. It now tests model on another new set of data (20%); if the test performance (the level to be indicated by the user) is not satisfactory, it then goes back and redevelops the model. When the test is satisfied, it takes another new set of data (balance 20%) and simulates or predicts the unknown and compares its simulated results with actual (experimental or field data), if it is satisfactory, the final model is generated;

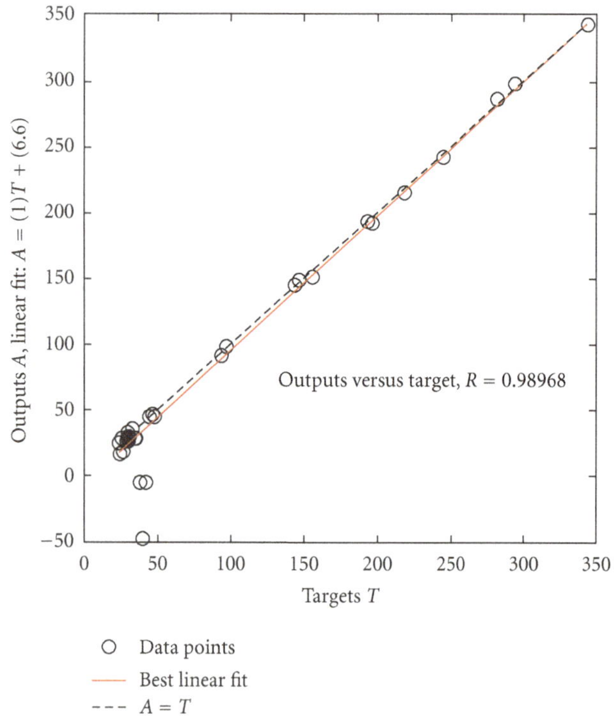

Figure 3: Regression line between output generated from the model and target of corrosion data obtained from experiments.

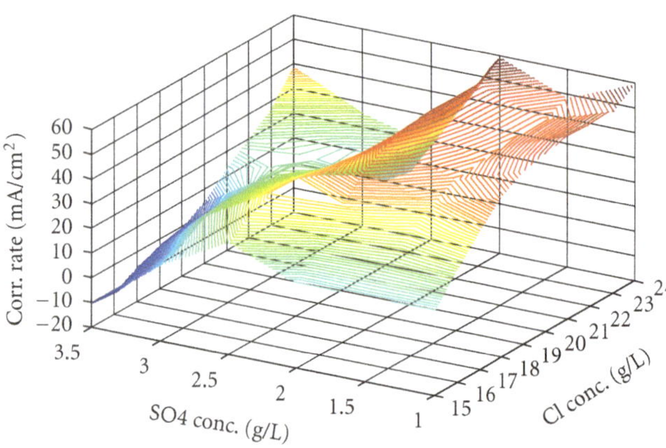

Figure 4: Corrosion behavior of MS in seawater with variation of Cl and SO_4 conc. at pH 8.

if not it goes back to the first step. So under so many steps of training, testing and validation, it has been found that the neural network can be very accurately predicted in most cases subject to the accuracy of experimental data and large number of data, covering full domain of the parameters variables.

The model is now used to simulate the unknown corrosion rates at any combinations of the five parameters to study the corrosion behavior of steel in any global marine environment. It also helps to find the corrosion rate of steel structure at a given geographical site in an ocean.

Figure 4 shows the effects of chloride and sulphate ion concentrations on the corrosion rate of mild steel in seawater. It is seen that, while the chloride ion aggravates the corrosion rate, the sulphate ion has marginal effect at lower concentration and decreases the rate sharply at higher concentration. This is due to the fact that the presence of sulphate ion helps in formation of calcareous deposit in alkaline region, with presence of Ca and Mg ions. Sulphate ions have an important role during calcareous deposition on a steel; they promote the formation of the Mg-containing porous layer. In seawater, the Mg-containing film forming on steel would then be Mg basic sulphate, $x\mathrm{Mg(OH)_2 \cdot MgSO_4} \cdot n\mathrm{H_2O}$ [20]. Secondly sulphate ions had participated directly to the corrosion process of iron in seawater by promoting the formation of an intermediate Fe(II-III) hydroxysulphate [21, 22] of the Green Rust by the following reactions:

$$6\mathrm{Fe^{2+}} + 12\mathrm{H_2O} + \mathrm{SO_4}^{2-} \longrightarrow \mathrm{Fe^{II}}_4\mathrm{Fe^{III}}_2\mathrm{(OH)_{12}SO_4} + 12\mathrm{H^+} + 2e \quad (1)$$

It is interesting to find that the corrosion rate enhances when both sulphate and chloride concentrations are very high. The effects of pH and chloride on corrosion rate have been illustrated in Figure 5. At higher level of chloride concentration, the variation of pH from 4 to 8 does not reflect much effect on corrosion rate. At lower level of Cl^- ions, the corrosion rate decreases at near neutral to alkaline region of the medium. Figure 6 is an interesting diagram of the conjoint effect of pH and sulphate at fixed chloride concentration. The corrosion surface shows a minima at near neutral pH and high sulphate concentration. The rate increases with acidity or alkalinity in seawater. The diagram clearly indicates the complex interrelationship that exits between the variables and the importance of conjoint effect of variables rather than with single variable. SO_4^{2-} ions seem to help in the formation of corrosion resistant deposit in near neutral seawater. The effects of temperature and dissolved oxygen content on corrosion rate have been illustrated in Figure 7. While both of the parameters aggravate the corrosion rate, the effect of temperature is pronounced only when appreciable amount of dissolved oxygen is present in seawater and vice versa. The electrochemical corrosion of carbon steel in aqueous environment like seawater takes place by the following anodic and cathodic reactions:

$$\begin{gathered} 2\mathrm{Fe} \longrightarrow 2\mathrm{Fe^{2+}} + 4e \quad \text{Anodic} \\ \mathrm{O_2} + 2\mathrm{H_2O} + 4e \longrightarrow 4\mathrm{OH^-} \quad \text{Cathodic} \end{gathered} \quad (2)$$

It is seen that the concentration of dissolved oxygen O_2 enhances the rate of cathodic reaction. It is known from the polarization diagram of low carbon steel in aerated aqueous environment [23] that the anodic line cuts at concentration polarization region of the cathodic polarization curve at $E_{corr} - I_{corr}$ point. The rate of cathodic reaction is controlled by I_L, the limiting current density, which is a strong function of concentration and diffusion of oxygen from the bulk solution to the interface. This explains the enhancement of the corrosion rate with increase of dissolved oxygen concentration. The limiting current density is also very

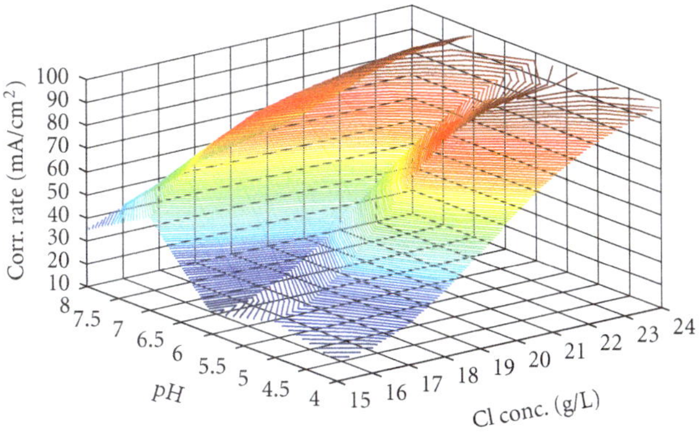

FIGURE 5: Corrosion behavior of MS in seawater with variation of Cl and pH at SO_4 conc. of 2 g/L.

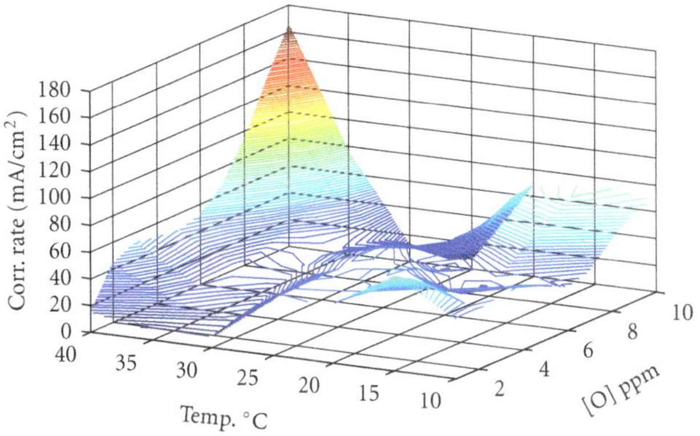

FIGURE 7: Corrosion behavior of MS in seawater with variation of temperature and dissolved Oxygen at Cl conc. of 15 g/L.

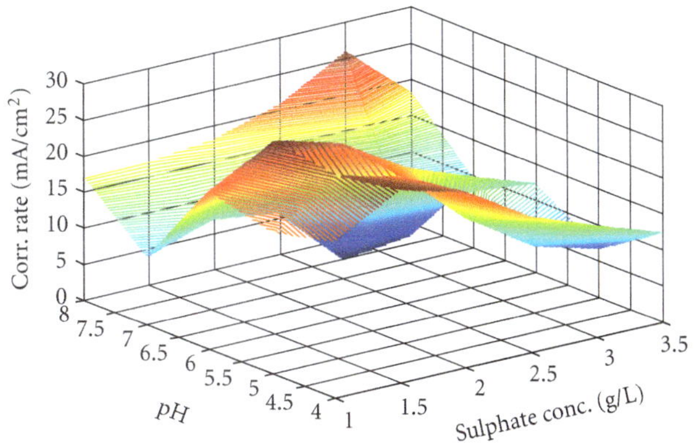

FIGURE 6: Corrosion behavior of MS in seawater with variation of pH and SO_4 at Cl conc. of 15 g/L.

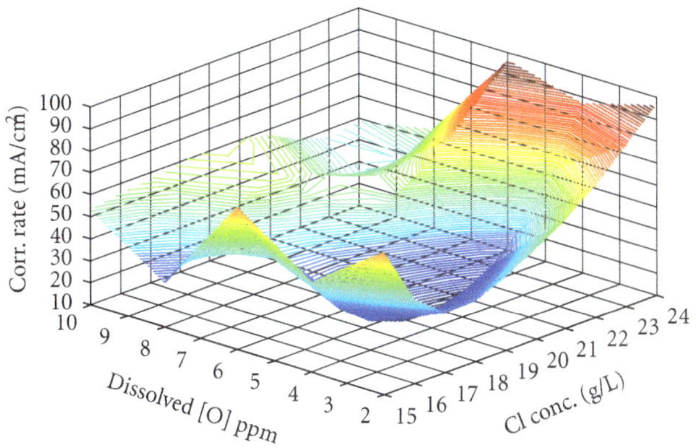

FIGURE 8: Corrosion behavior of MS in seawater with variation of dissolved oxygen and chloride.

much influenced by the temperature, which enhances the corrosion rate. Figure 8 shows the conjoint effect of dissolved oxygen and chloride concentration on corrosion rate. It is seen while both the parameters aggravate corrosion rate, the effect of chloride ion in enhancing corrosion rate is higher. The dissolved oxygen influences the cathodic reaction and the chloride ion controls the anodic reaction. Chloride ion increases the rate of 1st reaction (anodic) by forming metal chloride, which hydrolyzes to generate H^+ ions, increasing the local acidity that aggravates further the dissolution process. It is interesting to note from Figure 8 that at high chloride concentration and dissolved oxygen content, the corrosion rate is less than that at lower dissolved oxygen content. This might be due to the fact that excess oxygen can now act on anodic areas also and may form protective oxide layer, retarding the attack of chloride ion on anodic sites.

5. Conclusion

Effect of seawater parameters on corrosion rate of steel is very well studied by modeling with neural network. It is found from the 3D mapping that the corrosion rate is a complex non-linear function of the variables. Influence of one parameter on the rate is much dependent on concentration of other variable. While the chloride ion increases the corrosion rate, the sulphate ion decreases corrosion rate by helping to form calcareous deposit. This fact is dependent on the level of pH. At lower level of Cl^- ions, the corrosion rate decreases at near neutral to alkaline region of the medium. Temperature and dissolved oxygen, both parameters, aggravate the corrosion rate; the effect of temperature is pronounced only when appreciable amount of dissolved oxygen is present in seawater and vice versa. The model can predict the corrosion rate of steel structure at any location site of the ocean.

References

[1] *Dechema Handbook of Corrosion*, vol. 11, VCH Publishers, Weinheim, NY, USA, 1992.

[2] M. A. Yescas, H. K. D. H. Bhadeshia, and D. J. MacKay, "Estimation of the amount of retained austenite in austempered ductile irons using neural networks," *Materials Science and Engineering A*, vol. 311, no. 1-2, pp. 162–173, 2001.

[3] E. A. Metzbower, J. J. DeLoach, S. H. Lalam, and H. K. D. H. Bhadeshia, "Neural network analysis of strength and ductility of welding alloys for high strength low alloy shipbuilding steels," *Science and Technology of Welding and Joining*, vol. 6, no. 2, pp. 116–124, 2001.

[4] S. Malinov, W. Sha, and Z. Guo, "Artificial neural network modelling of crystallization temperatures of the Ni–P based amorphous alloys Material," *Materials Science and Engineering A*, vol. 283, pp. 1–218, 2000.
[5] S. Malinov, W. Sha, and J. J. McKeown, "Modelling the correlation between processing parameters and properties in titanium alloys using artificial neural network," *Computational Materials Science*, vol. 21, no. 3, pp. 375–394, 2001.
[6] O. Gundersen, A. O. Kluken, O. R. Myhr et al., *Mathematical Modeling of Weld Phenomena 5*, Edited by H. Cerjak, Institute of Materials, London, UK, 2001.
[7] S. McShane, S. Malinov, J. J. McKeown, and W. Sha, *Computational Modeling of Materials, Minerals, and Metals Processing*, Edited by M. Cross, J.W. Evans, C. Bailey, TMS, Easton, Pa, USA, 2001.
[8] S. Malinov and W. Sha, "Application of artificial neural networks for modelling correlations in titanium alloys," *Materials Science and Engineering A*, vol. 365, no. 1-2, pp. 202–211, 2004.
[9] V. D. Kiselev, S. M. Ukhlovtsev, A. N. Podobaev, and I. I. Reformatskaya, "Analysis of corrosion behavior of steel 3 in chloride solutions by using neural networks," *Protection of Metals*, vol. 42, no. 5, pp. 452–458, 2006.
[10] K. V. S. Ramana, T. Anita, S. Mandal et al., "Effect of different environmental parameters on pitting behavior of AISI type 316L stainless steel: experimental studies and neural network modeling," *Materials and Design*, vol. 30, no. 9, pp. 3770–3775, 2009.
[11] F. Yosserman, *Neirokomp'yuternaya tekhnika: Teoriya I Praktika (Neural-Computer Technique: Theory and Practice)*, Mir, Moscow, Russia, 1992.
[12] A. N. Gorban, *Obuchenie neironnykh setei (Learning of Neural Networks)*, SP ParaGraf, Moscow, Russia, 1991.
[13] K. Gurney, *An Introduction to Neural Network*, UCL Press Limited, 1997.
[14] L. Ileană, C. Rotar, and A. Incze, "The optimization of feed forward neural network structures using genetic algorithm," in *Proceedings of the International Conference on Theory and Applications of Mathematics and Informatics (ICTAMI '04)*, Thessaloniki, Greece, 2004.
[15] J. R. Scully, "Polarization resistance method for determination of instantaneous corrosion rates," *Corrosion*, vol. 56, no. 2, pp. 199–217, 2000.
[16] ASTM G 5, Potentiostatic and Potentiodynamic Anodic Polarization Measurements.
[17] ASTM G 59, Polarization Resistance Measurements.
[18] ASTMG 61, Cyclic Polarization Measurements for Localized Corrosion.
[19] N. G. Thompson and J. H. Payer, *DC Electrochemical Test Methods*, National Association of Corrosion Engineers, Houston, Tex, USA.
[20] C. Barchiche, C. Deslouis, O. Gil, S. Joiret, P. Refait, and B. Tribollet, "Role of sulphate ions on the formation of calcareous deposits on steel in artificial seawater; the formation of Green Rust compounds during cathodic protection," *Electrochimica Acta*, vol. 54, no. 13, pp. 3580–3588, 2009.
[21] Ph. Refait, J. B. Memet, C. Bon, R. Sabot, and J. M. R. Génin, "Formation of the Fe(II)–Fe(III) hydroxysulphate green rust during marine corrosion of steel," *Corrosion Science*, vol. 45, no. 4, pp. 833–845, 2003.
[22] S. Pineau, R. Sabot, L. Quillet et al., "Formation of the Fe(II–III) hydroxysulphate green rust during marine corrosion of steel associated to molecular detection of dissimilatory sulphite-reductase," *Corrosion Science*, vol. 50, no. 4, pp. 1099–1111, 2008.
[23] D. A. Jones, *Principles and Prevention of Corrosion*, Prentice Hall, 2nd edition, 1996.

17

Thermal Conductivity of Cu-Cr-Zr-Ti Alloy in the Temperature Range of 300–873 K

S. Chenna Krishna,[1] N. Supriya,[2] Abhay K. Jha,[1] Bhanu Pant,[1] S. C. Sharma,[1] and Koshy M. George[3]

[1] *Materials and Metallurgy Group, Materials and Mechanical Entity, Vikram Sarabhai Space Centre, Trivandrum 695022, India*
[2] *Analytical, Spectroscopy, and Ceramics Group, Propellants, Polymers, Chemicals, and Materials Entity, Vikram Sarabhai Space Centre, Trivandrum 695022, India*
[3] *Materials and Mechanical Entity, Vikram Sarabhai Space Centre, Trivandrum 695022, India*

Correspondence should be addressed to S. Chenna Krishna, chenna.sk@gmail.com

Academic Editors: R. Hebert, P. Markovsky, and T. Yue

In the present investigation, thermal conductivity of Cu-Cr-Zr-Ti alloy was determined as the product of the specific heat (C_p), thermal diffusivity (α), and density (ρ) in the temperature range of 300–873 K. The experimental results showed that the thermal conductivity of the alloy increased with increase in temperature up to 873 K and the data was accurately modeled by a linear equation. For comparison, thermal conductivity was also evaluated for OFHC copper in the same temperature range. The results obtained were discussed using electrical conductivity and hardness measurements made at room temperature. Transmission electron microscopy (TEM) was done to understand the microstructural changes occurring in the sample after the test. Wiedemann-Franz-Lorenz law was employed for calculating electronic and phonon thermal conductivity using electrical conductivity. On the basis of studies conducted it was deduced that in situ aging may be one of the reasons for the increase in thermal conductivity with temperature for Cu-Cr-Zr-Ti alloy.

1. Introduction

Cu-Cr-Zr alloy is a candidate material for high heat flux applications like the inner wall of a thrust chamber and the first wall of nuclear reactor owing to its high conductivity and strength [1, 2]. The high conductivity of the alloy is attributed to the low solubility of Cr and Zr in copper at room temperature [3], while the strength is due to the precipitation of Cr and Cu_5Zr in copper matrix [4, 5]. Zirconium plays an additional role of fixing elemental sulphur and suppresses dynamic embrittlement [6]. The alloy used in the present study is modified by the addition of titanium of Cu-Cr-Zr alloy. Titanium plays a role similar to that of zirconium in Cu-Cr-Zr alloy in suppressing dynamic embrittlement by fixing elemental sulphur as titanium sulphide [7]. Limited literature is available on mechanical properties, microstructure and thermal properties of Cu-Cr-Zr-Ti alloy. Lack of thermal conductivity data for the alloy at elevated temperature promoted us to design experiments to evaluate the same in the temperature of 300–873 K.

2. Experimental Procedure

Hot rolled and annealed plate of Cu-0.61Cr-0.038Zr-0.029Ti-0.003Fe (wt %) alloy was used in the present study. Small pieces cut from the plate were used for hardness, thermal, and electrical conductivity measurement. In addition, properties of OFHC copper (C10100) were evaluated in the annealed condition. OFHC copper represents the pure form of copper, and the effect of alloying elements (Cu-Cr-Zr-Ti) can be better understood by comparison. Thermal conductivity was measured using a Flashline 3000 laser flash apparatus as the product of thermal diffusivity (α), specific heat (C_p), and density (ρ). A disc of 12.7 mm diameter and 3 mm thickness was used for the measurement. The samples were made more opaque to the energy pulse by coating with

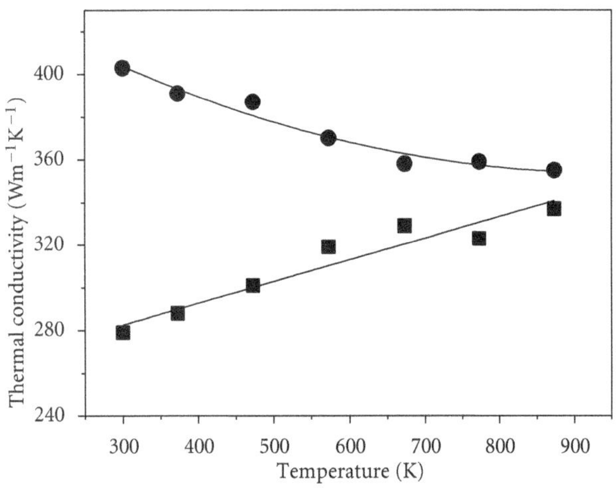

FIGURE 1: Effect of temperature on the thermal conductivity of OFHC copper and Cu-Cr-Zr-Ti alloy.

a thin layer of graphite spray. Three tests were conducted in the temperature range of 300–873 K, and average of three tests is reported. Hardness of the disc was measured at room temperature before and after thermal conductivity test using the Brinell Hardness Tester. An average of five hardness readings is reported. The electrical resistivity was determined using four-point technique by measuring the resistance of the specimens (50 mm (length) × 4 mm (width) × 6 mm (thickness)) utilizing meter calibrator-Wavetek 9100 and digital multimeter-HP 3458. Thin foils for TEM were prepared by mechanical polishing, dimpling, and ion milling. Discs of 3 mm diameter and 100 μm thick were punched, and one side was dimpled followed by ion milling with an incident beam angle of 4° using Gatan PIPS system. The ionmilled thin foils were examined under a JEOL transmission electron microscope operated at 200 kV.

3. Results and Discussion

Thermal conductivity of OFHC copper and Cu-Cr-Zr-Ti alloy as a function of temperature is shown in the Figure 1. OFHC copper shows a decreasing trend with an increase in temperature, and λ decreases from 397 Wm^{-1}K^{-1} at 300 K to 350 Wm^{-1}K^{-1} at 873 K. The observed trend is in agreement with previous works [8–11]. On the other hand, λ of Cu-Cr-Zr-Ti alloy increases with increase in temperature up to 873 K. The room temperature thermal conductivity was measured as 279 Wm^{-1}K^{-1}, while it reaches 337 Wm^{-1}K^{-1} at 873 K. The data obtained for OFHC copper and Cu-Cr-Zr-Ti alloy could be correlated to a second degree polynomial and linear equation, respectively, as shown in Table 1.

It was also found that the thermal conductivity of Cu-Cr-Zr-Ti alloy is lower than OFHC copper in the temperature range investigated. However, the gap between the curves was more at room temperature, and it diminished with increase in temperature. A similar increasing trend of thermal conductivity with temperature was reported in copper alloys [9, 10], platinum alloys [12], tungsten alloys [13], and magnesium alloys [14, 15]. The reason for lower thermal conductivity for alloys can be attributed to the alloying elements added to the matrix. These elements impede the movement of free electrons, which are the primary source for heat transport in metals and alloys by acting as scattering sites and decreasing the mean free path [11]. Nevertheless, the magnitude of drop in thermal conductivity depends on the position of an element in the periodic table and the concentration of the element [12, 16].

The thermal conductivity of the metals and alloys is assumed to compose of an electronic component (λ_e) and a phonon component (λ_p) [13]:

$$\lambda = \lambda_e + \lambda_p. \tag{1}$$

The electronic thermal conductivity can be approximately calculated using Wiedemann-Franz-Lorenz Law:

$$\lambda_e = L\sigma T, \tag{2}$$

where L is the Lorentz number (2.443×10^{-8} WΩK^{-2}), σ is the electrical conductivity (Ωm)$^{-1}$, and T is temperature (K)

$$\begin{aligned}\lambda_e&(\text{Cu-Cr-Zr-Ti})\\&= (2.443 \times 10^{-8}) \cdot (0.3386 \times 10^{8}) \cdot 300 \\&= 248 Wm^{-1}K^{-1}.\end{aligned} \tag{3}$$

λ_p (Cu-Cr-Zr-Ti) is calculated as 31 Wm^{-1}K^{-1} ($\lambda - \lambda_e$: 279 − 248). This indicates that the electron contribution is dominant for the alloy at room temperature.

It is noteworthy to mention that the addition of impurities largely reduces the electronic thermal conductivity (λ_e) of the alloy. This can be substantiated by comparison of λ_e for OFHC copper and Cu-Cr-Zr-Ti alloy at room temperature. The λ_e obtained for OFHC copper and Cu-Cr-Zr-Ti alloy was 398 Wm^{-1}K^{-1} and 248 Wm^{-1}K^{-1}, respectively, which clearly shows that λ_e for the alloy is suppressed due to alloying elements. This result is in line with that obtained for MgSc alloys, where it was reported that significant change has been noticed in λ_e with the concentration of the alloying element added to the alloy [14, 15].

It is seen from the Figure 1 that the thermal conductivity of Cu-Cr-Zr-Ti alloy increases with temperature up to 873 K. The reason for such increasing trend to a larger extent may be attributed to the increase in electronic thermal conductivity with temperature [13–15]. Nevertheless, some contribution can be ascribed to microstructural changes occurring in the specimen during the test. The alloy studied can attain age hardening in the temperature range of 673–823 K [3, 4]. In the present study, tests were conducted in the temperature range of 300–873 K and at each test temperature, and the sample was held for 30 min for attaining homogeneity. Therefore, there is a reason to believe that in situ aging might have occurred, which has contributed to the increase in thermal conductivity with temperature. This argument was supported by measuring the hardness of the disc after the

TABLE 1: Results of regression analysis for α, C_p, and λ and temperature for OFHC copper and Cu-Cr-Zr-Ti alloy.

Property	Material	Equation	Regression constant (R^2)	Valid range
Thermal conductivity (λ)	OFHC Cu	$\lambda = 461.1 - 0.22(T) + 1.20 \times 10^{-4}(T^2)$	0.9490	300–873 K
	Cu-Cr-Zr-Ti	$\lambda = 253.5 + 0.099(T)$	0.9123	300–873 K

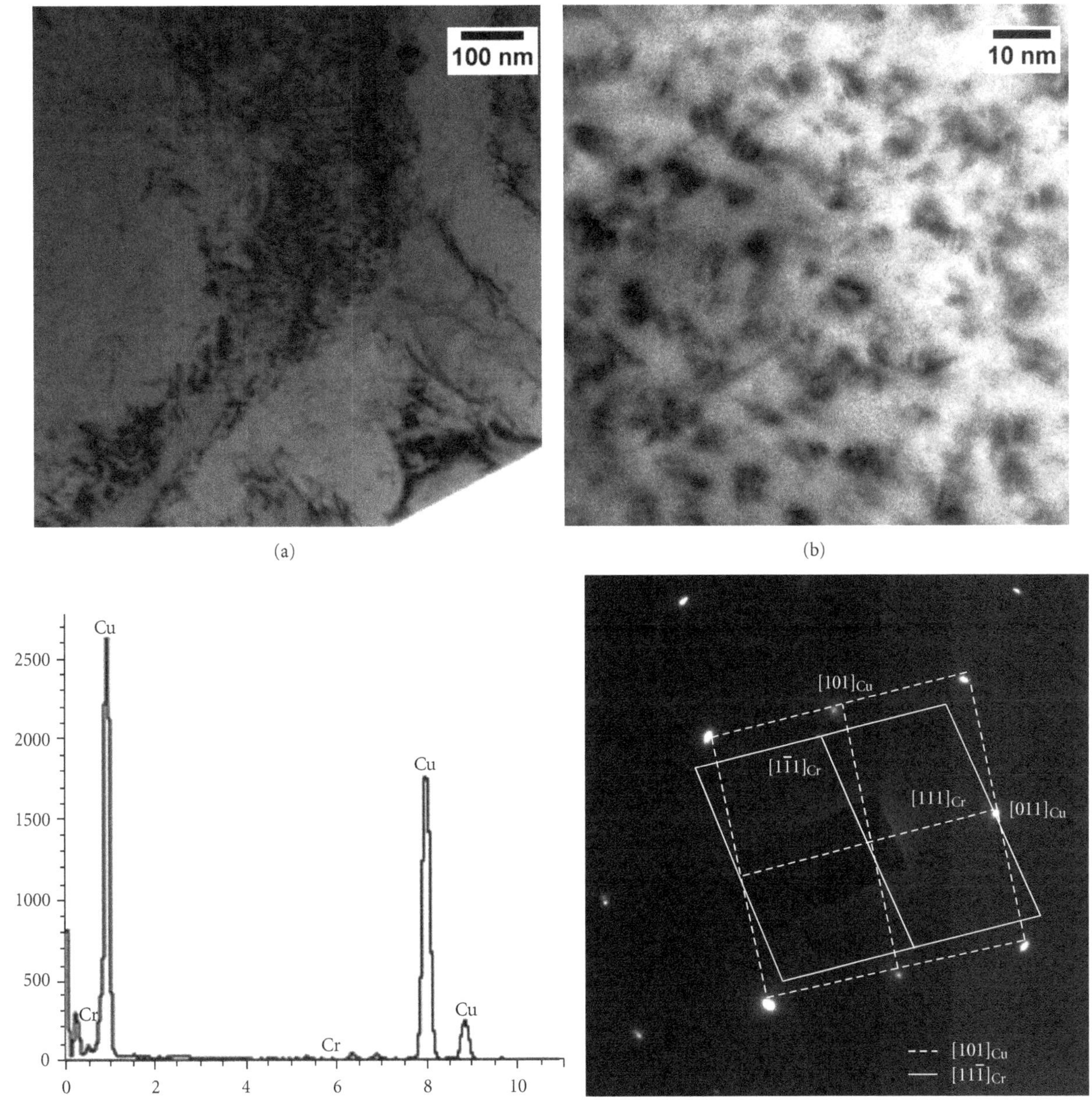

FIGURE 2: Transmission electron microscopy images of thermal conductivity specimens (a) before test, (b) after test, (c) EDS spectrum from (a), and (d) selected area diffraction from (b).

test. The hardness of the disc before the thermal conductivity test was 70 BHN, and after the test, it has increased to 91 BHN, which shows that hardening has occurred in the sample. The possible mechanism which enables hardening in the given temperature range of the alloy is in situ aging. In situ aging will result in the precipitation of excess element, which will make the matrix pure (rich in Cu), and increases the thermal conductivity.

The transmission electron microscopy images of the thermal conductivity specimens before and after the test are shown in Figure 2. The bright-field image of the specimen before the test is shown in Figure 2(a) with modulated structure free of any precipitates. Figure 2(b) shows the bright-field image of the specimen after the test where a uniform distribution of fine precipitates is observed in the matrix. The energy dispersive X-ray spectroscopy (EDS) analysis spectrum obtained from the sample before the test (Figure 2(a)) is shown in Figure 2(c). The spectrum shows prominent peaks of copper with few peaks of chromium. From the spectrum, it can be inferred that the sample is in the solution treated condition with excess chromium in the copper matrix. The selected area diffraction (SAD) pattern taken from Figure 2(b) is shown in Figure 2(d). The spot pattern obtained was solved and indexed for matrix and precipitate. The matrix had FCC crystal structure corresponding to copper, and precipitate has BCC crystal structure corresponding to chromium. The matrix and precipitate are indicated with subscripts Cu and Cr, respectively. From TEM studies, it is inferred that precipitates that form in the specimen after the test is chromium precipitates, and similar Cr precipitates were reported by Batra et al. [3, 4]. In addition to hardness measurement, TEM studies clearly confirm the occurrence of in situ aging in the specimen after the test.

4. Conclusions

The investigation of thermal conductivity of Cu-Cr-Zr-Ti alloy in the temperature range of 300 K–873 K showed that the thermal conductivity is sensitive to temperature. With an increase in temperature, thermal conductivity showed an increase up to 873 K. Thermal conductivity increased by 20 % at 873 K compared to that at room temperature. In addition to electronic thermal conductivity, one of the reasons for the increasing trend with temperature is in-situ aging of the sample which results in formation of precipitates and facilities precipitation of excess solute from the matrix. The curve obtained for thermal conductivity could be modeled to a linear equation. Electronic and phonon thermal conductivity of the alloy was calculated using Wiedemann-Franz-Lorenz law, and it was found that the phonon contribution was insignificant at room temperature.

Acknowledgment

The authors express their sincere gratitude to the director of VSSC for his kind permission to publish this paper.

References

[1] G. J. Butterworth and C. B. A. Forty, "A survey of the properties of copper alloys for use as fusion reactor materials," *Journal of Nuclear Materials*, vol. 189, no. 3, pp. 237–276, 1992.

[2] H. Groh, D. Ellis, and W. Loewenthal, "Comparison of GRCop-84 to other Cu alloys with high thermal conductivities," *Journal of Materials Engineering and Performance*, vol. 17, pp. 594–606, 2008.

[3] I. S. Batra, G. K. Dey, U. D. Kulkarni, and S. Banerjee, "Microstructure and properties of a Cu-Cr-Zr alloy," *Journal of Nuclear Materials*, vol. 299, no. 2, pp. 91–100, 2001.

[4] I. S. Batra, G. K. Dey, U. D. Kulkarni, and S. Banerjee, "Precipitation in a Cu-Cr-Zr alloy," *Materials Science and Engineering A*, vol. 356, pp. 32–36, 2003.

[5] H. Fuxiang, M. Jusheng, N. Honglong et al., "Analysis of phases in a Cu-Cr-Zr alloy," *Scripta Materialia*, vol. 48, no. 1, pp. 97–102, 2003.

[6] R. D. K. Misra and V. S. Prasad, "On the dynamic embrittlement of copper-chromium alloys by sulphur," *Journal of Materials Science*, vol. 35, no. 13, pp. 3321–3325, 2000.

[7] R. D. K. Misra, V. S. Prasad, and P. R. Rao, "Dynamic embrittlement in an age-hardenable copper-chromium alloy," *Scripta Materialia*, vol. 35, no. 1, pp. 129–133, 1996.

[8] P. Hanzelka, V. Musilova, T. Kralik, and J. Vonka, "Thermal conductivity of a CuCrZr alloy from 5 K to room temperatures," *Cryogenics*, vol. 50, no. 11-12, pp. 737–742, 2010.

[9] T. M. Tritt, *Thermal Conductivity: Theory, Properties, and Applications*, Plenum Press, New York, NY, USA, 2004.

[10] R. C. Smith, "Thermal conductivity of copper alloys.-I. Copper-zinc alloys," *Trans AIME Papers*, vol. 102, pp. 3–24, 1930.

[11] R. C. Smith and W. K. Palmer, "Thermal and electrical conductivities of copper alloys," *Trans AIME Papers*, vol. 221, pp. 225–241, 1935.

[12] D. E. Gray, *American Institute of Physics Handbook*, McGraw-Hill, New York, NY, USA, 1972.

[13] Y. Terada, K. Ohkubo, T. Mohri, and T. Suzuki, "Thermal conductivity in Pt-based alloys," *Journal of Alloys and Compounds*, vol. 285, no. 1-2, pp. 233–237, 1999.

[14] T. Tanabe, C. Eamchotchawalit, C. Busabok, S. Taweethavorn, M. Fujitsuka, and T. Shikama, "Temperature dependence of thermal conductivity in W and W-Re alloys from 300 to 1000 K," *Materials Letters*, vol. 57, no. 19, pp. 2950–2953, 2003.

[15] A. Rudajevová, M. Staněk, and P. Lukáč, "Determination of thermal diffusivity and thermal conductivity of Mg-Al alloys," *Materials Science and Engineering A*, vol. 341, no. 1-2, pp. 152–157, 2003.

[16] A. Rudajevová, F. von Buch, and B. L. Mordike, "Thermal diffusivity and thermal conductivity of MgSc alloys," *Journal of Alloys and Compounds*, vol. 292, pp. 27–30, 1999.

18

A Kinetic Model for Tribological Surface Transformations Occuring on the Railroads: Irreversible Near-Surface Metallurgical Transformations

G. Antoni

LMA UPR 7051, CNRS, Aix-Marseille Université, 13331 Marseille Cedex 03, France

Correspondence should be addressed to G. Antoni, antoni@lma.cnrs-mrs.fr

Academic Editors: A. Chrysanthou, C. Panagopoulos, and L. Takacs

During the manufacturing processes or the operational phases, some materials are liable to undergo metallurgical phenomena known as irreversible solid-solid phase transformations or Tribological Surface Transformations (TSTs). The treads of several rails in the French railroad network have been affected by TSTs. The kinetic model presented in this paper describes the initiation of TSTs and their development near the rail surface.

1. Introduction

For the last twenty years, the weight and velocity of freight trains have been increasing steadily, resulting in the development of "Tribological Surface Transformations" (TSTs) or "White Etching Layers" (WELs) [1–4] (see Figure 1(a)). TSTs are the irreversible solid-solid phase transformations which occur both at the surface and in depth, in the immediately vicinity of the wheels of passing trains. These processses, which mainly occur in straight parts of the rails, are induced by heavy corrugation [5–7] (see Figure 2), which generates an alternating pattern of troughs, consisting of the initial material (ferrito-pearlitic steel) and crests, forming the damaged part, where the material is transformed into "quasi-martensitic" steel (see Figure 1(b)) at depths ranging from several nanometers to more than 100 μm [8]. The values obtained in microhardness tests on TSTs in steel rails generally range from 700 to 1000 H_v (corresponding to a martensitic phase), as compared with the initial hardness of ferrito-pearlitic rails, which is about 300 H_v [9]. Rails with TSTs may, in the worst cases, develop cracks due to strong strain incompatibilities between the hard "white" phase (martensite) and the soft "untransformed" phase (ferrite-pearlite), see Figure 3 [10]. Although the origin of TSTs has not yet been clearly established, it seems to involve contact fatigue processes [11, 12]. The repeated mechanical loads applied (normal and shear stresses) combined with the highly localized thermal effects occurring near the surface at the wheel/rail contact points as the result of the friction may be largely responsible for these transformations [13].

Based on the above assumption, the kinetic model for TSTs [14] presented in Section 2 was developed. This model takes the thermo-mechanical state in the close vicinity of the wheel/rail contact area into account, that is, the fact that the high mechanical stress levels and the local temperature increase occurring near the surface may generate irreversible "quasi-surface" solid-solid phase transformations.

2. Metallurgical Behavior Model for the Irreversible Solid-Solid Phase Transformations and TSTs

In line with the principles of causality, whereby the future cannot influence the current state and those of localization, whereby the state of a material point depends only on its past history, and not on that of the neighboring material points, the metallurgical state of the material point X at a given time t is a function of both the temperature and the strain history, that is,

$$z(X,t) = \underset{\tau \le t}{\mathcal{F}}(T(X,\tau), \boldsymbol{\epsilon}(X,\tau)), \tag{1}$$

(a) (b)

Figure 1: (a) Macroscopic view on the tread of a rail (S54/900A steel) showing TSTs or "White Etching Layer". (b) Transverse cross-section of the upper part of a rail (a). The parent phase (grey part) is ferrite pearlite, which still remains in the rail when TSTs have developped and the daughter phase (white part), occurring at the surface, is "quasi-martensite" (extract from [8]).

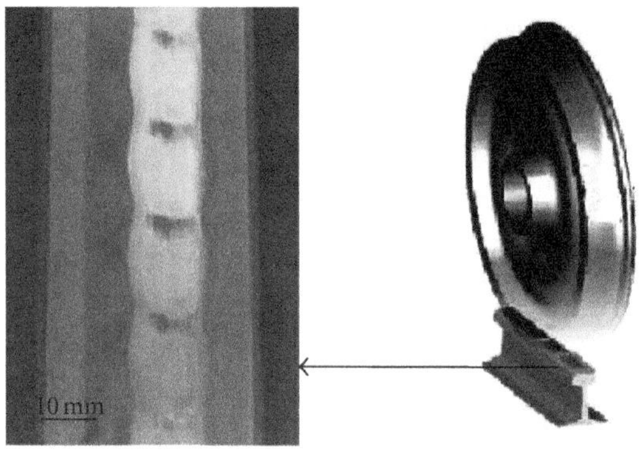

Figure 2: Rail corrugation: alternating of the troughs (grey part: ferrito-pearlitic phase) and crests (white part: "quasi-martensitic" phase) (extract from [7]).

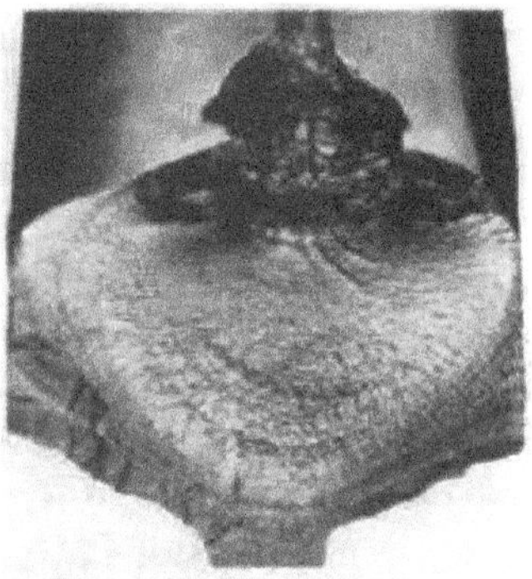

Figure 3: Fatigue defect due to wheel/rail contact: the "Squat" which corresponds to the local cracking and sagging of the rail tread. This defect occurs mainly in straight parts of railway tracks (extract from [10]).

where $z \in [0, 1]$ is the metallurgical variable, $\mathcal{F}$ denotes the metallurgical functional, T and $\boldsymbol{\epsilon}$ are the temperature and strain tensor, respectively. Note that when $z = 0$ (resp., $z = 1$), the material is composed of a single phase, that is, a ferrite-pearlite phase (resp., a martensite phase), and when $z \in]0, 1[$, it consists of two phases.

In the framework of Continuum Thermodynamics with internal variables [15], this functional dependence can be overcome by replacing the memory effects by additional internal variables. We assume that the equation of evolution of the metallurgical variable z can be written as

$$\dot{z} = f(\mathcal{S}, \dot{T}, \dot{\boldsymbol{\epsilon}}, \nabla T), \tag{2}$$

where $\mathcal{S} = (T, \boldsymbol{\epsilon}, \boldsymbol{\sigma}, z, \vartheta^i)$ denotes the thermomechanical state of the material point X at a given time t, $\boldsymbol{\sigma}$ is the stress tensor and ϑ^i are the additional internal variables (for example, the martensitic grain size); in a first approximation, ϑ^i are not taken account that is, $\vartheta^i = 0$.

The following kinetic model is adopted in order to describe the irreversible solid-solid phase transformations involved in TSTs:

$$\dot{z} = \kappa \dot{p} \tag{3a}$$

$$\text{with } \dot{p} = \frac{\left\langle \sigma^{\text{eq}} + \delta P - \left(\sigma_y(T, \mathbf{q}) + \beta(z/\kappa)\right)\right\rangle}{\overline{\sigma}_y \eta} H(1 - z) \geq 0, \tag{3b}$$

$$\sigma_y(T, \mathbf{q}) = \begin{cases} \overline{\sigma}_y \exp\left(\dfrac{T_i - T}{T_i}\right) \\ \text{or} \\ \overline{\sigma}_y \exp\left(-\dfrac{\|\mathbf{q}\|}{\|\mathbf{q}_c\|}\right), \end{cases} \tag{3c}$$

where $\overline{\sigma}_y$ denotes the TST's yield strength when $T = T_i$ (or $\mathbf{q} = 0$), T_i is the initial temperature, $\sigma^{\text{eq}} = ((3/2)\mathbf{s} : \mathbf{s})^{1/2}$ is the Von Mises equivalent stress, $\mathbf{s}$ is the deviatoric stress tensor, $P = -(1/3)\,\text{tr}(\boldsymbol{\sigma})$ is the pressure, σ is the Cauchy stress tensor, δ is a material parameter associated with pressure sensitivity, $\mathbf{q} = -k\nabla T$ is the heat flux (Fourier's law), k is the thermal conductivity, $\mathbf{q}_c$ is the critical heat flux, κ and β are material parameters associated with the irreversible solid-solid phase transformation, $\langle\cdot\rangle$ denotes the Macaulay brackets ($\langle x \rangle = x$ when $x \geq 0$ and $\langle x \rangle = 0$ when $x < 0$), $H(\cdot)$ denotes the Heaviside step function ($H(x) = 1$ when $x \geq 0$ and $H(x) = 0$ when $x < 0$), η is the characteristic time of the viscous effets associated with the TST process, and $\|\cdot\|$ is the Euclidean norm.

Note that: (i) Equation (3b) shows that the irreversible solid-solid phase transformation (or TST) is activated by the normal and shear stresses and accelerated with the temperature. The transformation occurs at stress levels which trigger dislocation motions due to the strain incompatibilities between the ferrite-pearlite and martensite phases. The triggering of these dislocation motions is facilitated by the increase in temperature at a given stress state. (ii) The irreversibility of the solid-solid phase transformation is expressed by $\dot{z}$, which cannot be negative. In addition, the eligibility field of the metallurgical variable $z \in [0,1]$ is also confirmed. (iii) Equation (3c) means that the TST's yield strength depends on either the temperature (T) or its gradiant (∇T). When $T \geq T_i$ (or $\mathbf{q} \geq \mathbf{q}_c$) in the immediate vicinity where the thermomechanical load is applied, the area is "thermally activated" and the original, nontransformed material ($z_0 = 0$) undergoes an irreversible solid-solid phase transformation at stress levels which are lower than the TST's yield strength $\overline{\sigma}_y$.

3. An Illustrative Example

The initiation and development of TSTs occur during the complex processes at work in a rail when a train is crossing over it: this situation is also known as the wheel/rail problem (see Figure 4). The repeated mechanical loading with friction applied to the rail due to the passing of trains creates a TST thickness (h_{TST}) in the immediate vicinity of the wheel/rail contact area, whereas the basic material constituting the rail remains unchanged beyond this area. In this section, the case of the structure studied does not resemble the wheel/rail problem, it follows therefore that the nonlinear contact and friction between wheel and rail are not taken into account. It is proposed here to solve a 1D problem by examining the rail in depth in order to confirm that the kinetic model presented in Section 2 can generate irreversible solid-solid transformations near the surface at which the thermomechanical loads are applied.

Taking the case of small perturbations, a thermoelastic bar (see Figure 5(a)) liable to undergo irreversible solid-solid transformations is subjected to quasi-static thermomechanical loading conditions, where the lateral surfaces of the bar are taken to be adiabatic in a first approximation.

The following constitutive equations are used for this purpose:

$$\sigma = E\left(\epsilon - \alpha(T - T_i) - \frac{z}{\kappa}\,\mathrm{sgn}(\sigma)\right), \quad (4)$$

$$\dot{z} = \kappa \frac{\left\langle |\sigma| - \left(\sigma_y(T,\mathbf{q}) + \beta(z/\kappa)\right)\right\rangle}{\overline{\sigma}_y \eta} H(1 - z), \quad (5)$$

$$\sigma_y(T,\mathbf{q}) = \begin{cases} \sigma_y^1 = \overline{\sigma}_y \exp\left(\dfrac{T_i - T}{T_i}\right) \\ \text{or} \\ \sigma_y^2 = \overline{\sigma}_y \exp\left(-\dfrac{\|\mathbf{q}\|}{\|\mathbf{q}_c\|}\right), \end{cases} \quad (6)$$

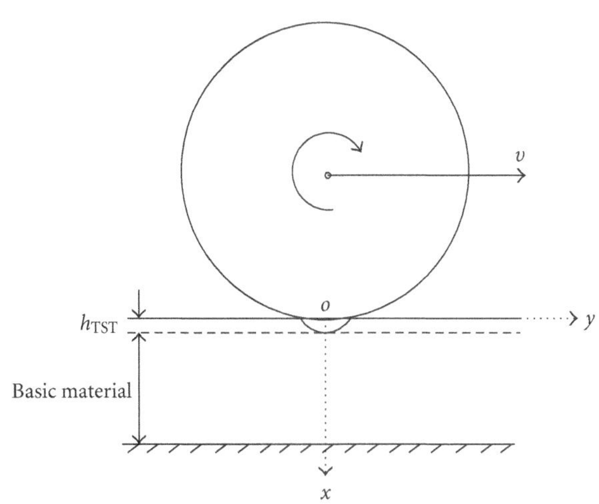

FIGURE 4: Wheel/rail problem with TSTs: presence of a TST thickness, h_{TST} (martensite), in the close vicinity of the wheel/rail contact area, below the basic material (ferrite pearlite) of the rail.

where E is the Young's modulus, α is the thermal expansion coefficient, $\mathrm{sgn}(\cdot)$ denotes the sign function ($\mathrm{sgn}(x) = -1$ when $x < 0$, $\mathrm{sgn}(x) = 0$ when $x = 0$ and $\mathrm{sgn}(x) = 1$ when $x > 0$). Note that (4) represents the thermoelastic mechanical behavior of the bar under the assumptions mentioned above.

Assuming no inertia and zero gravity, the quasi-static equilibrium can be written:

$$\frac{\partial \sigma}{\partial x} = 0, \quad (7)$$

where $\partial(\cdot)/\partial x$ denotes the first derivative with respect to x, $|\cdot|$ is the absolute value function ($|x| = x$ when $x \geq 0$ and $|x| = -x$ when $x < 0$).

Therefore (7) means that the stress field $\sigma(x,t)$ is uniform and equal to the mechanical loading applied $\sigma_0(t)$ throughout the bar for all $x \in [0, L]$.

Neglecting the source terms in the heat equation corresponding to the thermoelasticity and solid-solid phase transformations, this reduces to

$$\rho C_\epsilon \frac{T}{T_i}\frac{\partial T}{\partial t} - k\frac{\partial^2 T}{\partial x^2} = 0, \quad (8)$$

where ρ is the mass density, C_ϵ is the specific heat capacity, $\partial(\cdot)/\partial t$ and $\partial^2(\cdot)/\partial x^2$ denote the first and second partial derivatives with respect to t and x (temporal and spatial quantities).

The boundary conditions are summarized in Figure 5(b) and the constitutive thermomechanical parameters used are $C_\epsilon = 450\,\mathrm{J\cdot kg^{-1}\cdot K^{-1}}$; $E = 2\times 10^5$ MPa; $\rho = 78\times 10^{-7}\,\mathrm{kg\cdot mm^{-3}}$; $\alpha = 12\times 10^{-6}\,\mathrm{K^{-1}}$; $\kappa = 10^{-2}$; $\overline{\sigma}_y = 500$ MPa; $k = 5\times 10^{-2}\,\mathrm{W\cdot mm^{-1}\cdot K^{-1}}$; $\eta = 10^3$ s; $\beta = 5\times 10^3$ MPa; $q_c = 19\,\mathrm{W\cdot mm^{-2}}$.

The distribution (resp., temporal evolution) of the metallurgical variable z is plotted in Figure 6(a) (resp., Figure 6(b)). We can see from z-distribution obtained with both values of $\sigma_y(T,\mathbf{q})$ (σ_y^1 and σ_y^2) that the maximum phase transformation occurs near the surface and in its immediate

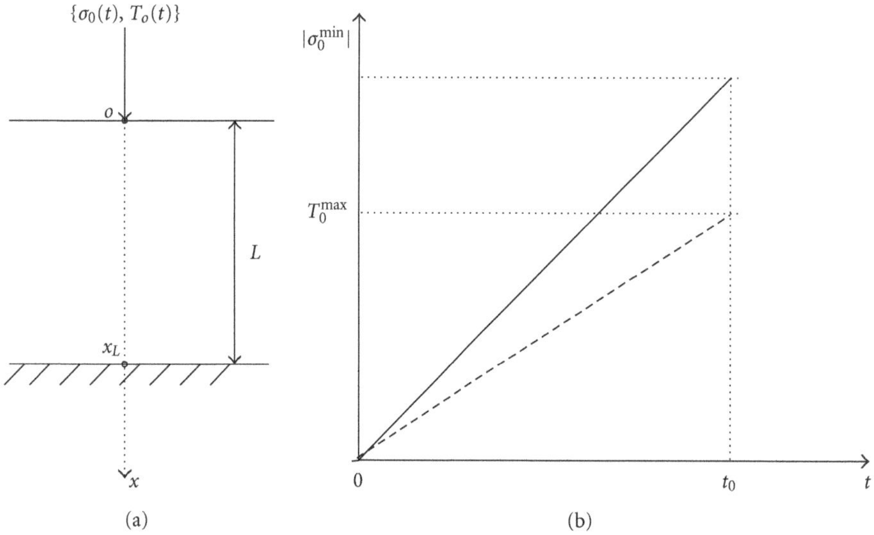

FIGURE 5: (a) 1D problem of TSTs: a bar subjected to a thermomechanical loading $\{\sigma_0(t), T_0(t)\}$ (where $L = 10\,\text{mm}$). (b) Boundary conditions: the loading function $\sigma_0(t)$ (resp., $T_0(t)$) is represented in solid lines with $|\sigma_0^{\min}| = 400$ MPa at time $t_0 = 10^{-1}$ s (resp., dashed lines, with $T_0^{\max} = 450$ K); At the point $x = x_L = L$, the temperature and the displacement are $T_L = T_i = 300$ K and $u_L = 0$, respectively.

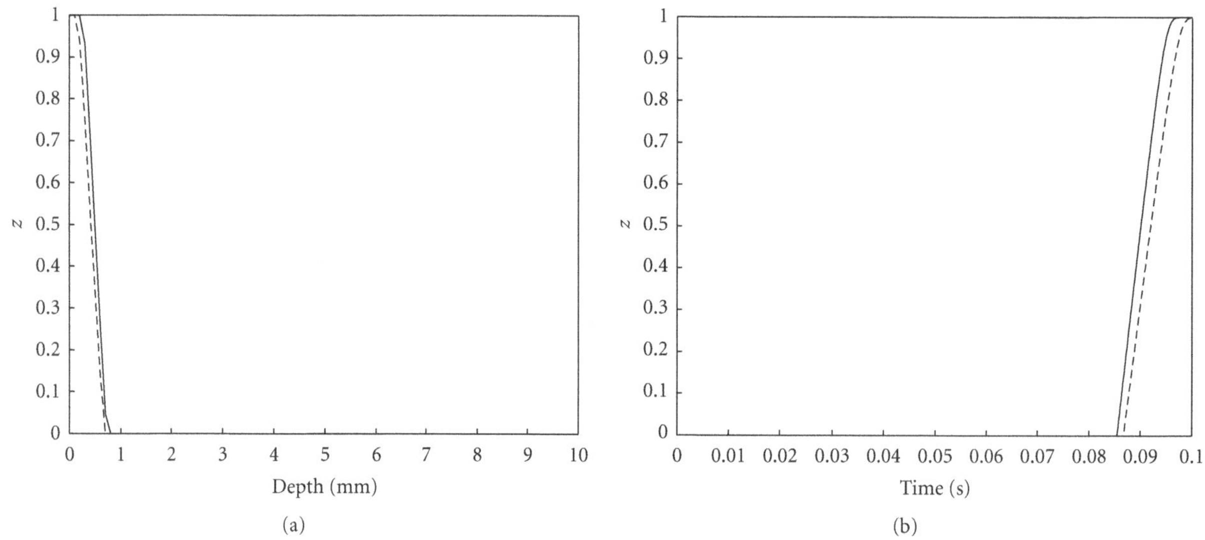

FIGURE 6: (a) Distribution of the metallurgical variable z in the bar at time $t = t_0 = 10^{-1}$ s where $\sigma_y(T, \mathbf{q}) = \sigma_y^1$ (solid line) and $\sigma_y(T, \mathbf{q}) = \sigma_y^2$ (dashed line). (b) Temporal evolution of the metallurgical variable z in the time $t \in [0, t_0]$ where $\sigma_y(T, \mathbf{q}) = \sigma_y^1$ (solid line) and $\sigma_y(T, \mathbf{q}) = \sigma_y^2$ (dashed line).

vicinity at the point where the thermomechanical loading is applied. A "new" phase ($z \neq 0$) develops near the left end of the bar where the load was applied: from $x = 0$ to about 8×10^{-1} mm in depth (see Figure 6(a)). At the end of the bar to about 2×10^{-1} mm (resp., 10^{-1} mm) with $\sigma_y(T, \mathbf{q}) = \sigma_y^1$ (resp., σ_y^2), a full phase transformation ($z = 1$) developed and there existed only one phase: the daughter phase (martensite). Beyond that depth, down to 8×10^{-1} mm, the partial transformation ($z < 1$) persisted, that is, the material was a mixture between the parent phase (ferrite pearlite) and the daughter phase (martensite). On the other hand, the temporal z-evolution (see Figure 6(b)) shows that at time $8.5 \times 10^{-2} \le t \le 9.6 \times 10^{-2}$ s (resp., $8.7 \times 10^{-2} \le t \le 9.8 \times 10^{-2}$ s) with $\sigma_y(T, \mathbf{q}) = \sigma_y^1$ (resp., σ_y^2), the solid-solid phase transformation continued from $z = 0$ until $z = 1$.

The distribution of the metallurgical variable z is plotted in Figure 7(a) (here only for $\sigma_y(T, \mathbf{q}) = \sigma_y^1$) for three different times: $t = \{9 \times 10^{-2}\text{ s}; 9.5 \times 10^{-2}\text{ s}; 10^{-1}\text{ s}\}$. It can be seen here that the irreversible metallurgical transformation was still confined near the left end of the bar, where the thermomechanical loading was applied. Figure 7(b) shows the distribution of the temperature in the bar for the same times as previously. The nonstationary effects in the heat (8)

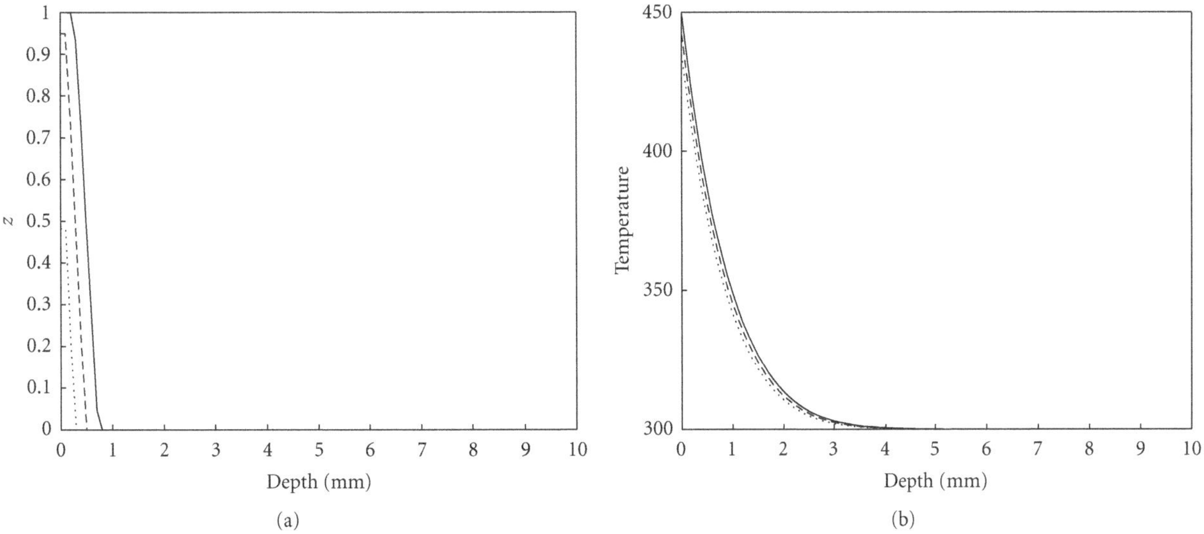

FIGURE 7: (a) Distribution of the metallurgical variable z in the bar at time $t = 9 \times 10^{-2}$ s (dotted line), $t = 9.5 \times 10^{-2}$ s (dashed line), and $t = t_0 = 10^{-1}$ s (solid line) where $\sigma_y(T,\mathbf{q}) = \sigma_y^1$. (b) Distribution of the temperature in the bar at time $t = 9 \times 10^{-2}$ s (dotted line), $t = 9.5 \times 10^{-2}$ s (dashed line), and $t = t_0 = 10^{-1}$ s (solid line) where $\sigma_y(T,\mathbf{q}) = \sigma_y^1$.

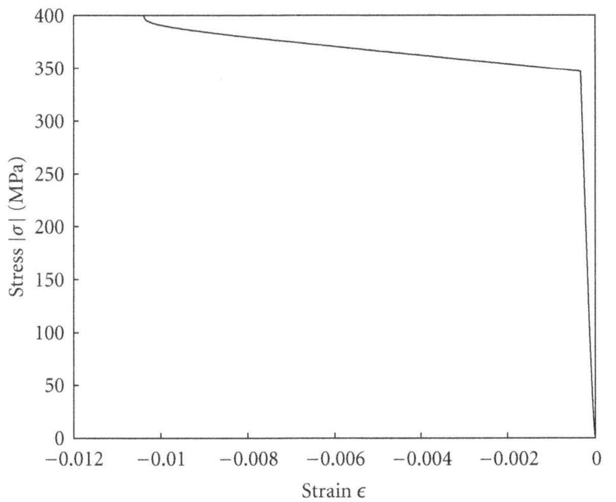

FIGURE 8: Stress-strain curve in $x = L/100$ at time $t = t_0 = 10^{-1}$ s (in this case, only in the case of $\sigma_y(T,\mathbf{q}) = \sigma_y^1$).

explain the nonlinear temperature distribution observed in the bar at the time considered.

The stress-strain curve is plotted in Figure 8 (in this case, only in the case of $\sigma_y(T,\mathbf{q}) = \sigma_y^1$). The response of the material is thermoelastic when $0 \leq |\sigma| \leq 340$ MPa, and inelastic behavior due to the irreversible solid-solid transformation then occurs when $340 < |\sigma| \leq 384$ MPa and lastly, a return to the thermoelastic response of the material occurs when $384 < |\sigma| \leq 400$ MPa.

The kinetics of the transformation process presented in Section 2 describe the initiation and development of TSTs in the immediate vicinity of the surface, where the thermomechanical loading was applied. In agreement with the experimental data available on TSTs in railway tracks [8], the results presented in this Section correspond to a realistic TST thickness, that is, at a depth of around one hundred micrometers. Therefore, this metallurgical behavior model will be tested on the wheel/rail problem in the future works.

4. Conclusion

The kinetic model presented here describes the initiation and development of Tribological Surface Transformations (TSTs) near the surface of several rails used to railway tracks. This model is based on the assumption that the thermomechanical state in the close vicinity of the wheel/rail contact area can generate TSTs. In agreement with experimental data on TSTs, the results obtained in the case of an illustrative example show the occurrence of a completely irreversible solid-solid phase transformation in the immediate vicinity of the surface at which the thermomechanical load was applied.

References

[1] Y. Jirásková, J. Svoboda, O. Schneeweiss, W. Daves, and F. D. Fischer, "Microscopic investigation of surface layers on rails," *Applied Surface Science*, vol. 239, no. 2, pp. 132–141, 2005.

[2] A. Pyzalla, L. Wang, E. Wild, and T. Wroblewski, "Changes in microstructure, texture and residual stresses on the surface of a rail resulting from friction and wear," *Wear*, vol. 250-251, no. 2, pp. 901–907, 2001.

[3] L. Wang, A. Pyzalla, W. Stadlbauer, and E. A. Werner, "Microstructure features on rolling surfaces of railway rails subjected to heavy loading," *Materials Science and Engineering A*, vol. 359, no. 1-2, pp. 31–43, 2003.

[4] H. W. Zhang, S. Ohsaki, S. Mitao, M. Ohnuma, and K. Hono, "Microstructural investigation of white etching layer on pearlite steel rail," *Materials Science and Engineering A*, vol. 421, no. 1-2, pp. 191–199, 2006.

[5] M. Hiensch, J. C. O. Nielsen, and E. Verheijen, "Rail corrugation in the Netherlands—measurements and simulations," *Wear*, vol. 253, no. 1-2, pp. 140–149, 2002.

[6] H. G. Feller and K. Walf, "Surface analysis of corrugated rail treads," *Wear*, vol. 144, no. 1-2, pp. 153–161, 1991.

[7] A. Saulot, *Analyse tribologique du contact roue-rail—modélisation exérimentations [Ph.D. thesis]*, INSA Lyon, Lyon, France, 2005.

[8] W. Österle, H. Rooch, A. Pyzalla, and L. Wang, "Investigation of white etching layers on rails by optical microscopy, electron microscopy, X-ray and synchroton X-ray diffraction," *Materials Science and Engineering A*, vol. 303, no. 1-2, pp. 150–157, 2001.

[9] E. Wild, L. Wang, B. Hasse, T. Wroblewski, G. Goerigk, and A. Pyzalla, "Microstructure alterations at the surface of a heavily corrugated rail with strong ripple formation," *Wear*, vol. 254, no. 9, pp. 876–883, 2003.

[10] Revue Générale des Chemins de Fer, Editions Delville, no. 144, 2005.

[11] R. I. Carroll and J. H. Beynon, "Rolling contact fatigue of white etching layer: part 1. Crack morphology," *Wear*, vol. 262, no. 9-10, pp. 1253–1266, 2007.

[12] R. I. Carroll and J. H. Beynon, "Rolling contact fatigue of white etching layer. Part 2. Numerical results," *Wear*, vol. 262, no. 9-10, pp. 1267–1273, 2007.

[13] G. Baumann, H. J. Fecht, and S. Liebelt, "Formation of white-etching layers on rail treads," *Wear*, vol. 191, no. 1-2, pp. 133–140, 1996.

[14] G. Antoni, *Transformation tribologique de surface: une approche thermo-mécanique [Ph.D. thesis]*, Université de Provence, Marseille, France, 2010.

[15] J. Lemaitre and J. L. Chaboche, *Mécanique des Matériaux Solides*, Dunod, Paris, France, 1996.

19

Aluminium Alloy-Based Metal Matrix Composites: A Potential Material for Wear Resistant Applications

Rupa Dasgupta

CSIR-Advanced Materials and Processes Research Institute (AMPRI), Hoshangabad Road, Madhya Pradesh, Bhopal 462064, India

Correspondence should be addressed to Rupa Dasgupta, rupa_dasgupta@rediffmail.com

Academic Editors: R. Asthana and P. Lukac

Aluminium alloy-based metal matrix composites (AMMCs) have been by now established themselves as a suitable wear resistant material especially for sliding wear applications. However, in actual practice engineering components usually encounter combination of wear types. An attempt has been made in the present paper to highlight the effect of dispersing SiC in 2014 base alloy adopting the liquid metallurgy route on different wear modes like sliding, abrasion, erosion, and combinations of wear modes like cavitation erosion, erosion abrasion, sliding abrasion, and the results obtained compared with the base alloy. It is found that there are a number of contributing factors for the resulting wear and all are not necessarily derogatory in nature. The limits within which the AMMCs can exhibit superior performance over the base alloy have been discussed. Worn surface and subsurface studies have been carried out to understand the mechanism of material removal and the role of the different contributing factors to material removal. Wear mechanisms that have been prevalent have been suggested and the possibility of making wear resistant components from the MMCs is discussed based on the experimental results obtained.

1. Introduction

Material scientists and researchers in this area have been fulfilling the demand of the engineering sector since decades in synthesizing materials to attain the demanded properties to enhance efficiency and cost savings in the manufacturing sector. In fulfilling this demand, a certain trend has been followed, the materials presently been used is tried for improvement through known methods of alloy additions, heat treatment, grain modification, and the like. Once the limit is reached through these methods, either due to economic constraint, difficulty in mass production, or further improvement is ruled out, a different line of thought emerges in further improving the properties or decreasing cost and increasing efficiency. At times, a completely new system takes over, like was done around three decades back when metal matrix composites (MMCs) were thought of. Since a few years, the economically feasible routes, dispersoids, and alloy system that can give meaningful improvement have been narrowed down. Among MMCs, Al-alloy-based composites were always on the forefront of research. Parallel areas of research had then emerged but after about two decades of research in various disciplines to further enhance the properties to satisfy the ever increasing demand of the engineering sector, composites took a lead compared to the other processes when the cost and ease of fabrication were compared. The other methods changed track and chose for themselves different areas of application and Al-based metal matrix composites (AMMCs) remained as the most potential candidate to be researched on for making engineering components viable [1–5].

Worldwide research in the area of AMMCs established beyond doubt the advantages of Al-based metal matrix composites (AMMCs) over the base alloy in the laboratory scale. With time, the demands moved ahead and engineering components were demanded from AMMCs. When the industrial application of these were targeted, the question of viability arose and from the different methods adopted for making AMMCs in the laboratory scale, the liquid metallurgy route stood ahead due to its ease of fabrication. Although the other routes were more efficient regarding property attainment and microstructural features, still the ease of fabrication added to the cost economics made the liquid metallurgy route a competitive and viable method for bigger sized components. It can be easily concluded from the different results published from across the globe for two decades in the area of AMMCs that the end use needs to be kept in mind before adopting a material and a process of fabrication. This is because the cost factors, ease of fabrication, second phase's nature, shape, size, distribution, wetability between matrix alloy and second phase all have to be considered for making a final choice depending on the component selected, mainly its working conditions for the engineering component to be fabricated.

In the last decade though research continues in this field yet the results obtained were a mere confirmation of the previously attained results, and a plateau in the possible improvement has been reached. Also attempts are being made to make engineering components from the AMMCs but still there is no mass utility of these composites especially those made from the liquid metallurgy route. Some areas where research is still needed before these versatile materials see the light of the day as engineering components include machining, joining, repeatability of properties on mass production, and effect of secondary processing. The difficulty in meeting the industry's demand can be easily seen from the lesser number of research publications since the millennium, and old researchers were finding it difficult to answer questions related to the engineering potential of these composites. The material with better property needs to be converted to a component with better properties than presently available and here lies the challenge, like (i) cast AMMCs will seldom find use as engineering components, (ii) advantages accrued in cast composites do not always hold good when processed, (iii) moreover secondary processing of composites has its own set of challenges, and (iv) only selected alloying elements and second phase dispersions can withstand the rigours and give meaningful results.

To date, the literature at least does not pose a very rosy picture regarding availability of engineering components been used which are made from AMMCs. Maybe this would need a few years more, or may be since the development of engineering components is difficult to publish, the details are not openly available. However, a few components that have seen the light of the day from AMMCs include Swash plate, tennis racket, transfer bench, foot rest, and a few automobile components like brake drum, casings, head light combos, mining equipments, and a few components for structural applications, with an intention to replace either cast iron products or Al-alloy products presently used especially aimed at weight reduction, strength, stiffness, and energy efficiency. However, the aerospace industries, automobile and railway sectors are keenly observing the composite research development.

The enormous amount of R&D that has gone into Al-based MMCs of every possible alloy with different dispersoids establishing beyond doubt the usefulness of making composites but a choice has to be made with both the base alloy selection and dispersoid size and volume percentage for making engineering components. Another point is that all engineering components will undergo somewear related degradation with use even in an otherwise nonwear environment. The Al-Cu and Al-Zn-based alloys are commonly used for making engineering components used in environments like those mentioned above.

Literature review of Al-Cu-based composites show that this system has been widely investigated both in terms of understanding the mechanism of formation and property evaluation of composites [6–8]; the main dispersoids being SiC and Al_2O_3. Although investigations with varying volume fraction of dispersoids have been done, meaningful results have been obtained when the dispersoid percentage has been restricted to 15. Again the size of the dispersoids is dependent on/or restricted to the processing route adopted. The Al-Cu-based MMCs have been evaluated for their sliding, and abrasive wear properties in most cases have reported improvement in the wear resistance of composites over the control alloy; however, the degree of improvement depends to a large extent on the experimental conditions and microstructure of the MMCs. In some cases, however the wear resistance of MMCs was found to be inferior to the alloy; this could be correlated to extreme experimental conditions or improper formation of composites leading to detachment of the dispersoids. The wear resistance of MMCs indicated that both the reinforcement size and matrix strengthening have a significant effect on the wear properties of the composites. Sliding wear resistance of the composites increases with increasing SiC particulate size. For the composite containing various particulate sizes, the wear properties of the composites cannot be directly correlated to the bulk hardness (HRB) of the composites. Under lubricated conditions (paraffin oil) the wear rates were about 1/10 those under dry conditions. Wear mechanism has been explained on the nature of wear debris, wear cracks, its growth, propagation, and detachment. These are the factors which contribute to wear loss. No direct relationship has been found between hardness and wear resistance. Maximum wear resistance has been reported after over ageing [9–12]. The improvement of sliding wear resistance of particulate MMCs over the control alloy is more evident at lower sliding speeds; at higher speeds the wear rate increases which is in contrast to the control alloy which shows better wear resistance (i.e., lower wear rate) at higher speeds. However the seizure pressure of composites is better than the control alloy as reported by several investigators. Reports indicate that the wear resistance of composites increases with increasing volume fraction and size of the particulates. The excellent wear resistance of the composites is mainly attributed to the effective resistance of the SiC particle to penetration, grinding, and cutting by the ceramic slider. The effect of SiC size is greater than that of volume fraction. The improvement in the wear resistance of composites containing small particles over the unreinforced alloy is attributed to matrix strengthening of the particles as well as wear resistance of the particles, whereas the higher wear resistance of the composites containing large particles compared with those containing small particles is mainly accounted for by the excellent wear resistance of the large SiC particles. Abrasive wear tests have been carried out by investigators under high stress conditions. In general, wear rate of composites are less than the control alloy. The nature of variation with load, however, remained the same for the control alloy and composite [13–15]. The nature and volume of material removal are controlled by factors like dispersoid size, penetration depth (dependent on load), size of abrasive particles, work hardening of the subsurface (matrix) regions, microcracking tendency of material, reduction in cutting efficiency of the abrasive, blunting of abrasives, fragmentation, and so forth. Detailed investigations have demonstrated an increase in wear resistance with increasing volume fraction of particulates (at constant particle size) and also with increasing particle size (with constant volume fraction).

The present paper details highlights the effect of dispersing SiC in different alloy matrices adopting the liquid metallurgy route on different wear modes normally encountered in non wear/mild environments by engineering components like sliding, abrasion, combinations of sliding-abrasion, and the results obtained compared with the base alloy. It is found that there are a number of contributing factors for the resulting wear and all are not necessarily derogatory in nature. The limits within which the AMMCs can exhibit superior performance over the base alloy have been discussed. Worn surface and subsurface studies have been carried out to understand the mechanism of material removal and the role of the different contributing factors to material removal. Wear mechanisms that have been prevalent have been suggested and the possibility of making wear resistant components from the MMCs is discussed based on the experimental results obtained.

2. Experimental Details

2.1. Synthesis of Composites. The composites used in the present study were prepared by the liquid metallurgy technique. For preparing the composites, constituents in appropriate proportions conforming to 2014 and 7075 alloy system consisting of Al-4.4 Cu-0.5 Mg-0.8 Si-0.8 Mn and 1.6 Cu, 2.5 Mg, 5.6 Zn, 0.3 Cr, respectively, were melted in a gas-fired furnace in a graphite crucible. Weighed and preheated dispersoids of size between 15 and 20 μm and 10 weight% as optimized earlier [16] was poured into the melt after passing through a sieve. During insertion of the dispersoid, the melt was stirred constantly by means of a stirrer placed in the melt operated by a motor. The stirrer speed was controlled as required. After complete insertion of the dispersoids in the molten alloy, the alloy was simultaneously stirred and heated for some time for uniform mixing and temperature. The melt with the dispersoids were then poured into preheated permanent moulds of required size and shape. The chemical composition of the alloy was analysed using SPARKMET optical spectrometer (Model: SPECTRA) for confirmation.

2.2. Ageing and Tempering. In the case of 2014 alloy system, it was found that the optimised ageing conditions are different for the alloy and composite. The composite attains peak hardness at 180°C compared to 210°C for the alloy [17].

The alloys and composites of 7075 system were aged under T6 tempering and adopting retrogression and reageing

Table 1: Conditions and hardness of alloys and composites tested for sliding wear behaviour.

Sample	Condition	Hardness (HB)
2014 Alloy system		
Alloy	As cast	86.6 ± 6
Alloy	Aged at 210°C for 3 hours	114.2 ± 4
Alloy	Aged at 240°C for 3 hours	109.4 ± 3
Composite	As cast	83.4 ± 2
Composite	Aged at 180°C for 3 hours	124.2 ± 7
7075 Alloy system		
Alloy	As cast	96.6 ± 3
Alloy	T6 ageing condition	137.6 ± 4
Alloy	Retrogressed and reaged	71.0 ± 3
Composite	As cast	101.2 ± 4
Composite	T6 ageing condition	99.2 ± 3
Composite	Retrogressed and reaged	72.1 ± 3

Table 2: Tensile properties of alloys and composites.

System	Condition	UTS, MPa
2014 Alloy	Cast	176
	Aged under optimized condition	259
2014 Composite	Cast	93
	Aged under optimized condition	170
7075 Alloy	Cast	87
	Aged under optimized condition	158
7075 Composite	Cast	63
	Aged under optimized condition	112

method (RRA). In the RRA method, the samples were quenched and aged as per the following steps [18].

(i) The alloys and composites were aged at 490°C for 8 hours and water quenched. The quenched samples were heat treated at 120°C for 24 hours and were furnace cooled. This corresponds to T6 ageing.

(ii) The T6 aged samples were retrogressed at 200°C for 45 minutes.

(iii) The retrogressed samples were reaged at 120°C for 24 hours.

2.3. Test Specimens. The ageing conditions which exhibited maximum hardness were used for studying the microstructure, sliding and two body abrasive wear behaviour are given in (Table 1).

2.4. Microstructure. Samples around 20 mm diameter cut from the cast and aged specimens were metallographically polishing and etching in Kellar's reagent for metallographic studies. The microstructures were observed in the optical microscope and/or JEOL scanning electron microscope operating at 20 kV in both the secondary mode of electron transmission and the back scattered electron emission mode, although the features observed were the same in both modes of electron emission, in some cases where the secondary mode of electron emission could not resolve the microconstituents properly, the features observed in the back scattered electron emission mode have been reported.

2.5. Sliding Wear Test. Dry sliding wear tests were carried out on a pin-on-disc wear testing machine in which the specimen was held against a rotating steel disc and this sliding motion results in wear of the pin. The weight of the specimen was taken after every 500 m, sliding distance. The weight loss was calculated from the difference in weight between the initial weight and weight after a specified number of rotations. The volume loss calculated by dividing the weight loss by the density of the material (calculated by Archimedes' principle) or the wear rate calculated by dividing the volume loss by the total distance traversed is used for comparison between the different conditions. Experiments were carried out at two loads of 1 and 3 MPa and at two speeds of rotation corresponding to 200 and 750 rpm up to sliding distance of 2500 m. Weight loss measurements were taken after every 500 m sliding distance approximately. The weight loss was divided by the density of the sample to obtain the volume loss. Volume loss divided by the actual distance traversed was used to determine the wear rate of the sample. The volume loss (m^3) was used to compare the wear behaviour of the alloy and composites. The sliding wear properties of the alloys and composites were determined under dry sliding conditions for comparison of properties between the alloys and composites in the cast and aged conditions under similar experimental conditions.

2.6. High Stress Abrasive Test. High-stress (Two Body) abrasive wear tests were carried out using a Suga make abrasion tester (Model: NUS 1, Japan). Rectangular shaped specimens having dimensions $40 \times 35 \times 4$ mm (thickness) were cut from the alloys and composites cast in discs of 70 mm diameter and 6 mm thickness, and ground to obtain a smooth surface before carrying out the tests. Emery paper embedded with the desired size of SiC particles was used as abrasive medium. The emery paper was fixed on rotating wheel (diameter of the wheel is 50 mm and thickness is 12 mm). The specimen was fixed with a locking arrangement against the abrasive medium. Load was applied on the specimen with a cantilever mechanism. The sample was made to slide over the abrasive paper fixed on the rotating wheel. The abrasive changed its position after each stroke corresponding to a sliding distance of 0.065 m. Weight loss of the specimen was measured after it completed 400 strokes, that is, corresponding to a sliding distance of 26 m. A Mettler make microbalance was used for weighing the specimens. The specimen was cleaned with acetone prior to and after the wear test. Fresh abrasive paper was used after every 400 strokes. Tests were carried out up to a distance of 260 m. Loads applied were 3 and 5 N with the abrasive size of 220 grit. The weight loss was obtained from the difference in weight of the sample before and after the test after every 26 m sliding distance. The volume loss and wear rate attained (calculated as mentioned before) were used to compare the attained properties.

The same set of samples, as for sliding wear tests mentioned above, were tested for their abrasive wear behaviour.

2.7. Worn Surface Studies. The worn surface and subsurface was studied under the scanning electron microscope (SEM) to understand the nature of material removal and surface damage and to compare between material removal mechanism between the different types of wear and also between the different conditions and the effect of dispersoid addition if any on the mechanism of material removal.

3. Results

3.1. Microstructural Features. In the back scattered electron image of the as cast 2014 alloy (Figure 1(a)) clear grain boundaries are observed with some precipitation in them. The important micro constituents of the cast structure correspond to primary α-dendrites surrounded by eutectic α-$CuAl_2$ (marked A & B resp. in Figure 1(a)). The main features on solutionising and ageing under optimised conditions are fine precipitates along the grain boundaries which grow from discontinuous to continuous precipitates without coarsening (Figure 1(b)). At longer duration of time there is also coarsening at lower ageing temperatures (Figure 1(c)). The back scattered electron image of the cast composite show a near uniform distribution of the SiC particulates (Figure 2(a)). A good bonding between the alloy and particulate is also observed. The microstructural features observed are definite grain boundaries that are more continuous as compared to the alloys and the aged composites (Figure 2(b)) show similar features as the aged alloy.

In the as cast state, the 7075 alloy shows heterogeneous precipitation along the grain boundaries as well as inside the grain boundaries (Figure 3(a)). The alloy exhibits a black filigree of MgZn. Under RRA conditions dissolution of grain boundaries was observed (Figure 3(b)) and showed a granular structure with grains of varying size and distribution of very fine phases in the grains in the Al-matrix resulting from nonuniform solidification rates during casting. The cast composite exhibits (Figure 3(c)) a near uniform distribution of the particulates within the matrix and clear dendrites are seen in the matrix of Aluminium; grain boundaries are clearly defined with some precipitates in the grains. On ageing, dissolution of grain boundaries was observed together with coagulation of precipitates around the SiC dispersoids (Figure 3(d)).

3.2. Sliding Wear Behaviour. In order to compare the behaviour between the alloy and compositefor aparticular experimental condition, the volume loss against sliding distanceis plotted as in Figures 4–18. A general observation for both the alloy systems both the composite and alloy in the cast and aged conditions, the alloy samples seized at the load of 3 MPa before traversing any meaningful distance; other experimental factors, namely, the speed of rotation and sliding distance did not report seizure for any specimen condition till the limit of the tests, namely, 750 rpm speed

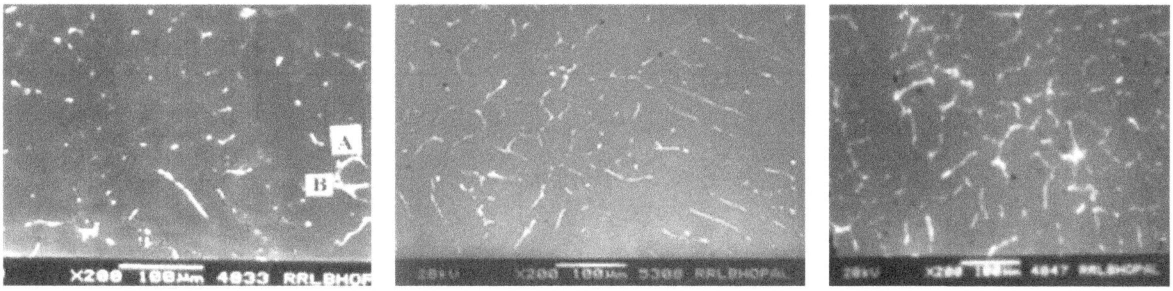

(a) (b) (c)

FIGURE 1: SEM photographs of alloy. (a) As cast alloy (b) under optimised ageing conditions (c) aged alloy at longer duration exhibiting coarsening.

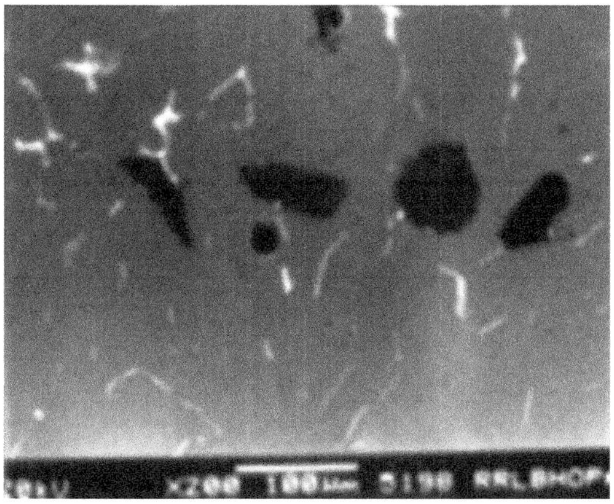

(a)

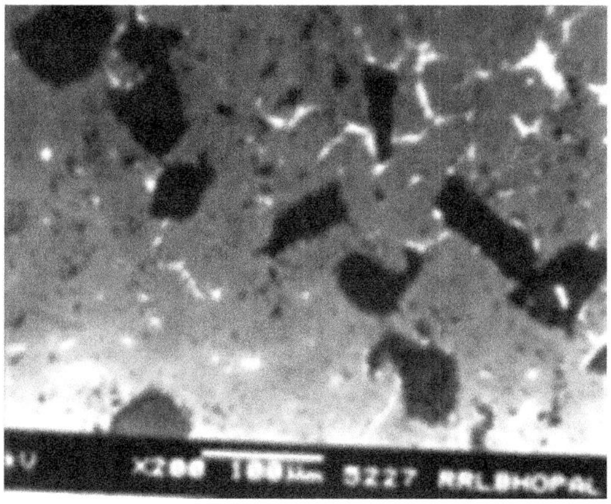

(b)

FIGURE 2: SEM photographs of composite. (a) As cast (b) under optimised ageing conditions.

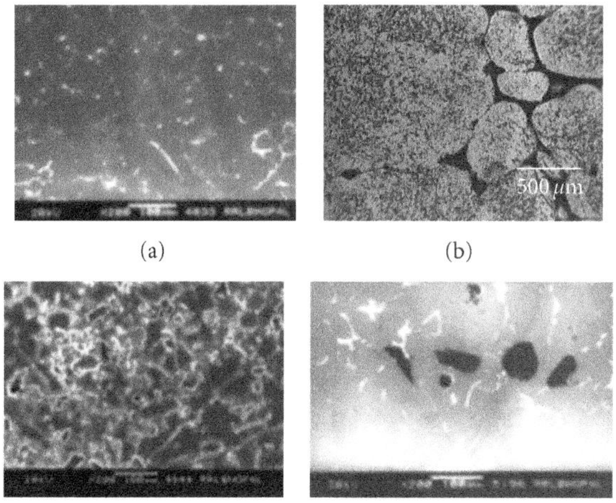

(a) (b)

(c) (d)

FIGURE 3: Microstructures of 7075 alloy. (a) Cast alloy (b) aged alloy (c) cast composite (d) aged composite.

and 2500 m sliding distance. However, the composite made with 7075 alloy both in the cast and aged conditions can withstand the pressure of 3 MPa till completion of the tests.

Again making composites reduces the volume loss in all the cases as compared to their alloy counterparts. Beneficial effect of ageing is more pronounced in the 2014 alloy system (Figure 4). Increases speed decreases wear rate probably due to the less time of contact, this is reflected in the worn surface's degradation also in the form of shallow ridges.

Ageing improves the wear rate in both the alloy and composite of the 2014 system, this can be attributed to the strengthening attained on ageing (Table 2 showing improved UTS). In some cases, the wear rate (calculated as volume loss divided by the actual sliding distance traversed) increases at longer sliding distance, which could be due to softening of the matrix on prolonged exposures.

The behaviour of volume loss as a function of the sliding distance for the 7075 system under the different experimental conditions for the cast and aged condition is plotted in Figure 5. It is observed that ageing improves the wear resistance inverse of volume loss significantly in the case of composite; for alloys there is no appreciable effect (Figures 5(a) and 5(b)). However the effect is most pronounced in both the alloy and composite when tested at higher load of 3 MPa.

Increasing speed of rotation decreases the volume loss for both the alloy and composite (Figure 6(a)); its effect is more pronounced than the effect of load (Figure 6).

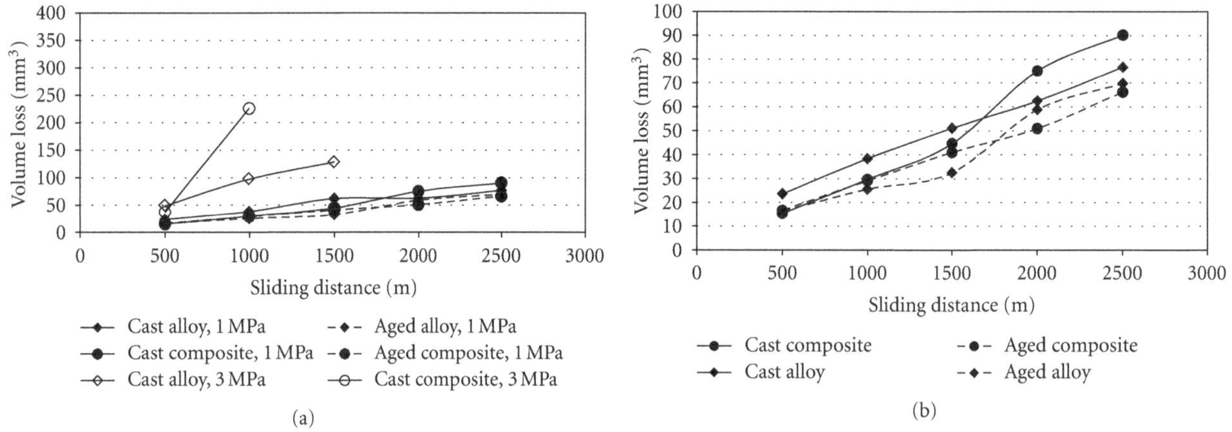

FIGURE 4: Sliding wear of 2014 alloy and composite at (a) 200 rpm (b) 750 rpm.

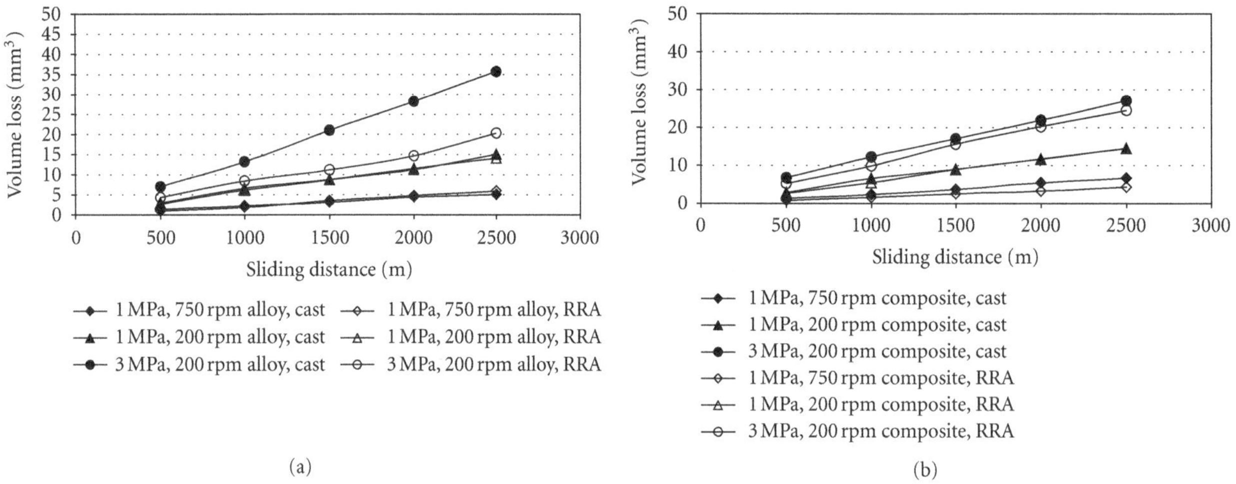

FIGURE 5: Effect of ageing on the sliding wear behaviour of 7075 alloy and composite (a) alloy and (b) composite.

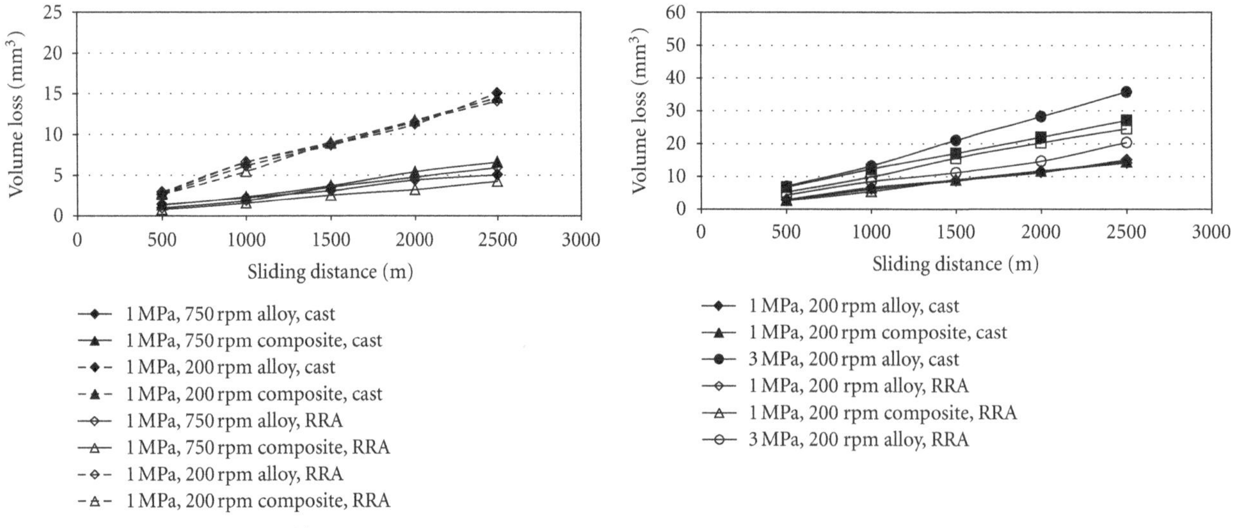

FIGURE 6: Effect of experimental parameters on the sliding wear behaviour of 7075 alloy and composite (a) at different speeds of rotation and (b) at different loads.

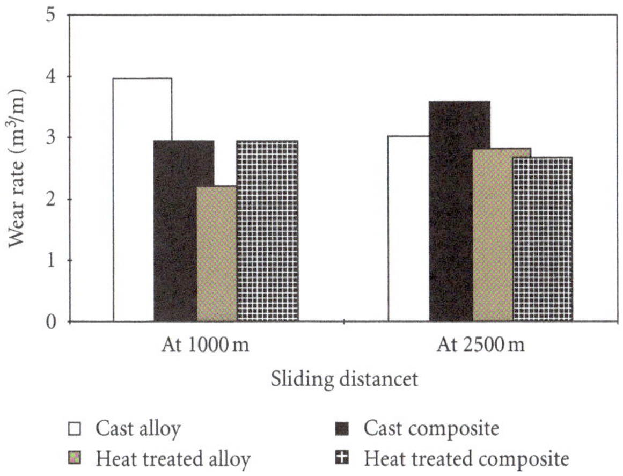

FIGURE 7: Sliding wear behaviour of 2014 alloy and composite at 750 rpm sliding speed and load of 1 MPa.

Moreover, ageing of alloy or composite in the case of 7075 system does not have any significant bearing in reducing the volume loss and this is probably due to softening of the matrix as a result of homogenisation and volume loss is directly dependant on the hardness and stiffness of the material. In this alloy system like in the case of 2014 system, it is interesting to note that higher speeds of rotation results in lesser volume loss which is probably due to less time that the surfaces get to slide against each other due to the enhanced speed; the material damage is also less in such cases as seen from comparing its worn surfaces at a particular load at different speeds.

The final wear rate for the samples after completion of the tests shows significant decrease at higher speeds for both the cast and aged conditions of the composites as compared to the base alloy (Figures 7 and 8). This is to say that if these composites were used for making engineering components, they would perform better at higher speeds and lower loads than at other combinations.

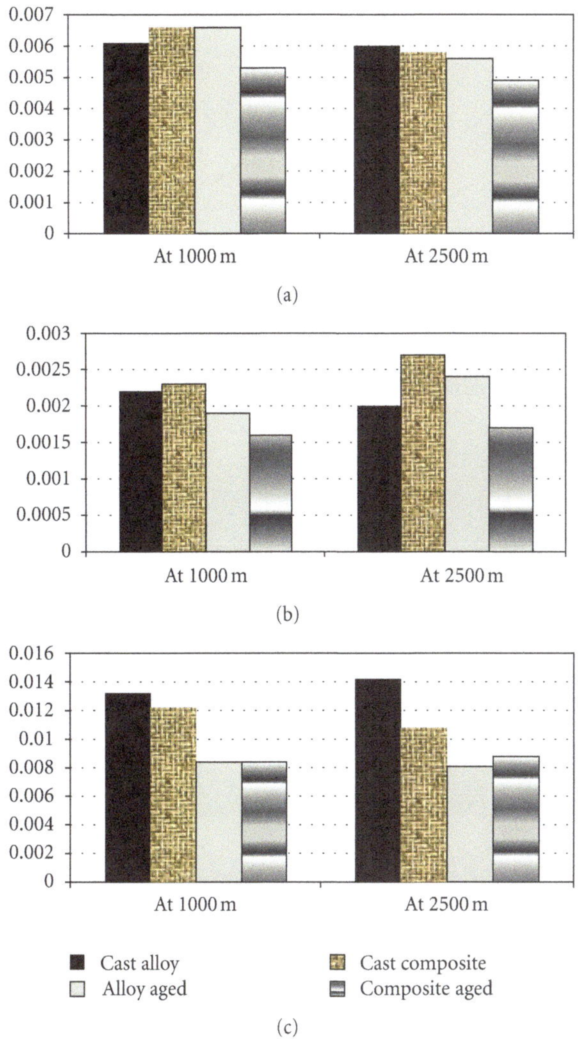

FIGURE 8: Wear rate variation for 2014 alloy and composite at sliding distance 1000 at and 2500 m (a) 200 rpm, 1 MPa and (b) 750 rpm, 1 MPa (c) 200 rpm, 3 MPa.

3.3. Worn Surface Studies. The worn surface features are in accordance with the material loss observed. The SEM photographs exhibit the features and give a relative idea of the damage due to sliding wear. The depth of the wear grooves is also a measure of the material loss. The worn surface of both the alloy systems is similar. Those for 2014 are reported here. In the cast condition at a load of 1 MPa (Figure 9(a)) is characterised by large, continuous dimples and uniform wear tracks; at higher sliding distances breaking of the surface is also observed, suggestive of brittle nature of the alloy and its inability to withstand long sliding distance. The worn surface at a load of 3 MPa is suggestive of seizure (Figure 9(b)) with severe metal flow. In the aged alloy, the worn surface at 1 MPa load is characterized by uniform wear tracks and small dimples (Figure 9(c)).

The worn surface of the composite under cast and aged conditions is shown in Figure 9. The worn surface of the cast alloy (Figure 10(a)) is characterized by continuous wear grooves and discontinuously scattered small dimples. This mechanism of material removal in the case of composite appears to be from the surface only, whereas for the alloy under similar experimental conditions removal is from beneath the sample surface. The cast composite is shown to suffer maximum material loss at times (Figure 10(b)) by peeling off of the dispersoids resulting in sudden weight (volume) loss. The worn surface of the aged composite does not show very clear wear tracks (Figure 10(c)) and is characterized by small dimples and pitting indicating less material loss and damage.

The worn surface shows only shallow wear tracks under high speeds of rotation (sliding speed) as compared with the slower speed at the same load for the alloy (Figures 11(a) and 11(b)). The worn tracks are deeper when the pressure is high, the damaged surface is characterized by deep wear grooves and material removal is through micro ploughing. The depth of the grooves is indicative of the quantity of material removal. Deeper wear grooves are characteristics

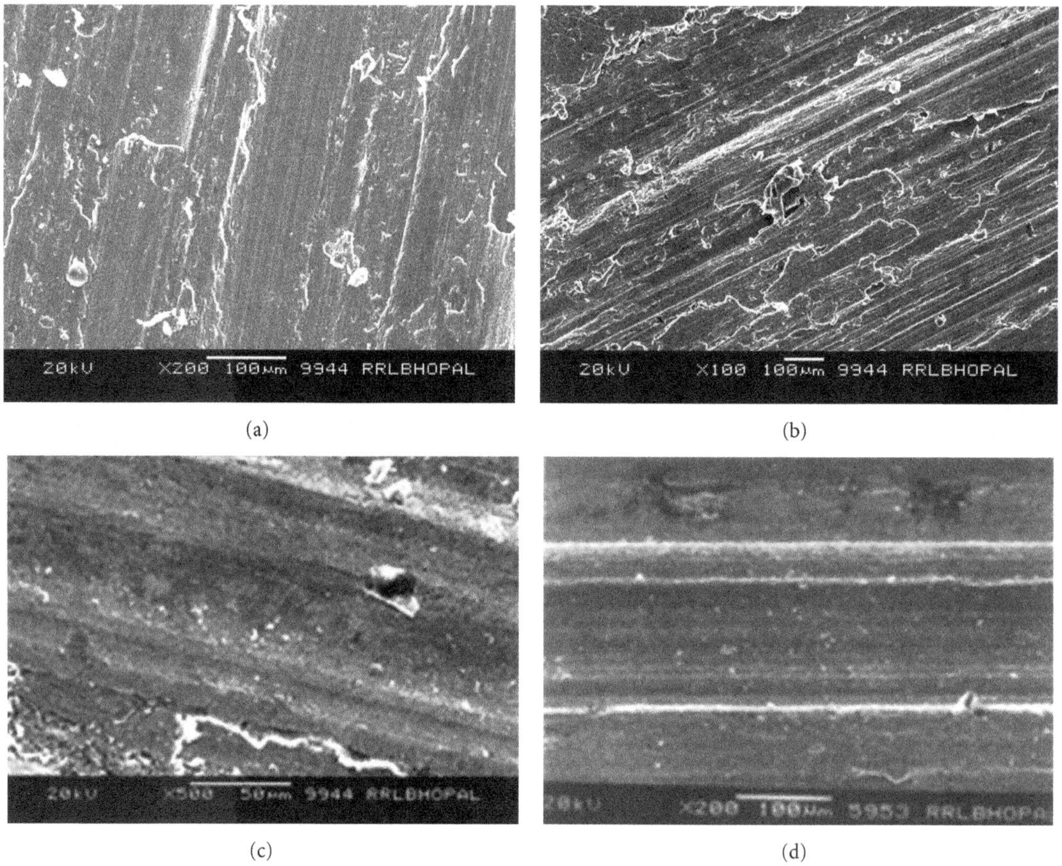

FIGURE 9: Worn surface of alloy. (a) at a pressure of 1 MPa at 200 rpm (b) at 3 MPa suggestive of seizure (c) aged alloy (d) at higher speeds showing less damage.

of material removal of micro ploughing, whereas material removal by oxidation of the surface is the cause for material removal from the surface only.

Seizure of the material can be understood from the machine stopping and also from the surface being tested which shows material movement; the scanning electron micrographs of seized samples (Figure 12) shows complete overlapping of material and material movement cutting across wear tracks (Figure 12(a)); the seized material in turn does not show any structure and looks like a lump of mass on the surface adhered to the surface and material from the inner surface are also exposed (Figure 12(b)).

The resistance offered by the SiC particulates influences sliding wear resistance of the composites. It is interesting to note that the wear rate or (volume loss) variation with hardness is different for the alloy and composite. A direct proportionality between hardness and wear rate would be increase in hardness resulting in decrease wear rate. Contrary to this, in the case of composite (for all conditions of experimental factors and material conditions) for increase in hardness there is an increase in wear rate; however in the case of alloy, wear rate decreases with increase in hardness. This behaviour is probably due to the fact that the soft surface can accommodate larger damage during sliding than harder samples which trend to be brittle resulting in more material loss. Addition of SiC to form composites increases brittleness of the alloy, hence the variation between wear rate and hardness are different for the alloy and composite. Higher material loss is reported for composites when the SiC particulates are removed due to sliding, prevalent under conditions of high pressure and low sliding speeds (Figures 9(b) and 10(b)). Under high speeds of sliding, no damage is seen of the SiC particulates (Figure 9(d)) and lower material loss (volume loss) is reported; the wear mechanism is similar to that of the alloy under same experimental conditions, that is, from the surface only.

3.4. High Stress Abrasive Wear Behaviour. The high stress abrasive wear properties of alloys and composites of the 2014 system in the cast and heat treated under optimised conditions reveal the effectiveness of both heat treatment and formation of composite (Figure 13). The aged samples exhibit lower wear rate (improved wear resistance) as compared to the cast counterparts under identical experimental conditions. Again the composites in both the aged and cast condition exhibit significantly reduced wear rates (which would result in much superior wear performance) over the counterpart alloys under similar experimental conditions. Higher loads as is expected resulted in higher wear rates. It

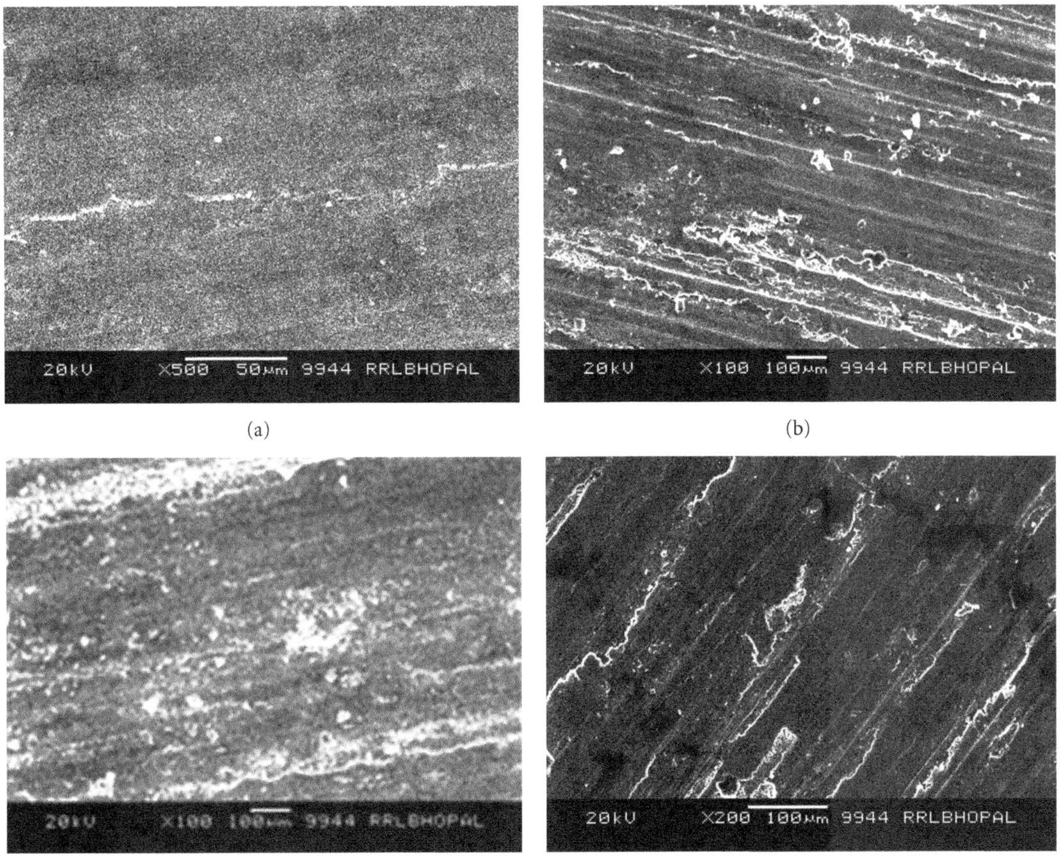

(a) (b)

(c) (d)

FIGURE 10: Worn surface of composite. (a) Cast at a pressure of 1 MPa at 200 rpm (b) peeling off of the dispersoids resulting in sudden weight loss at high loads (c) aged composite do not show very clear wear tracks (d) at higher pressures.

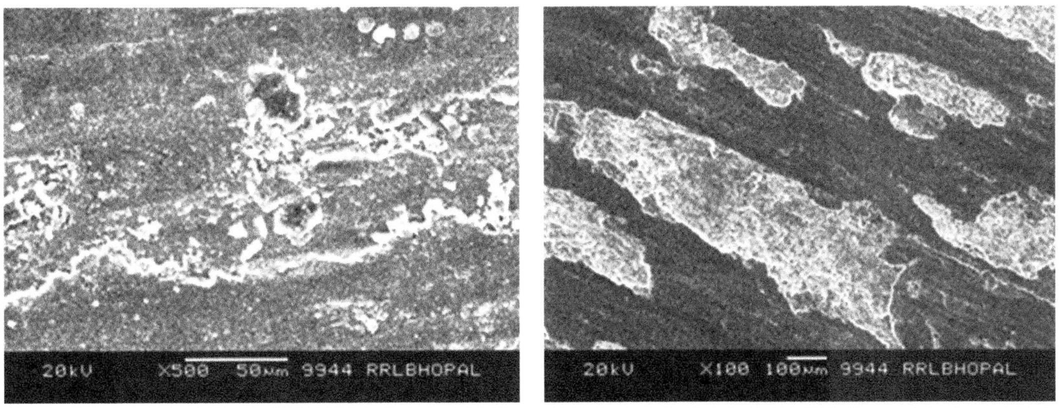

(a) (b)

FIGURE 11: Increase in depth of wear tracks with pressure (a) at 1 MPa (b) at 3 MPa.

can be undoubtedly be said that the composites from these alloys would perform much better than the alloys where there is a demand of superior abrasive wear resistance. The improved wear resistance of the composites is attributed to the resistance offered by the SiC dispersoids uniformly distributed in the matrix to the abrading media which as a result of the resistance offered loses its sharpness and eventually breaks.

The high stress abrasive wear behaviour of the alloy and composite in the cast and aged conditions were carried out at two loads of 3 and 5 N; the abrasive used was emery paper of 220 grit size.

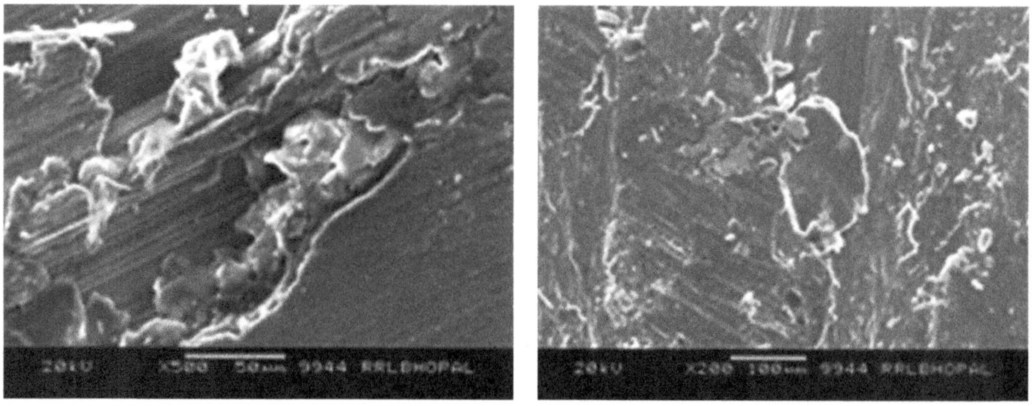

(a) (b)

FIGURE 12: Micrographs of seized samples.

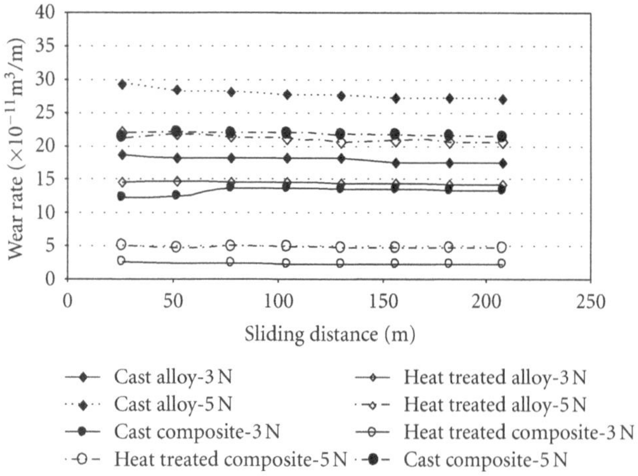

FIGURE 13: High stress abrasive wear behaviour of 2014 alloy and composite.

Composite making improves the high stress abrasive wear rate under all the experimental conditions for the 7075 series of alloys, which shows a steady rate with distance mostly in all the cases (Figure 14).

However, the high stress abrasive wear behaviour in the aged condition exhibits higher wear rates as shown in Figure 15. This could be attributed to the softening of the aged sample under loading conditions as prevalent under high stress abrasive wear conditions, which is more pronounced in the presence of abrasives. In the case of composite also, there is no beneficial effect of ageing, that is, the wear rate of the annealed composites is higher than the cast counterparts under identical experimental conditions.

3.5. Subsurface Features. The microstructural features of the longitudinal cross section of the worn pin for the 2014 alloy and composite are shown in Figure 16. In the case of cast alloy, (Figure 16(a)) a white layer is formed but the microstructural features are not well pronounced. In the case of aged alloy (Figure 16(b)), the thickness of the white layer is much less and the microstructural features are well resolved. The damaged region is also clearly seen and the region below the damaged region shows a tendency of plastic deformation in the sliding direction (Figure 16(c)). The subsurface features of the cast and aged composite (Figure 17) shows similar features as in the case of alloy. The plastic deformation appears to have been stopped/reversed on encountering the reinforced SiC particulates. The damaged region as measured by the thickness beneath the subsurface is also appreciably lesser in the case of composites as compared to the alloy. The subsurface of the aged composite shows cracking till a substantial depth from the affected region (Figure 17(c)). This indicates that metal flow when arrested by the presence of SiC particulates results in stress accumulation and hence in cracking.

4. Discussion

The mechanism of wear is in general very complex mainly due to the number of contributing factors to material removal [19–24]. The complexity and prediction of wear to be encountered in any system is increased as in actual conditions under any mode of wear, the real area of contact between two solid surfaces compared with the apparent area of contact is invariably very small, being limited to points of contact between surface asperities. The load applied to the surfaces will be transferred through these points of contact and the localised forces can be very large. The material intrinsic surface properties such as hardness, strength, ductility, work hardening, and so forth are very important factors for wear resistance, but other factors like surface finish, lubrication, load, speed, corrosion, temperature and properties of the opposing surface, and so forth are equally important. The gradual removal of a material(s) from one or both surfaces in contact increases the clearance between the surfaces. This spatial freedom induces a vibration that often leads to fatigue failures of machinery components. In many cases, the presence of a wear product during the process of wear like the wear debris even in small amounts

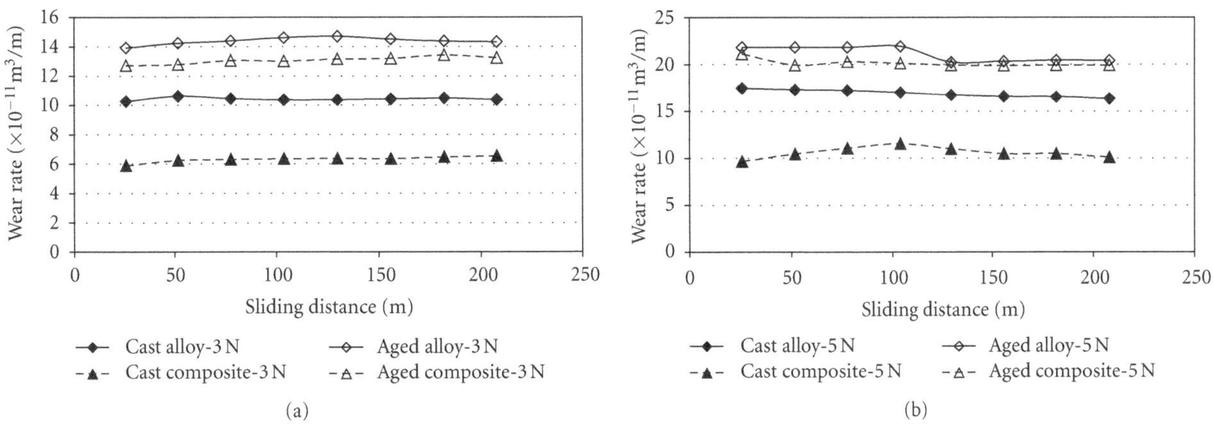

Figure 14: High stress abrasive wear rate behaviour between cast and aged conditions for 7075 alloy system (a) alloy (b) composite.

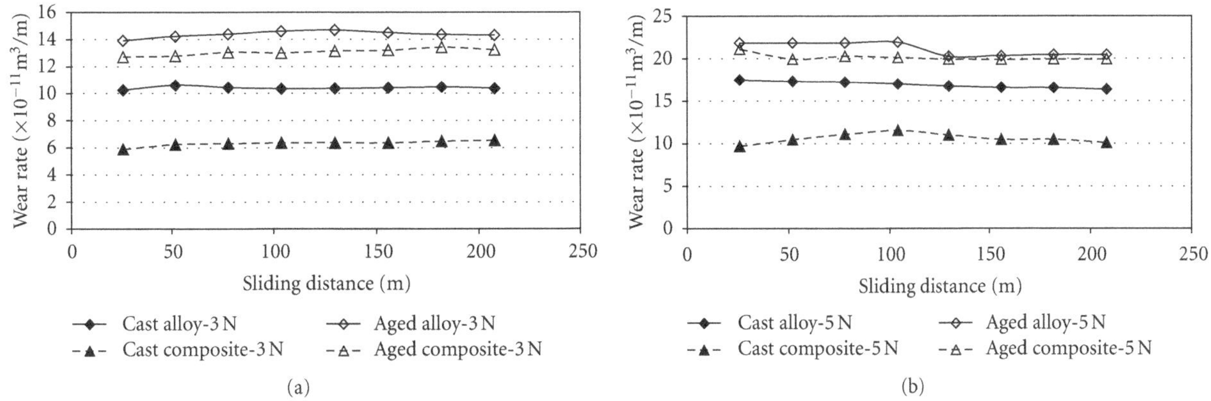

Figure 15: High stress abrasive wear rate behaviour between alloy and composite (a) load = 3 N (b) load = 5 N.

can cause catastrophe and even before the breakdown affects the performance of the parts undergoing wear. Thus, even if wear cannot be done away with, the wear debris during wear should be attempted to be minimised either by removing them or designing of materials that have better wear resistant properties.

Under conditions where mild sliding wear conditions prevail, in this case at higher speeds, wear is generated by fracture and removal of the oxide layer which is formed in aluminium alloys, as a result of continued sliding action (Figure 18(a)). Higher material loss is associated with micro ploughing action, observed under low speed conditions at both the loads (Figure 18(b)). It may be said that the benefits of composite formation is seen from the improved wear resistance as observed in shallower wear tracks under the same conditions when compared with the alloy (Figure 18(c)); also the beneficial effects of ageing is evident in all the cases (alloy and composite) under identical experimental conditions. The improved wear resistance in composites is mainly attributed to the effective resistance of the SiC particle to penetration, grinding and cutting by the ceramic particulates. When the particle sizes are small, matrix strengthening of the particles as well as wear resistance of the particles contributes to improving the wear resistance (inverse of wear rate). Ageing improves the toughness of the material hence inspite of lower hardness the wear rate is decreased (improved wear resistance).

There are many possible wear mechanisms: adhesion, abrasion, erosion, and so on. Generally, more than a single mechanism occurs at the same time both in actual operation of machines and even in experimental conditions, this cannot be completely ruled out. However, there is always a primary mechanism that determines the material removal rate. Whenever there is a possibility of abrasive wear, it is the most important problem to be solved. The wear rate of abrasive wear is at least one to two orders of magnitude larger than those of other mechanisms. The abrasive wear mechanism is basically the same as machining, grinding, polishing, or lapping that we use for shaping materials. Two body abrasive wear occurs when one surface (usually harder than the second) cuts material away from the second, although this mechanism very often changes to three body abrasion as the wear debris then acts as an abrasive between the two surfaces. Abrasives act as in grinding where the

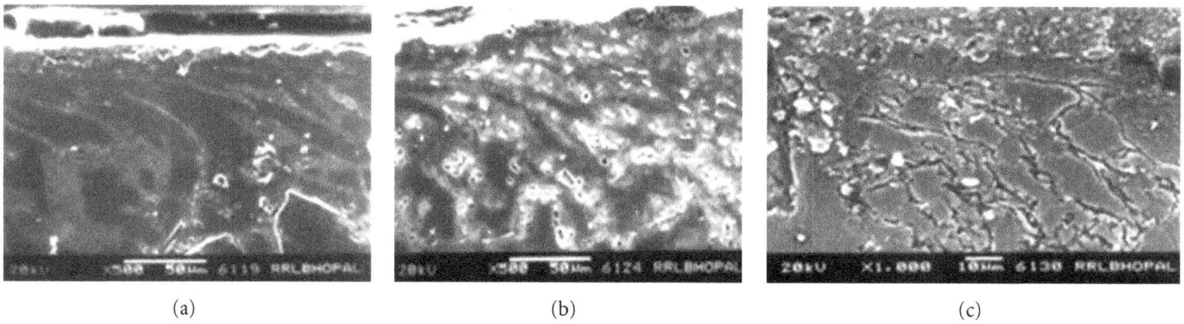

Figure 16: Subsurface features of 2014 alloy. (a) Cast alloy showing white layer formation (b) aged alloy showing well resolved microstructural features and thin white layer (c) damaged region show a tendency of plastic deformation in the sliding direction.

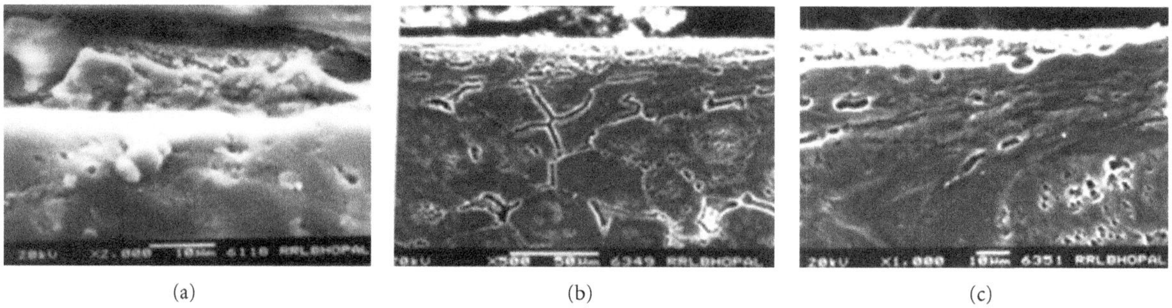

Figure 17: Subsurface features of 2014 composite. (a) cast composite (b) aged composite showing behaviour of plastic deformation on encountering the reinforced SiC particulates (c) aged composite show cracking till a substantial depth from the affected region.

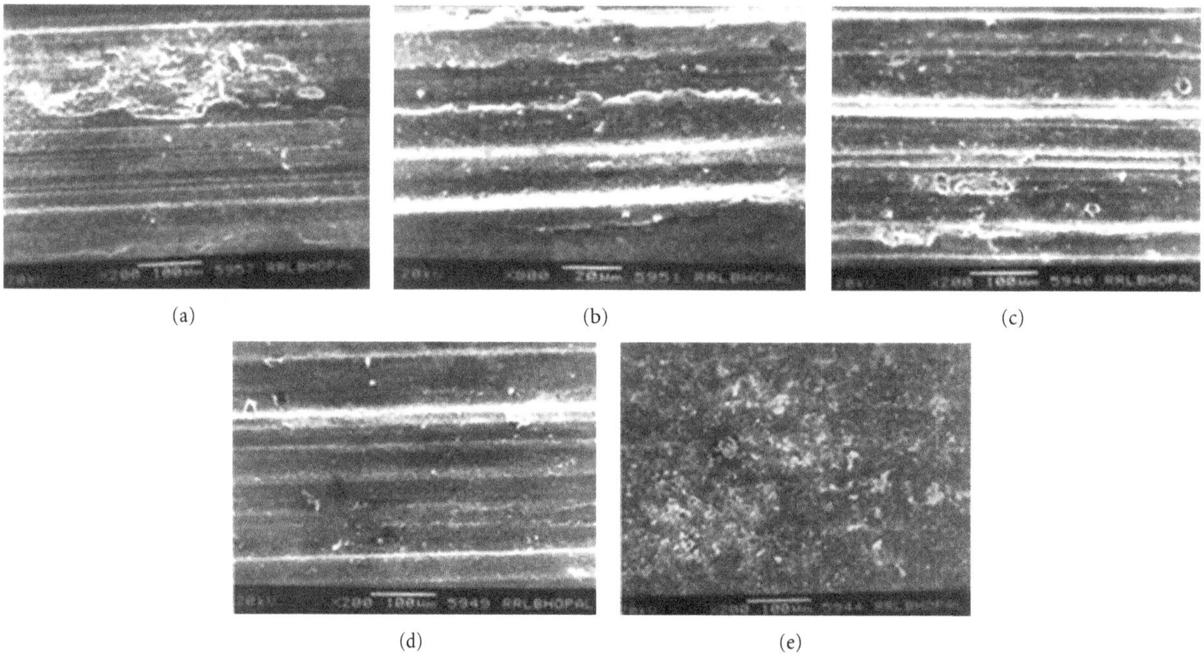

Figure 18: Worn surfaces under different experimental conditions (a) cast alloy tested at low load (b) cast alloy tested at higher load (c) aged alloy tested at higher load (d) cast composite under same experimental conditions as in "a" (e) aged composite tested at higher load.

abrasive is fixed relative to one surface or as in lapping where the abrasive tumbles producing a series of indentations as opposed to a scratch.

Simple predictive models developed for wear loss in abrasive wear of ductile metals, assuming that mass loss resulted from repeated indentation from the corners and asperities of angular particles rolling across the surface of the wear specimen does not hold good under most, as it does not take into account wear through sliding and cutting by the particles. Examination of wear surfaces suggested that individual particles traversed the wear surface in a mixed slide (scratch)—roll manner. Under most of the test conditions examined, sliding/cutting was the dominant metal removal mechanism, although wear by indentation was generally not insignificant. Abrasive wear happens when the particular material is being removed from one surface, by another harder material, as a result, hard particles of the debris between the two surfaces are formed. The terms scratching, gouging, or scoring are used depending on the degree of abrasive wear. In simple words, the harder material scratches the softer material. There are 2 conditions in which abrasive wear occurs. The 2 body abrasion and 3 body abrasion. In the former, the harder surface rubs the other. Examples of these include mechanical grinding, cutting, and machining. The latter involves a third body. This third body is generally a small particle of abrasive between the 2 softer rubbing surfaces. Microscopically, abrasive wear causes asperities of the harder surface to press into the softer surface, with plastic flow of the softer surface formed around the harder asperities. These result in "micro ploughing" and "micro cracking" where tangential motion is imposed [24, 25].

Abrasive wear resistance of a monolithic material is usually poor as it offers very little resistance to the harder surface sliding against it during wear modes classified as abrasive. Making multiphase material cannot improve the abrasive wear resistance of such materials. The technique is by introducing a secondary phase(s) into the matrix material; hence, composites have invariably shown much better wear resistant properties under abrasive conditions over the base alloys. The mechanism of material removal during two body abrasive wear of Al-alloy composites with SiC dispersoids are primarily cutting and plowing, which lead to formation of continuous wear grooves. The advantage of making composites can be seen from the significantly lower wear rate in them as compared to the base alloy; in the former the SiC particles act as protrusions over the surface and protect the matrix from wear (Figures 18(a), 18(d), 18(b), and 18(e)). But at higher applied load, coarser abrasive size, and larger sliding distances, some of the SiC particles get fractured into fine particles and scooped off from the wear surface leading to a higher wear rate and since the abrasives are now loose and if not removed from the surface the behaviour changes to that of three body abrasion. The subsurface studies show severe plastic deformation and finally formation of a mechanically mixed layer (MML) over the plastically deformed zone. The MML gets fractured during the wear process and finally removed by the formation of lateral and transverse cracking. The cracks are generally initiated at the interface of MML and the plastically deformed zone and propagate along the weaker region in MML [26].

This is a scraping, grinding mechanism that rubs away metal surfaces and can be caused by the scouring action of the abrasives in the abrasive paper. The scouring action is a result of both abrasion and sliding and can be seen microscopically as scratches under low load and small sliding distance but as the load and/or sliding distance increases, the scratches become continuous and parallel grooves are observed and later the depth increases. The abrasive paper is degraded and the loose abrasives fall into the created grooves; however, these are removed when fresh emery paper is used after a certain traversal distance. Where lubricants are used, the oil serves as a bed for accumulation of abrasives and thus needs to be changed at regular intervals for smooth operation, else the contaminated oil at times results in more wear loss that under dry/unlubricated conditions.

A number of wear models and mathematical expressions of wear rate have been predicted; however for reliable predictions, it is better to conduct fresh experiments as the number of variables and their effects on the wear rate are always sketchy and depend on experimental conditions [27].

5. Conclusions

Al-alloy based composites were thought of about a quarter of a century back when the possibilities of improvement in Al-alloys by the then conventional methods of heat treatment and microstructural modification had reached its limit. Composites took a lead as compared to the other processes when the cost and ease of fabrication were compared. The other methods changed track and chose for itself different areas of application and Al-based metal matrix composites (AMMCs) remained as the most potential candidate to be researched on for making engineering components viably. When the question of viability arose, from the different methods adopted for making AMMCs the liquid metallurgy route stood ahead due to its ease of fabrication; although the other routes were more efficient regarding property attainment and microstructural features, still the ease of fabrication added to the cost economics made the liquid metallurgy route a competitive and viable method. Worldwide research in the area of AMMCs established beyond doubt the advantages of these materials over the base alloy. The demands moved ahead and engineering components were demanded from AMMCs. However, engineering products could not make a dent when this technology was adopted. One of the properties that showed marked improvement was the wear resistance, the contributing factor being the better wear resistance of the dispersoids. However, attaining uniform dispersion was an area of specialisation.

The present paper highlights the improvements in different types of wear normally encountered by engineering components during service and the improvements attained in them by making composites through the liquid metallurgy route and also ageing. Two alloy systems generally used by the engineering sector amongst the different aluminium based alloys have been reported in the present study. The interesting point to be noted is that an otherwise nonwear resistant alloy like the 7075 system also exhibits significant improvement in wear resistance when composites are fabricated from them. This opens up additional avenues for this alloy system.

The wear mechanism has been explained and conditions where composites can perform better than alloys have been arrived at. Engineering components where the working conditions are similar to the conditions under which experiments have been conducted can be fabricated from the composites, and it is expected to be performed with ease and be better than the presently used alloys.

References

[1] M. S. Zedalis, P. S. Cilman, and S. K. Das, "High temperature aluminum-base composites," in *High Performance Composites for the 1990's*, S. D. Das et al., Ed., pp. 61–81, The Metallurgy Society of AIME, Warrendale, Pa, USA, 1990.

[2] P. K. Rohatgi, S. Das, and T. K. Dan, "Cast Al-graphite particle composites, a potential engineering material," *Journal of the Institution of Engineers*, vol. 67, no. 2, pp. 77–83, 1987.

[3] P. K. Rohatgi, S. Das, and R. Asthana, "Science, technology and industrial potential of cast metal ceramic particle composites," in *Materials Science and Technology in the Future*, pp. 123–184, CSIR, Bhopal, India, 1985.

[4] *Engineered Materials Handbook: Composites*, vol. 1, ASM International, Materials Park, Ohio, USA, 1989.

[5] M. K. Surappa, "Aluminium metal matrix composites: challenges and opportunities," *Sadhana*, vol. 28, no. 1-2, pp. 319–334, 2003.

[6] U. Hecht and S. Rex, "On the transition from pushing to engulfment during directional solidification of the particle-reinforced aluminum-based metal-matrix composite 2014 + 10 Vol Pct Al_2O_3 ," *Metallurgical and Materials Transactions A*, vol. 28, no. 13, pp. 867–874, 1997.

[7] R. Asthana, "Reinforced cast metals: part I Solidification microstructure," *Journal of Materials Science*, vol. 33, no. 7, pp. 1679–1698, 1998.

[8] C. P. Hong, H. F. Shen, and I. S. Cho, "Prevention of macrosegregation in squeeze casting of an Al-4.5 wt pet Cu alloy," *Metallurgical and Materials Transactions A*, vol. 29, no. 1, pp. 339–349, 1998.

[9] L. H. Chen and D. A. Rigney, "Transfer during unlubricated sliding wear of selected metal systems," *Wear*, vol. 105, no. 1, pp. 47–61, 1985.

[10] R. L. Deuis, C. Subramanian, and J. M. Yellup, "Abrasive wear of aluminium composites—a review," *Wear*, vol. 201, no. 1-2, pp. 132–144, 1996.

[11] F. M. Hosking, F. F. Portillo, R. Wunderlin, and R. Mehrabian, "Composites of aluminium alloys: fabrication and wear behaviour," *Journal of Materials Science*, vol. 17, no. 2, pp. 477–498, 1982.

[12] S. Turenne, Y. Châtigny, D. Simard, S. Caron, and J. Masounave, "The effect of abrasive particle size on the slurry erosion resistance of particulate-reinforced aluminium alloy," *Wear*, vol. 141, no. 1, pp. 147–158, 1990.

[13] K. H. Z. Gahr, "Wear by hard particles," *Tribology International*, vol. 31, no. 10, pp. 587–596, 1998.

[14] P. J. Blau, "Fifty years of research on the wear of metals," *Tribology International*, vol. 30, no. 5, pp. 321–331, 1997.

[15] A. T. Alpas and J. D. Embury, "Sliding and abrasive wear behaviour of an aluminum (2014)-SiC particle reinforced composite," *Scripta Metallurgica et Materiala*, vol. 24, no. 5, pp. 931–935, 1990.

[16] D. Rupa and M. Humaira, "High stress abrasive wear behaviour of Al-2014 alloy and its SiC composite," *International Journal of Microstructure and Materials Properties*, vol. 3, no. 2-3, 2008.

[17] R. Dasgupta, H. Meenai, and S. Das, "Age hardening of Al 201-alloy and 2014-SiC composites," *Journal of Materials Science Letters*, vol. 22, no. 5, pp. 391–394, 2003.

[18] D. Rupa, "The stretch, limit and path forward for particle reinforced metal matrix composites of 7075 Al-Alloys," *Engineering*, vol. 2, no. 4, pp. 237–256, 2010.

[19] S. Das, D. P. Mondal, S. Sawla, and S. Dixit, "High stress abrasive wear mechanism of LM13-SiC composite under varying experimental conditions," *Metallurgical and Materials Transactions A*, vol. 33, no. 9, pp. 3031–3044, 2002.

[20] M. S. Bingley and S. Schnee, "A study of the mechanisms of abrasive wear for ductile metals under wet and dry three-body conditions," *Wear*, vol. 258, no. 1–4, pp. 50–61, 2005.

[21] K. Kato, "Classification of wear mechanisms/models," *Proceedings of the Institution of Mechanical Engineers J*, vol. 216, no. 6, pp. 349–356, 2002.

[22] X. F. Zhang, G. Y. Lee, D. Chen, R. O. Ritchie, and L. C. De Jonghe, "Abrasive wear behavior of heat-treated ABC-silicon carbide," *Journal of the American Ceramic Society*, vol. 86, no. 8, pp. 1370–1378, 2003.

[23] R. I. Trezona, D. N. Allsopp, and I. M. Hutchings, "Transitions between two-body and three-body abrasive wear: influence of test conditions in the microscale abrasive wear test," *Wear*, vol. 225–229, no. I, pp. 205–214, 1999.

[24] J. A. Williams and A. M. Hyncica, "Mechanisms of abrasive wear in lubricated contacts," *Wear*, vol. 152, no. 1, pp. 57–74, 1992.

[25] F. M. Hosking, F. F. Portillo, R. Wunderlin, and R. Mehrabian, "Composites of aluminium alloys: fabrication and wear behaviour," *Journal of Materials Science*, vol. 17, no. 2, pp. 477–498, 1982.

[26] B. Venkataraman and G. Sundararajan, "Correlation between the characteristics of the mechanically mixed layer and wear behaviour of aluminium, Al-7075 alloy and Al-MMCs," *Wear*, vol. 245, no. 1-2, pp. 22–38, 2000.

[27] Y. L. Gun, C. K. H. Dharan, and R. O. Ritchie, "A physically-based abrasive wear model for composite materials," *Wear*, vol. 252, pp. 322–331, 2002.

20

Study of Amorphous-Crystalline Phase Transformations by DSC and Dilatometer in the Case of Al-Based Amorphous Alloys

Éva Fazakas,[1] Béla Varga,[2] and Lajos Károly Varga[1]

[1] *Wigner Research Centre for Physics, Institute for Solid State Physics and Optics, Hungarian Academy of Sciences, P.O. Box 49, Budapest 1525, Hungary*
[2] *Faculty of Materials Science and Engineering, Transylvania University, Bulevardul Eroilor, No. 29, 500036 Brasov, Romania*

Correspondence should be addressed to Éva Fazakas, efazakas@szfki.hu

Academic Editors: R. Hebert and A. Tuissi

The amorphous-crystalline transformation is studied by DSC and dilatometer. From the DSC signal the characteristic temperatures and the heat evolved during each transformation step can be determined. The dilatometer depicts the thermal contractions accompanying the changes in topological short range order. Although the characteristics temperatures determined by DSC and dilatometer, two methods are slightly different their dependences on the heating rate are very similar, where from the activation energies of each transformation steps can be determined using the Kissinger plots. As an example, two aluminum-based amorphous ribbon samples ($Al_{88}Mm_5Ni_5Co_2$ and $Al_{85}Y_8Ni_5Co_2$) will be measured and compared.

1. Introduction

20 years have been passed since the discovery of the first ductile Al-based amorphous alloy ribbons with superior mechanical properties compared to the conventional aluminum alloys [1, 2]. These ribbons have been obtained by melt spinning technique starting from a multicomponent ingot contained as a rule rare earth (RE) metal and one or two late transition metals (TMs) [3]. It has been found that alloys with high glass forming ability (low critical cooling rate for glass formation or large critical thickness) are the so-called strong metallic glasses in the Angell plot [4]. Unfortunately, the Al-based glasses belong to the fragile glasses with low glass forming ability and consequently no bulk metallic glass (BMG) exits on Al-base. The bulk amorphous samples can be obtained only by hot consolidation of the amorphous ribbon flakes, which should be performed below the transformation temperature. Therefore, the thermal stability of these ribbons (in amorphous or nanocrystalline state) marks a limit of their applicability, and the study of the kinetics of crystallization of Al-based alloys is an important task to which several studies have been devoted so far (for a review, see [5]).

In order to reduce the price of the alloy in this paper we have studied a composition containing Mischmetal instead of pure RE element. Mischmetal (Mm) is a natural mixture of rare earth metals; typical composition contains 48% cerium, 25% lanthanum, 17% neodymium, and 5% praseodymium, with the balance being the other lanthanides. Many studies were devoted to compare the behavior of Mm-substituted alloys with those with pure rare earth elements [5–7]. In the present paper the kinetics of crystallization are studied for two alloys of nominal compositions $Al_{85}Y_8Ni_5Co_2$ and $Al_{88}Mm_5Ni_5Co_2$ by DSC and dilatometer. We have chosen these two compositions because they show big differences during dilatometer investigations although both show three transformation steps in their DSC patterns. The dilatometer is sensitive to the volume changes accompanying the amorphous-crystalline transformation, that is, to the change in topological short range order (TSRO). In contrast with dilatometer, the DSC is sensitive to the heat evolved during the formation of new phases, that is, to the changes in chemical short range order (CSRO). In addition, Vickers microhardness data are presented to compare their mechanical properties.

2. Experimental

The $Al_{88}Mm_5Ni_5Co_2$ (Mm5) and $Al_{85}Y_8Ni_5Co_2$ (Y8) alloy ingots were prepared by induction melting under argon atmosphere [8]. High-purity (>99.9 at% purity) elements were placed in a water-cooled copper boat and held above the melting point to homogenize the ingot. The as-prepared ingot was used to prepare ribbon samples (thickness was about 20–22 μm and the width 2.5–3 mm of the ribbon) by a single-roller melt-spinning technique under protective atmosphere at a wheel speed of 25°m/s. Bulk amorphous Mm5 and Y8 samples can be produced by hot consolidation of the flakes produced by cryogenic chopping of melt-spun amorphous ribbons. The evolution of the microstructure of the as-quenched samples was examined by X-ray diffraction (Bruker D8). Differential scanning calorimeter (Perkin-Elmer TGA7) and dilatometer measurements (Linseis L75H) under argon atmosphere were used to study the structural transformations up to the melting point.

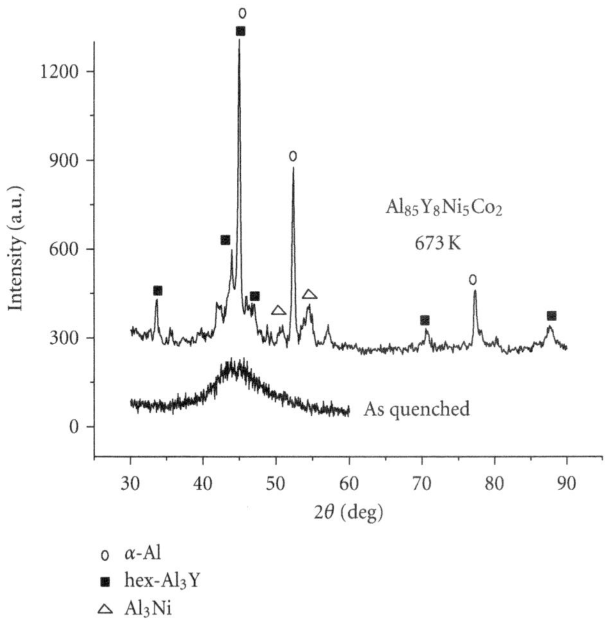

FIGURE 1: XRD patterns for the samples of the as-quenched Y alloy being heated up to 673 K temperatures at 10 K/min.

3. Results and Discussion

Figures 1 and 2 show the XRD patterns obtained from samples annealed up to the end of crystallization sequences. Amorphous patterns obtained for the as-cast ribbons are also shown. The peaks were identified to correspond to fcc α-Al as the main component; hexagonal Al_3Y and orthorhombic Al_3Ni for $Al_{85}Y_8Ni_5Co_2$ and orthorhombic $Al_{11}Mm_3$ and orthorhombic Al_3Ni for $Al_{88}Mm_5Ni_5Co_2$, respectively. Measuring the XRD pattern after different crystallization stages one can remark a narrowing of the fcc α-Al peak indicating that grain growth of this phase takes place during the three crystallization stages.

The two glasses show very different crystallization sequences. For Y8 an endothermic prepeak (see Figure 3) just before the main crystallization stage can be observed as an evidence for the glass transition. Whereas for the Mm5 sample a small exothermic prepeak (see Figure 4), more than 100 K degree separated from the main crystallization stage, denotes the primary crystallization of the fcc α-Al in a residual amorphous matrix. The crystallization transformation followed on the dilatometer scans (Figures 5 and 6) shows significant differences compared to the DSC scans: for the Y8 samples (Figure 5) there is apparently only one crystallization step on the derivative of the length change, ΔL, contradicting the DSC scan (Figure 3), whereas for the M5 sample all the 3 crystallization stages are well marked (Figure 6) in accordance to the DSC scans (Figure 4). However, enlargement of the details around the transformation peak (see the inset in Figure 5) reveals the existence of the three transformation steps at relative low heating rates for the Y8 sample as well. For 10 K/min heating rate one can observe a change in the volume of the sample even before the onset of the glass transition marked by the endothermic peak in DSC pattern. More investigations are necessary to separate the free volume elimination caused contraction during the relaxation of the amorphous state and the contraction caused by the glass transition into the supercooled liquid state. At the moment we concentrate on the devitrification process, which depends mostly on the RE content. For high Al content (88 at % Al in Mm5 sample) we observe the precrystallization peak of the nanosized α-Al, characteristic for the hypoeutectic compositions at the edge

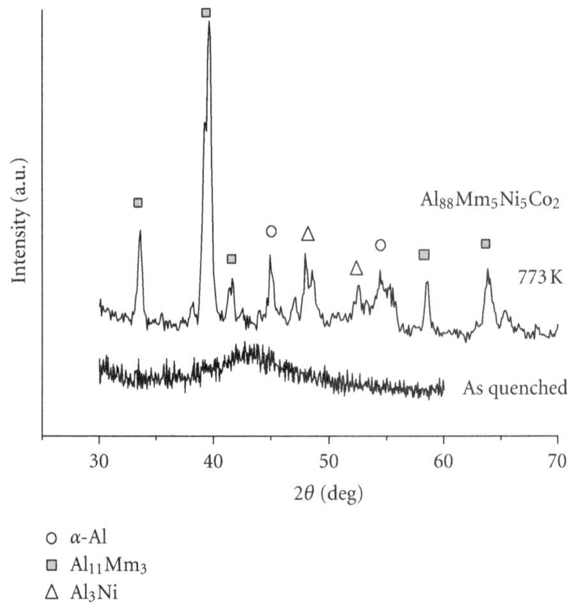

FIGURE 2: XRD patterns for the samples of the as-quenched Mm alloy being heated up to 773 K temperatures at 10 K/min.

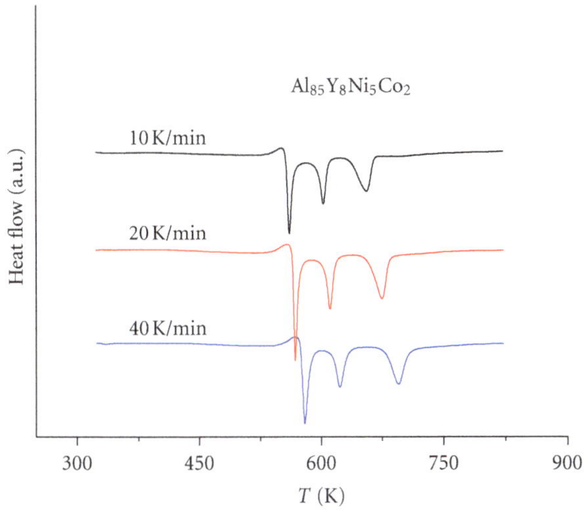

FIGURE 3: DSC scans at different heating rate for as-cast samples of the $Al_{85}Y_8Ni_5Co_2$ alloys.

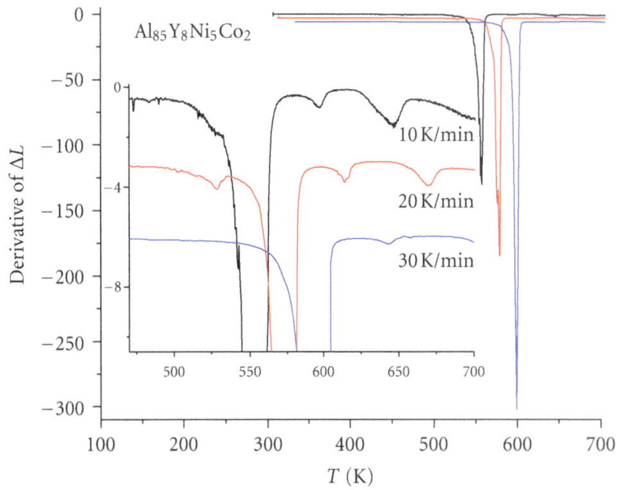

FIGURE 5: Dilatometer scans at different heating rate for as-cast $Al_{85}Y_8Ni_5Co_2$ ribbons. The inset shows a magnified detail around the main crystallization peak.

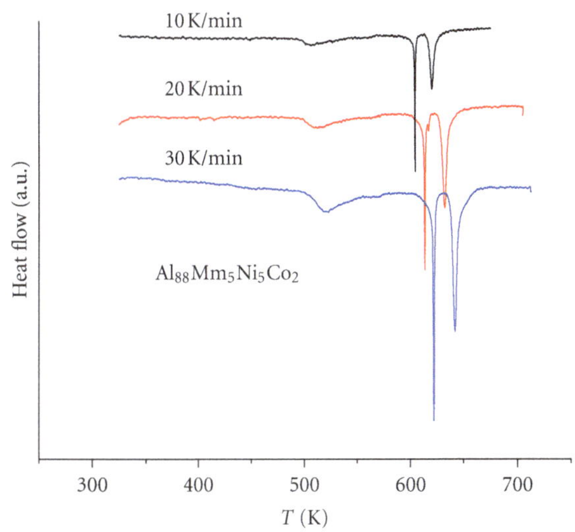

FIGURE 4: DSC scans at different heating rate for as-cast samples of the $Al_{88}Mm_5Ni_5Co_2$ alloys.

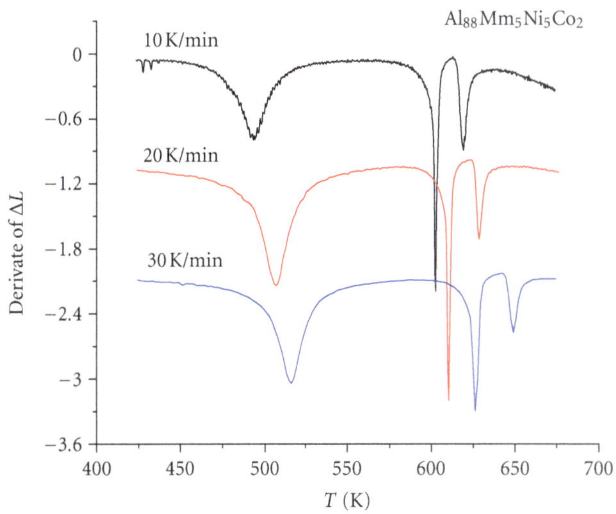

FIGURE 6: Dilatometer scans at different heating rate for as-cast samples of the $Al_{88}Mm_5Ni_5Co_2$ alloys.

of glass forming ability. The devitrification proceeds on by successive decomposition of the residual amorphous phase into compound phases at higher temperatures. For lower Al content (85 at % Al in Y8 sample) the amorphous state is more stable showing even glass transition into supercooled liquid state which stability is usually measured by the difference between the first crystallization peak and the Tg (about 30 K for Y8 sample). This supercooled liquid nucleates as a whole at the onset of crystallization into α-Al and metastable compound phases [9], that is, the majority of the changes in the topological short range order take place. As a result a huge contraction peak appears for the Y8 sample and the subsequent small contractions belong to the small volume changes accompanying the changes in CSRO during the diffusion rearrangement of the metastable crystalline phases into stable stoichiometric compound phases. The change of the chemical short range order with increasing temperature produces similar heat effect for the Y8 sample as the decomposition of the residual amorphous phase for the Mm5 sample. The coincidence of the number of the main crystallization peaks is just fortuitous, the devitrification process is completely different for the two compositions. For the Mm5 sample the transformation steps are dominated by the subsequent decomposition of the residual amorphous matrix, whereas for the Y8 sample the transformation steps are determined by the evolution of the metastable compound phase nucleated in the first step. The three crystallization steps are as follows. For the Mm5 sample: Am → Am′ + supersaturated nano-α-Al → A″ + α-Al + Al_3Ni + Al_3Co → α-Al + Al_3Ni + Al_3Co + $Al_{11}Mm_3$. For the Y8 sample: Am → supersaturated nano-α-Al + metastable Xcryst → α-Al + X′ +Al_3Ni +Al_3Co → α-Al +Al_3Ni +Al_3Co+ Al_3Y. The three contraction peaks on the dilatometer scan of

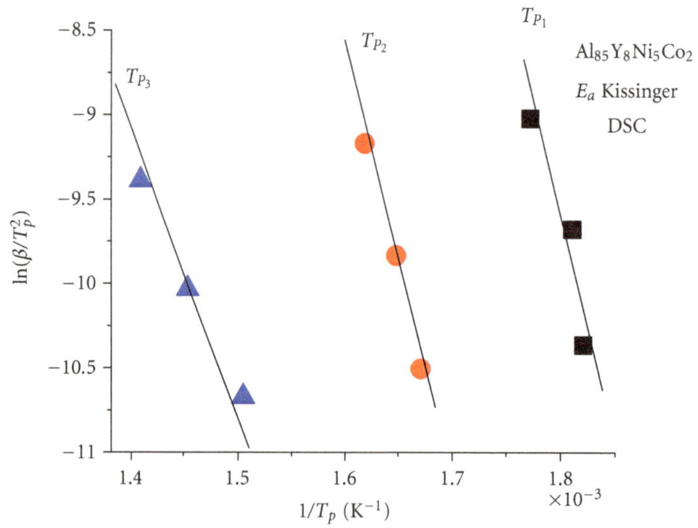

FIGURE 7: Kissinger plots for the glass transition and crystallization peaks for the Y8 alloy.

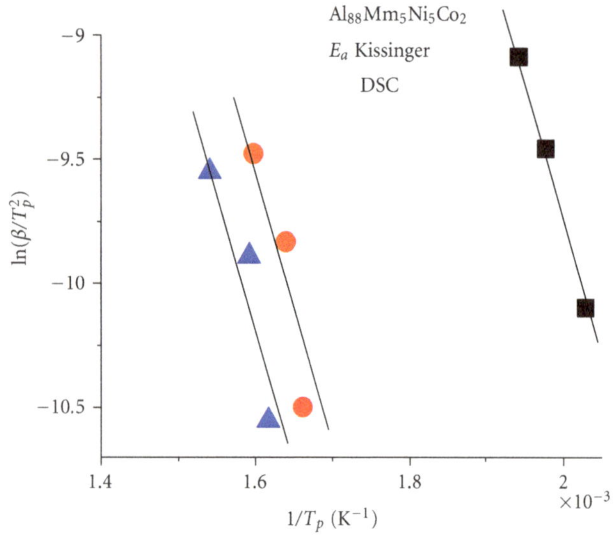

FIGURE 8: Kissinger plots for the crystallization peaks for the Mm5 alloy.

the Mm5 sample (Figure 6) mark the TSRO changes in the initial amorphous (Am) and subsequent residual amorphous (Am′ and Am″) phases. The accompanying CSRO changes produce the three DSC exothermic peaks (Figure 4). The apparently single dilatometer peak for the Y8 marks the decomposition of the majority of the initial amorphous phase in the first crystallization step, and the subsequent diffusion rearrangement of the metastable crystalline phase produces hardly observable volume changes. These decompositions of the metastable phase give rise to the second and third exothermic peaks in the DSC scan of the Y8 sample (Figure 3). The parameters of the amorphous-crystalline transformation are collected in the Tables 1 and 2 for the Mm5 and Y8 samples, respectively.

TABLE 1: Crystallization parameters for $Al_{85}Y_8Ni_5Co_2$ alloy obtained by DSC and dilatometer.

$Al_{85}Y_8Ni_5Co_2$							
	Dilatometer			DSC			
β (K/min)	T_{p1} (K)	T_{p2} (K)	T_{p3} (K)	T_g (K)	T_{p1} (K)	T_{p2} (K)	T_{p3} (K)
10	551	591	640	532	560	603	654
20	568	607	662	534	568	611	674
30	574.5	—	—	—	—	—	—
40				545	579	623	695
E_a (eV)	2.3			2.25	2.64	2.7	1.43

TABLE 2: Crystallization parameters for $Al_{88}Mm_5Ni_5Co_2$ alloys obtained by DSC and dilatometer.

$Al_{88}Mm_5Ni_5Co_2$						
	Dilatometer			DSC		
β (K/min)	T_{p1} (K)	T_{p2} (K)	T_{p3} (K)	T_{p1} (K)	T_{p2} (K)	T_{p3} (K)
10	493	602	618.5	506	604	621
20	506	610	628	512	612	631
30	515	626	649	520	622	641
E_a (eV)	1.43	2.01	1.51	1.38	2.03	1.79

The activation energies (E_a) of the 3 crystallization steps have been determined from the Kissinger plots based on the shift of the peaks as a function of the heating rate, β:

$$\ln\left(\frac{\beta}{T_p^2}\right) = -\frac{E_a}{RT_p} + C, \ln\left(\frac{\beta}{T_p^2}\right) = -\frac{E_a}{RT_p} + C, \quad (1)$$

where R is the universal gas constant.

These activation energies are collected also in the Tables 1 and 2.

In the Figures 7 and 8 the collected Kissinger plots for all the peaks are shown for the samples Y8 and Mm5, respectively. From the scatter of the data the error of the activation energy is estimated to be ±0.2 eV.

Perusal of the Tables 1 and 2 shows that the DSC and the dilatometer methods give similar values for the activation energies within the error of the E_a determinations. The values of the crystallization peaks are in general lower for the dilatometer method compared to the DSC data. The dilatometer data for the peak temperatures of the second and third crystallization steps deviates more and more with increasing heating rate from the reference DSC data, but these deviations remain below 10 degree for the Mm5 sample. The glass transition appeared only for the Y8 sample (see Table 1), which shows no primary crystallization. The determination of the glass transition temperature, T_g, is somehow ambiguous. In this work, we have adopted the intersection of the steepest slope with the baseline, which might be the glass transition onset [10]. However, other authors [9] define T_g as the inflexion point of the curve denoting the endothermic effect. The latter definition gives higher T_g values, and, consequently, the super cooled liquid region (defined as $\Delta T = T_x - T_g$) decreases (in our case by about 10 K). The T_g is also shifting to higher values as a function of heating rate and from these shifted T_g data an activation energy can be determined for the glass transition as well, similar to the crystallization data. The corresponding Kissinger plot for the Y8 sample is presented in Figure 7, and the activation energy determined from the slope of the fitted line is a little bit smaller than the activation energy for the first step of the crystallization.

The thermal stability of the amorphous state measured by the first crystallization temperature is higher for the Y8 sample than for the Mm5 sample. As a consequence, the hot pressure consolidation can be performed at higher temperatures for the Y8 sample.

Vickers microhardness data were determined obtaining $HV_{0.01} = 337\ kg/mm^2$ for the Mm5 and $309\ kg/mm^2$ for Y8 as cast ribbon samples. Using the $HV = 3\sigma$ relationship, from these hardness data very similar tensile strength values can be estimated in the as cast state: $\sigma = 1120$ and 1030 MPa for the samples Mm5 and Y8, respectively.

4. Conclusions

Both the DSC and dilatometer methods are suitable to study the crystallization processes of the amorphous ribbons and for the determination of the activation energy. Based on the different sensitivity of these two methods to the changes in the TSRO and CSRO, we have established that the crystallization steps are evaluated through the decomposition of the residual amorphous phase and of the metastable crystalline compound phase for the Mm5 and Y8 samples, respectively.

The higher thermal stability and the observed wide glass transition region ($\Delta T \sim 30$ K) for the Y8 composition facilitate the future hot pressure consolidation experiments in order to prepare bulk amorphous sample.

Acknowledgments

This work was supported by Hungarian Grant OTKA 73451 and by Romanian Academy of Sciences.

References

[1] Y. He, S. J. Poon, and G. J. Shiflet, "Synthesis and properties of metallic glasses that contain aluminum," *Science*, vol. 241, no. 4873, pp. 1640–1642, 1988.

[2] A. Inoue, K. Ohtera, A. P. Tsai, and T. Masumoto, "New Amorphous Mg-Ce-Ni Alloys with High Strength and Good Ductility," Japanese Journal of Applied Physics," *JapaneseJournal of Applied Physics*, vol. 27, p. L289, 1988.

[3] C. A. Angell, "Formation of glasses from liquids and biopolymers," *Science*, vol. 267, no. 5206, pp. 1924–1935, 1995.

[4] A. Inoue, "Amorphous, nanoquasicrystalline and nanocrystalline alloys in Al-based systems," *Progress in Materials Science*, vol. 43, no. 5, pp. 365–520, 1998.

[5] D. V. Louzguine, A. R. Yavari, and A. Inoue, "Mischmetal as an alloying addition to amorphous materials and glass formers," *Journal of Non-Crystalline Solids*, vol. 316, no. 2-3, pp. 255–260, 2003.

[6] I. T. H. Chang and R. R. Botten, "Heat treatment of rapidly solidified Al-Ni-Mm alloys," *Materials Science and Engineering A*, vol. 226–228, pp. 183–186, 1997.

[7] H. E. Kissinger, "Reaction kinetics in differential thermal analysis," *Analytical Chemistry*, vol. 29, no. 11, pp. 1702–1706, 1957.

[8] "Vacuum and Technologies for Metallurgy," http://web.ald-vt.de/cms/fileadmin/pdf/prospekte/ALD_Bibel_full.pdf.

[9] D. V. Louzguine and A. Inoue, "Influence of a supercooled liquid on crystallization behaviour of Al-Y-Ni-Co metallic glass," *Materials Letters*, vol. 54, no. 1, pp. 75–80, 2002.

[10] J. S. Blázquez, E. Fazakas, H. Dimitrov, J. Latuch, L. K. Varga, and T. Kulik, "Effect of substitution of rare earth by mischmetal on the devitrification process of Al-X-Ni-Co (X = Y, Ce, Mm) alloys," *Journal of Non-Crystalline Solids*, vol. 351, no. 2, pp. 158–166, 2005.

A Model for Conjoint Shape Memory and Pseudo-Elastic Effects during Martensitic Transformation

K. Boubaker

Unité de Physique des Dispositifs à Semi-Conducteurs, Faculté des Sciences de Tunis, Tunis El Manar University, 2092 Tunis, Tunisia

Correspondence should be addressed to K. Boubaker, mmbb11112000@yahoo.fr

Academic Editors: Y.-D. Wang and Y. Xu

Shape memory alloys (SMA) are metals which can restore their initial shape after having been subjected to a deformation. They exhibit in general both nonlinear shape memory and pseudoelastic effects. In this paper, shape memory alloy (SMA) and its constitutive model with an empirical kinetics equation are investigated. A new formulation to the martensite fraction-dependent Young modulus has been adopted and the plastic deformation was taken into account. To simulate the variations, a one-dimensional constitutive model was constructed based on the uniaxial tension features.

1. Introduction

Recently, smart metals and alloys have been extensively used in several metallurgical applications, due to their great potential in updated structures and design [1–10]. Among these materials, shape memory alloys (SMA) have attracted more attention, due to their ability to develop extremely large recoverable strains and great forces in the field of biomedical, metallurgy, aerospace, and civil structures [5–10]. In SMA matrices, pseudoplastic effect creates different stress strain behavior resulting in a stress strain curve which lies on the curve produced by the initial linear elastic response during loading. Consecutive and continued unloading may produce linear elastic behavior that eventually returns the structure to the zero stress strain state.

In the present work, an attempt is made to model typical martensitic transformations occurring in shape memory alloys, taking into account pseudoplasticity patterns. In this martensitic transformation, austenite undergoes transformation to form different variants of martensite under a controlled mechanical loading. The formation of martensite in the material is monitored through the coexistence of the initial austenite phases and martensite inside periodic units. Solutions for the implemented governing equations are obtained numerically via explicit numerical protocols and compared to some records presented in the recent related literature [11–18].

2. Model Patterns

2.1. Governing Equations. The studied system is a mono-dimensional rod subjected to axial solicitation (Figure 1). The phase transformations in this considered structure occur by nucleation and growth of platelet inclusions perpendicular to x-axis (Figure 1). In this configuration, elastic modulus local expression can be obtained considering the medium as a succession of austenite-martensite periodic units (Figure 1).

The main assumptions of the present model consists of setting one scalar internal variable η which represents the martensite fraction, along with linear kinetic rules in terms of the uniaxial stress T.

Young modulus E relative to the periodic unit can be computed via the elongation (x) of the periodic cell and the local strain ε:

$$x = \varepsilon w = x_A + x_M = \varepsilon_A w_A + \varepsilon_M w_M,$$
$$\varepsilon = \varepsilon_A \frac{w_A}{w} + \varepsilon_M \frac{w_M}{w}, \tag{1}$$

where w, w_A, and w_M represent the total length of the periodic cell, the length of the austenite, and martensite fractions, respectively, (Figure 1). Subscripts A and M indicate austenite and martensite.

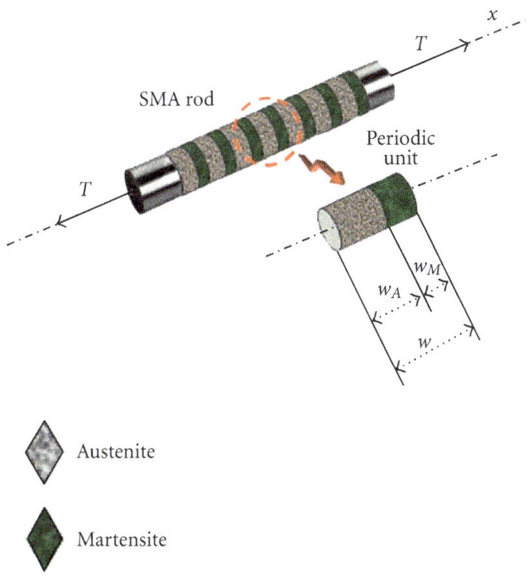

Figure 1: The studied model.

Since the martensite fraction η is defined as

$$\eta = \frac{w_M}{w}, \tag{2}$$

it gives

$$\varepsilon = \varepsilon_A (1 - \eta) + \varepsilon_M \eta. \tag{3}$$

The η-dependent boundary condition concerning Young modulus $E(\eta)$ are:

$$\begin{aligned} E(\eta)\big|_{\eta=0} &= E_A, \\ E(\eta)\big|_{\eta=1} &= E_M. \end{aligned} \tag{4}$$

The expression of $E(\eta)$ has been discussed by several authors. Tanaka-Mori [19, 20] scheme proposed the expression:

$$\begin{aligned} E(\eta) &= \frac{E_A E_M}{2}\left[\frac{(1-\eta)+\eta\Pi^{MA}}{E_M(1-\eta)+E_A\eta\Pi^{MA}} + \frac{(1-\eta)\Pi^{AM}+\eta}{E_M(1-\eta)\Pi^{AM}+E_A\eta}\right], \\ \Pi^{MA} &= \frac{2E_M}{E_A+E_M}, \\ \Pi^{AM} &= \frac{2E_A}{E_A+E_M}, \end{aligned} \tag{5}$$

whereas Voigt [21–25] proposed a simple linear model:

$$E(\eta) = E_A(1-\eta) + E_M(\eta). \tag{6}$$

Both expressions verify the imposed condition for Young modulus, but present the disadvantage of lack of control on the first derivatives of $E(\eta)$, which are either sophisticated (Tanaka-Mori scheme [19, 20]) or constant (Voigt [21–25]).

As per the model of derived by Tanaka and Nagaki [26], the main equation deduced from the first and second laws of thermodynamics can be formulated as

$$T(\varepsilon,\eta) = \rho\frac{\partial\varphi(\varepsilon,\eta)}{\partial\varepsilon}, \tag{7}$$

with

T: Second Piola-Kirchhoff stress,

ρ: Density,

φ: Helmholtz free energy,

ε: Green strain,

η: Martensite fraction.

In the case of pure mechanical constraint, and taking into accounts the presumptions of Chen et al. [27], the main equation becomes

$$\begin{aligned} T(\varepsilon,\eta) - T_0 &= \rho\frac{\partial^2\varphi}{\partial\varepsilon^2}(\varepsilon-\varepsilon_o) + \rho\frac{\partial^2\varphi}{\partial\varepsilon\partial\eta}(\eta-\eta_0), \\ E(\eta) &= \rho\frac{\partial^2\varphi}{\partial\varepsilon^2}, \end{aligned} \tag{8}$$

where $E(\eta)$ is Young modulus and $T_0\varepsilon_o$ and η_0 refer to initial condition for stress, strain and martensite fraction, respectively.

In the actual model, the transformation tensor $\rho(\partial^2\varphi/\partial\varepsilon\partial\eta)$ is considered as constant, while the Young modulus $E(\eta)$ is expressed as

$$E(\eta) = E_M + \frac{E_M - E_A}{N_0}\sum_{k=1}^{N_0}\xi_k \times B_{4k}(\eta \times r_k), \tag{9}$$

where B_{4k} are the $4k$-order Boubaker polynomials [28–44], r_k are B_{4k} minimal positive roots, N_0 is a prefixed integer, and $\xi_k|_{k=1,\dots,N_0}$ are unknown pondering real coefficients.

This expression refers to the Boubaker polynomials expansion scheme (BPES) [29–31]. This scheme is a resolution protocol which has been successfully applied to several applied-physics and mathematics problems. The BPES protocol ensures the validity of the related boundary conditions regardless of main equation features. The BPES is mainly based on Boubaker polynomials first derivatives properties:

$$\begin{aligned} \sum_{q=1}^{N} B_{4q}(x)\Bigg|_{x=0} &= -2N \neq 0, \\ \sum_{q=1}^{N} B_{4q}(x)\Bigg|_{x=r_q} &= 0, \end{aligned}$$

$$\sum_{q=1}^{N} \left.\frac{dB_{4q}(x)}{dx}\right|_{x=0} = 0,$$

$$\sum_{q=1}^{N} \left.\frac{dB_{4q}(x)}{dx}\right|_{x=r_q} = \sum_{q=1}^{N} H_q,$$

$$\text{with } H_n = B'_{4n}(r_n) = \left(\frac{4r_n\left[2 - r_n^2\right] \times \sum_{q=1}^{n} B_{4q}^2(r_n)}{B_{4(n+1)}(r_n)} + 4r_n^3\right). \tag{10}$$

Several solutions have been proposed through the BPES in many fields such as numerical analysis [28–31], theoretical physics [31–34], mathematical algorithms [35], heat transfer [36], homodynamic [37, 38], material characterization [39], fuzzy systems modeling [40–42], and biology [43, 44].

The main advantage of this formulation (9) is the verification of boundary conditions, expressed in (4), in advance to resolution process, along with affording controllable and easy-to-access first derivatives of $E(\eta)$. In fact, thanks to the properties expressed in (10), these conditions are inherently verified.

$$E(\eta)\big|_{\eta=0} = E_M + \frac{E_M - E_A}{N_0} \sum_{k=1}^{N_0} \xi_k \times (-2) = E_A,$$
$$E(\eta)\big|_{\eta=1} = E_M + \frac{E_M - E_A}{N_0} \sum_{k=1}^{N_0} \xi_k \times (0) = E_M. \tag{11}$$

Plots of the η-dependent Young modulus $E(\eta)$ are gathered in Figure 2, for both actual and referred models [19–25].

2.2. Numerical Results and Discussion. Numerical simulation was achieved for the values of parameters gathered in Table 1. The system was taken as insulated thermally and pinned-end mechanically. A variable mechanical load T, expressed as follows, has been considered [45]

$$T = \begin{cases} 3500t, & t \in \left[0, \frac{\theta}{6}\right[, \\ 3500(4 - t), & t \in \left[\frac{\theta}{6}, \frac{\theta}{3}\right[, \\ 0, & t \in \left[\frac{\theta}{3}, \frac{\theta}{2}\right[, \\ 3500(6 - t), & t \in \left[\frac{\theta}{2}, \frac{2\theta}{3}\right[, \\ 3500(t - 10), & t \in \left[\frac{2\theta}{3}, \frac{5\theta}{6}\right[, \\ 0, & t \in \left[\frac{5\theta}{6}, \theta\right[. \\ \theta = 12. \end{cases} \tag{12}$$

Consequently, the dynamics were simulated using (7)-(8) with the given load (12) and the BPES-related expression (9). Finite element simulations in simple uniaxial solicitation

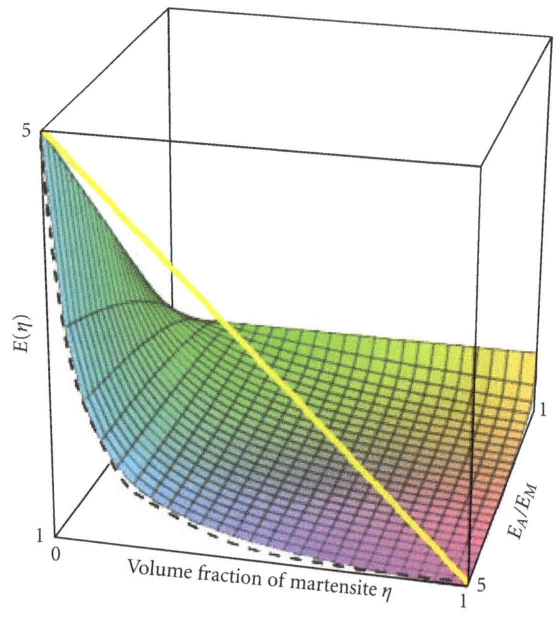

FIGURE 2: Young modulus $E(\eta)$ versus martensite fraction for different values of E_A/E_M.

TABLE 1: Values of d_{hkl} of relevant parameters.

Parameters	Value	Unit
E_A	6.0 10^5	MPa
E_M	7.5 10^3	MPa
Rod diameter	1.49	mm

mode were carried out with the preset load using load gradient and the preset η-dependent Young modulus expression as input. The resulting stress-strain response was plotted in Figure 3, along with some records from the related literature [46, 47].

3. Discussion and Perspectives

Numerical stress-strain plots obtained from the actual model confirm that the patterns of hysteresis loop generated in the positive quadrant (Figure 3) are in good agreement with the profiles presented by Auricchio [45], Motahari and Ghassemieh [46], and Sayyadi et al. [47].

Strain span shows also a good agreement with the model performed by Motahari and Ghassemieh [46]. The unique divergence lies in the upper zone ($\varepsilon > 5\%$) and may be attributed to the linear approximation of the η-dependent Young modulus $E(\eta)$ (Voigt-type) instead of the polynomial form of evolution equations assumed in our study. In this context, it can been stated that Tanaka-Mori [19, 20] and Voigt [21–25] models, cannot be efficiently used for predicting the shape memory effect behavior of SMA. This is due to the fact that in the constitutive equations used in those

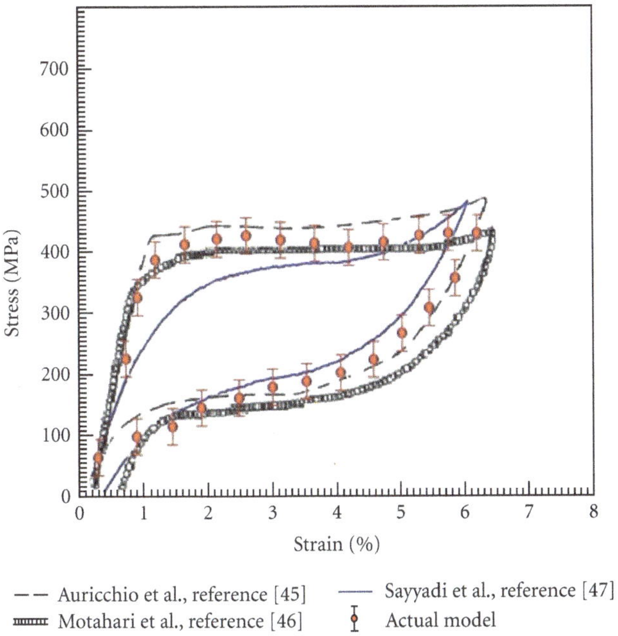

FIGURE 3: Solution plots represented in the Strain-Stress plane.

models, the transformational component is proportional to both martensite volume fraction and its derivatives. Indeed, these models, oppositely to the actual one, introduce a prefixed and uncontrollable derivative-dependence.

4. Conclusion

In summary, we have implemented constitutive model for shape-memory alloys capable of undergoing austenite to martensite phase transformation using fundamental thermodynamic laws and the principle of martensite fraction-dependence. The stress-strain plots obtained from uniaxial monitored load were predicted using the finite-element simulations. A key parameter of the performed model consists of avoiding avoids singularity of the main equilibrium equation during the transition since the derivative of the preset η-dependent Young modulus expression are controllable. η-dependent boundary conditions were also inherently taken into account in the model.

Although applied to a particular geometry, the model should be suitable to study other configurations since it was based on a single scalar internal variable: the martensite fraction. This model may be extended to 2D and 3D, while other possible future developments are the inclusion of permanent inelastic effects, the prediction of coupled thermomechanical behavior, and the nonlinear hardening mechanisms.

References

[1] G. B. Olson and M. Cohen, "A mechanism for the strain-induced nucleation of martensitic transformations," *Journal of The Less-Common Metals*, vol. 28, no. 1, pp. 107–118, 1972.

[2] T. Angel, "Formation of martensite in austenitic stainless steels: effects of deformation, temperature and composition," *Journal of The Iron and Steel Institute*, vol. 1977, pp. 165–174, 1954.

[3] G. W. Greenwood and R. H. Johnson, "The deformation of metals under stresses during phase transformation," *Proceedings of the Royal Society A*, vol. 283, no. 1394, pp. 403–422, 1965.

[4] I. Tamura, "Deformation-induced martensitic transformation and transformation-induced plasticity in steels," *Journal of Materials Science*, vol. 16, no. 5, pp. 245–253, 1982.

[5] J. R. Patel and M. Cohen, "Criterion for the action of applied stress in the martensitic transformation," *Acta Metallurgica*, vol. 1, no. 5, pp. 531–538, 1953.

[6] R. G. Stringfellow, D. M. Parks, and G. B. Olson, "A constitutive model for transformation plasticity accompanying strain-induced martensitic transformations in metastable austenitic steels," *Acta Metallurgica et Materialia*, vol. 40, no. 7, pp. 1703–1716, 1992.

[7] Y. Tomita and T. Iwamato, "Constitutive modeling of trip steel and its application to the improvement of mechanical properties," *International Journal of Mechanical Sciences*, vol. 37, no. 12, pp. 1295–1305, 1995.

[8] J. G. Cook and F. P. Koffyberg, "Solar thermal absorbers employing oxides of Ni and Co," *Solar Energy Materials*, vol. 10, no. 1, pp. 55–67, 1984.

[9] I. Doghri and A. Ouaar, "Homogenization of two-phase elasto-plastic composite materials and structures study of tangent operators, cyclic plasticity and numerical algorithms," *International Journal of Solids and Structures*, vol. 40, no. 7, pp. 1681–1712, 2003.

[10] I. Doghri and C. Friebel, "Effective elasto-plastic properties of inclusion-reinforced composites. Study of shape, orientation and cyclic response," *Mechanics of Materials*, vol. 37, no. 1, pp. 45–68, 2005.

[11] K. Tanaka and T. Mori, "The hardening of crystals by non-deforming particles and fibres," *Acta Metallurgica*, vol. 18, no. 8, pp. 931–941, 1970.

[12] J. C. Simo and T. J. R. Hughes, *Computational Inelasticity*, Springer, New York, NY, USA, 1998.

[13] S. Miyazaki, T. Imai, Y. Igo, and K. Otsuka, "Effect of cyclic deformation on the pseudoelasticity characteristics of Ti-Ni alloys," *Metallurgical and Materials Transactions A*, vol. 17, no. 1, pp. 115–125, 1986.

[14] J. D. Eisenwasser and L. C. Brown, "Pseudoelasticity and the strain-memory effect in Cu-Zn-Sn alloys," *Metallurgical and Materials Transactions B*, vol. 3, no. 6, pp. 1359–1363, 1972.

[15] S. Miyazaki, S. Kimura, F. Takei, T. Miura, K. Otsuka, and Y. Suzuki, "Shape memory effect and pseudoelasticity in a TiNi single crystal," *Scripta Metallurgica*, vol. 17, no. 9, pp. 1057–1062, 1983.

[16] J. Ortin and A. Planes, "Thermodynamic analysis of thermal measurements in thermoelastic martensitic transformations," *Acta Metallurgica*, vol. 36, no. 8, pp. 1873–1889, 1988.

[17] G. B. Olson and M. Cohen, "Thermoelastic behavior in martensitic transformations," *Scripta Metallurgica*, vol. 9, no. 11, pp. 1247–1254, 1975.

[18] H. C. Tong and C. M. Wayman, "Thermodynamic considerations of "solid state engines" based on thermoelastic martensitic transformations and the shape memory effect," *Metallurgical and Materials Transactions*, vol. 6, no. 1, pp. 29–32, 1975.

[19] K. Tanaka and T. Mori, "The hardening of crystals by non-deforming particles and fibres," *Acta Metallurgica*, vol. 18, no. 8, pp. 931–941, 1970.

[20] K. Tanaka, "Thermomechanical sketch of shape memory effect: one-dimensional tensile behavior," *Research Mechanicanical*, vol. 18, no. 3, pp. 251–263, 1986.

[21] J. G. Boyd and D. C. Lagoudas, "Thermomechanical response of shape memory composites," *Journal of Intelligent Material Systems and Structures*, vol. 5, no. 3, pp. 333–346, 1994.
[22] J. G. Boyd and D. Lagoudas, "A thermodynamical constitutive model for shape memory materials. Part I. The monolithic shape memory alloy," *International Journal of Plasticity*, vol. 12, no. 6, pp. 805–841, 1996,.
[23] J. G. Boyd and D. C. Lagoudas, "A thermodynamical constitutive model for shape memory materials. Part II. The SMA composite material," *International Journal of Plasticity*, vol. 12, no. 7, pp. 843–873, 1996.
[24] L. C. Brinson and R. Lammering, "*Finite element analysis of the* behavior of shape memory alloys and their applications," *International Journal of Solids and Structures*, vol. 4, no. 2, pp. 229–242, 1993.
[25] L. C. Brinson, "One dimensional constitutive behavior of shape memory alloys: thermo-mechanical derivation with non-constant material functions," *Journal of Intelligent Material Systems & Structures*, vol. 4, no. 2, pp. 229–242, 1993.
[26] K. Tanaka and S. Nagaki, "A thermomechanical description of materials with internal variables in the process of phase transitions," *Ingenieur-Archiv*, vol. 51, no. 5, pp. 287–299, 1982.
[27] L. Chen, *Constitutive modeling of shape memory alloys [Ph.D. thesis]*, VPI, Blacksburg, Va, USA, 1990.
[28] A. Milgram, "The stability of the Boubaker polynomials expansion scheme (BPES)-based solution to Lotka-Volterra problem," *Journal of Theoretical Biology*, vol. 271, no. 1, pp. 157–158, 2011.
[29] J. Ghanouchi, H. Labiadh, and K. Boubaker, "An attempt to solve the heat transfer equation in a model of pyrolysis spray using 4q-order m-boubaker polynomials," *International Journal of Heat and Technology*, vol. 26, no. 1, pp. 49–53, 2008.
[30] S. Lazzez, K. B. Ben Mahmoud, S. Abroug, F. Saadallah, and M. Amlouk, "A Boubaker polynomials expansion scheme (BPES)-related protocol for measuring sprayed thin films thermal characteristics," *Current Applied Physics*, vol. 9, no. 5, pp. 1129–1133, 2009.
[31] T. Ghrib, K. Boubaker, and M. Bouhafs, "Investigation of thermal diffusivitymicrohardness correlation extended to surface-nitrured steel using Boubaker polynomials expansion," *Modern Physics Letters B*, vol. 22, no. 29, pp. 2893–2907, 2008.
[32] S. Fridjine, K. B. Ben Mahmoud, M. Amlouk, and M. Bouhafs, "A study of sulfur/selenium substitution effects on physical and mechanical properties of vacuum-grown ZnS1-xSex compounds using Boubaker polynomials expansion scheme (BPES)," *Journal of Alloys and Compounds*, vol. 479, no. 1-2, pp. 457–461, 2009.
[33] C. Khélia, K. Boubaker, T. Ben Nasrallah, M. Amlouk, and S. Belgacem, "Morphological and thermal properties of β-SnS2 sprayed thin films using Boubaker polynomials expansion," *Journal of Alloys and Compounds*, vol. 477, no. 1-2, pp. 461–467, 2009.
[34] M. Dada, O. B. Awojoyogbe, and K. Boubaker, "Heat transfer spray model: an improved theoretical thermal time-response to uniform layers deposit using Bessel and Boubaker polynomials," *Current Applied Physics*, vol. 9, no. 3, pp. 622–624, 2009.
[35] S. Tabatabaei, T. Zhao, O. B. Awojoyogbe, and F. O. Moses, "Cut-off cooling velocity profiling inside a keyhole model using the Boubaker polynomials expansion scheme," *International Journal of Heat and Mass Transfer*, vol. 45, no. 10, pp. 1247–1251, 2009.
[36] A. Belhadj, J. Bessrour, M. Bouhafs, and L. Barrallier, "Experimental and theoretical cooling velocity profile inside laser welded metals using keyhole approximation and Boubaker polynomials expansion," *Journal of Thermal Analysis and Calorimetry*, vol. 97, no. 3, pp. 911–915, 2009.
[37] A. Belhadj, O. F. Onyango, and N. Rozibaeva, "Boubaker polynomials expansion scheme-related heat transfer investigation inside keyhole model," *Journal of Thermophysics and Heat Transfer*, vol. 23, no. 3, pp. 639–640, 2009.
[38] P. Barry and A. Hennessy, "Meixner-type results for Riordan arrays and associated integer sequences," *Journal of Integer Sequences*, vol. 13, no. 9, pp. 1–34, 2010.
[39] M. Agida and A. S. Kumar, "A Boubaker polynomials expansion scheme solution to random Love's equation in the case of a rational Kernel," *Electronic Journal of Theoretical Physics*, vol. 7, no. 24, pp. 319–326, 2010.
[40] A. Yildirim, S. T. Mohyud-Din, and D. H. Zhang, "Analytical solutions to the pulsed Klein-Gordon equation using Modified Variational Iteration Method (MVIM) and Boubaker Polynomials Expansion Scheme (BPES)," *Computers and Mathematics with Applications*, vol. 59, no. 8, pp. 2473–2477, 2010.
[41] A. S. Kumar, "An analytical solution to applied mathematics-related Love's equation using the Boubaker polynomials expansion scheme," *Journal of the Franklin Institute*, vol. 347, no. 9, pp. 1755–1761, 2010.
[42] M. Benhaliliba, C. E. Benouis, K. Boubaker, M. Amlouk, and A. Amlouk, "The Amlouk-boubaker optothermal expansivity ψab in," in *The Book: Solar Cells-New Aspects and Solutions*, L. A. Kosyachenko, Ed., pp. 27–41, 2011.
[43] H. Rahmanov, "A solution to the non linear korteweg-de-vries equation in the particular case dispersion-adsorption problem in porous media using the spectral Boubaker polynomials expansion scheme (BPES)," *Studies in Nonlinear Sciences*, vol. 2, no. 1, pp. 46–49, 2011.
[44] C. Khélia, K. Boubaker, and M. Amlouk, "Boubaker polynomials expansion scheme (BPES)-related optical properties of β-sns2 sprayed layers," *Fizika*, vol. 18, no. 2, pp. 81–88, 2009.
[45] F. Auricchio, *Shape-memory alloys: applications, micromechanics, macro-modelling and numerical simulations [Ph.D. thesis]*, Department of Civil Engineering, University of California at Berkeley, Berkeley, Calif, USA, 1995.
[46] S. A. Motahari and M. Ghassemieh, "Multilinear one-dimensional shape memory material model for use in structural engineering applications," *Engineering Structures*, vol. 29, no. 6, pp. 904–913, 2007.
[47] H. Sayyadi, M. R. Zakerzadeh, and H. Salehi, "A comparative analysis of some one-dimensional shape memory alloy constitutive models based on experimental tests," *Scientia Iranica B*, vol. 19, no. 2, pp. 249–257, 2012.

22

Overview of Intermetallic Sigma (σ) Phase Precipitation in Stainless Steels

Chih-Chun Hsieh and Weite Wu

Department of Materials Science and Engineering, National Chung Hsing University, 250 Kuo-Kuang Road, Taichung 402, Taiwan

Correspondence should be addressed to Weite Wu, wwu@dragon.nchu.edu.tw

Academic Editors: M. Gjoka and Y. Yamabe-Mitarai

The σ phase which exists in various series of stainless steels is a significant subject in steels science and engineering. The precipitation of the σ phase is also a widely discussed aspect of the science and technology of stainless steels. The microstructural variation, precipitation mechanism, prediction method, and effects of properties of σ phase are also of importance in academic discussions. In the first section, a brief introduction to the development and the precipitation characteristics (including morphologies and precipitation sites) of σ phase in stainless steels is presented. In the second section, the properties effect, prediction method, processing effect, elemental addition, retardation method and Thermo-Calc simulation of the σ phase in stainless steels are highlighted.

1. Motivation

The precipitation of the σ phase, which is often observed in various series of stainless steels, is one of the main reasons for the deterioration of stainless steels' properties, for example, mechanical property, corrosion resistance, and weldability. The σ phase can be precipitated under an elevated temperature environment, for example, casting, rolling, welding, forging, and aging. It is difficult to prevent the precipitation of the σ phase when the Cr content is above a certain level (above 20 wt.%) in stainless steels. The addition of a strong ferrite stabilizer into the stainless steels (Cr, Si, or Mo) rapidly leads to the formation of the σ phase. This means that the transformation from δ-ferrite to the σ phase can be accelerated when the Cr, Si, or Mo diffuse efficiently in δ-ferrite. Hence, understanding the mechanism of the $\delta \rightarrow \sigma$ phase transformation is crucial to the prediction and control of stainless steels' properties. This review paper presents a comprehensive overview of the precipitation of the σ phase in stainless steels, and it is hoped that it will result in some scientific contributions in the materials science and engineering of stainless steels.

2. Review of Sigma Phase (σ)

2.1. Development Progress of the σ Phase.

(1) In 1907, Treitschke and Tammann found that the σ phase in the Fe-Cr binary system was an intermetallic compound of 30 wt.% Cr~50 wt.% Cr [1].

(2) In 1927, Bain and Griffiths observed the successful σ phase in the Fe-Cr-Ni ternary system. They found that the σ phase was a very hard and brittle compound which impacted the toughness of the steels. At that time, the σ phase was called the "B constituent" [2].

(3) In 1936, this Fe-Cr compound was called the "σ phase" by Jett and Foote, which became its official name [3].

(4) In 1951, the crystal structure of the σ phase in the Fe-Cr binary system was examined by Yano and Abiko They pointed out that the σ phase exhibited slower precipitation kinetics in the Fe-Cr alloy system than in the Fe-Cr-Mo and Fe-Cr-Si ternary systems [4].

Table 1: Chemical composition and lattice constant of the σ phase [10].

Alloy	Lattice parameter (Å)	Composition of phase (wt%) Fe	Cr	Ni	Mo	Si	Formula
Fe-Cr	$a_0 = 8.799$, $c_0 = 4.544$						Fe-Cr
Fe-Mo	$a_0 = 9.188$, $c_0 = 4.812$						Fe-Mo
17Cr-11Ni-2Mo-0.4Ti	—		30	4.3	9	0.8	
17Cr-11Ni-0.9Mo-0.5Ti	—		33	4.5	5.4	0.7	
Type 316	$a_0 = 8.28\sim8.38$, $c_0 = 4.597\sim4.599$	55	29	5	11	—	$(FeNi)_x(CrMo)_y$
Type 316L	$a_0 = 9.21$, $c_0 = 4.78$						
20Cr-25-34Ni-6.5-8Mo	$a_0 = 8.87$, $c_0 = 4.61$	35/37	17/26	15/21	21/28	—	
25Cr-20Ni	—	40	46	9.4	—	3	

(5) In 1966, the σ phase was observed by Hattersley and Hume-Rothery [5] and Hall and Algie [6] in austenitic stainless steels.

(6) By 1966, the σ phase had been found in over 50 transition alloys [6], including Cr-Ni, Fe-Cr-Ni, Fe-Cr-Mo, Fe-Cr-Mn, Fe-Cr-Ni-Mo, Fe-Cr-Si, Fe-V, Re-Cr, Mo-Re, Ta-Al, W-Te, Ta-V, Zr-Ir, Nb-Pd, Ti-Mn, and Nb-Fe.

2.2. Precipitation Characteristic of the σ Phase. The σ phase is a tetragonal crystal structure, and its precipitation temperature is between 600°C and 1000°C [7–9]. The crystal information and chemical composition of the σ phase are listed in Table 1 [10]. The σ phase can increase the hardness and decrease the toughness, as well as the elongation, of steels [10]. The σ phase issues from the phase transformation of $\delta \rightarrow \sigma$ (δ-ferrite to σ phase). When the $\delta \rightarrow \sigma$ phase transformation occurs, the σ phase will precipitate in the high Cr-concentrated region of δ-ferrite and is formed directly in δ-ferrite particles. When the Cr content is below 20 wt.%, the precipitation of the σ phase is not readily observable in austenitic stainless steels [11]. However, the σ phase can be formed quickly when the Cr content is 25 wt.%~30 wt.%. Furthermore, the σ phase can also precipitate from γ-austenite when there is no δ-ferrite in the stainless steels [12]; albeit this is unlikely. The δ-ferrite is a Cr-rich region, compared to the γ-austenite, and it is a body-centered crystal (BCC) structure. Cr is easy to diffuse in δ-ferrite, and it is also a ferritic stabilizer. Hence, δ-ferrite is a beneficial site for the precipitation of the σ phase [13].

Shinohara et al. [14] pointed out that the precipitation of the σ phase was a diffusion-controlled phase transformation in 25% Cr-20% Ni stainless steel, and that the Cr atom had an important effect on the acceleration of the precipitation rate of the σ phase. The precipitation rate of the σ phase in δ-ferrite was higher (by about 100 times) than that of γ-austenite [15]. This meant that the diffusion velocity of the σ phase in δ-ferrite was faster than that in γ-austenite. The contents of Cr and Mo of δ-ferrite decreased and the Ni contents increased simultaneously after the σ phase precipitated. The above result led to the formation of secondary austenite phases (γ_2) via the δ-ferrite [16].

Figure 1 shows the Time-Temperature-Transformation (TTT) curve of the σ phase in various series of stainless steels [17]. It can be seen that the nose of the TTT curve is located at a temperature range of 800°C~850°C, which means that this temperature range has the fastest precipitation rate. Furthermore, the σ phase precipitates at 0.1 h in AISI 309S stainless steel, which has a higher precipitation rate than other stainless steels because of its high Cr content, low C content, and high δ-ferrite (15%).

Barcik [18] reported on the precipitation behavior of the σ phase in Cr-Ni austenitic stainless steel. His study found that the precipitated order of the σ phase was in triple point: δ-ferrite, incoherent twin boundary, and inclusion grain. Sikka pointed out that the σ phase was easy to precipitate at the δ/γ interphase boundaries [19] because the δ/γ interphase was a high boundary energy place and so a beneficial site for the heterogeneous nucleation of the σ phase. Sato et al. [20] found that a high lattice coherent degree and low δ/γ boundary energy could suppress the precipitation of the σ phase. Na et al. [21] found that the σ phase formed in γ-austenite when the aging time was high; they constructed a schematic diagram to explain the precipitation behavior of the σ phase, as shown in Figure 2. For Case 1, the σ phase was formed in γ-austenite, but needed a higher aging temperature (above 1000°C), as shown in Figure 2(a). For Case 2, the σ phase precipitated at the δ/γ interphase boundary because this boundary was a region of high boundary energy, as displayed in Figure 2(b). For Case 3, the σ phase precipitated in δ-ferrite particle and was attributed to the high Cr region of δ-ferrite. This precipitation behavior was the least difficult condition to accomplish, as represented in Figure 2(c). Padilha and Rios [22] pointed out that the σ phase precipitating directly in γ-austenite was very slow (about several thousand hours) [23]. Three main reasons for this were as follows.

(1) The σ phase was insoluble from C and N elements so that the precipitation time of the carbide and nitride was earlier than that of the σ phase.

(2) The σ phase was a second phase of a rich substitutional element. The diffusion of the substitutional element was very slow in γ-austenite so that the nucleation of the σ phase in γ-austenite was difficult.

(3) The crystallography of the σ phase and γ-austenite was incoherent.

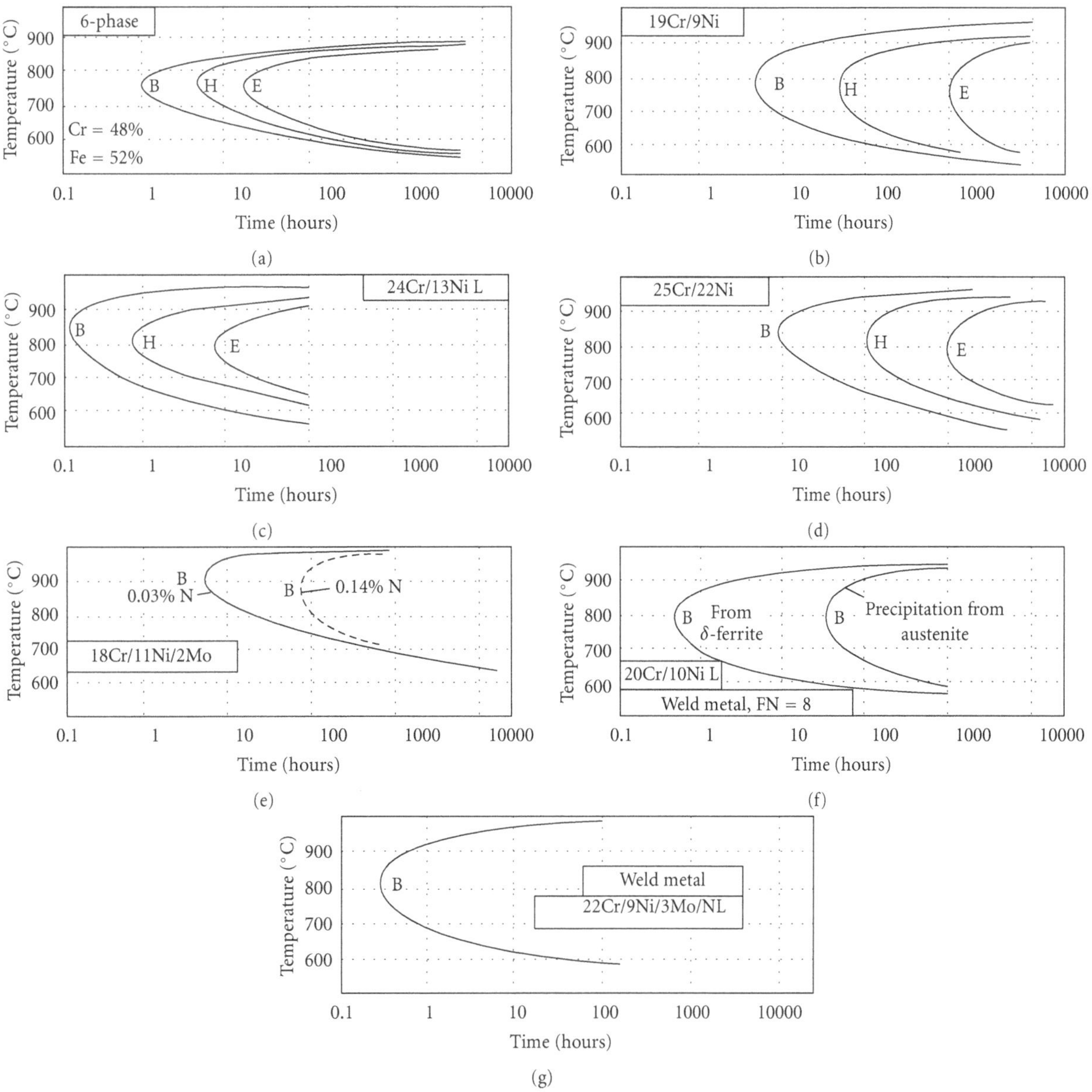

FIGURE 1: TTT curves of the σ phase in various series of stainless steels (B: initial precipitation, H: intermediate stage, E: final precipitation stage [17].

2.3. Microstructural Characteristic of the σ Phase. In 1978, Gray et al. [24] used an advanced magnetic etching technique to observe the $\delta \rightarrow \sigma + \gamma_2$ phase transformation. Figures 3 and 4 show the microstructures of AISI 304 stainless steel aged at 595°C and 650°C for 31000 h, respectively, using magnetic etching. The plate σ phase and γ_2 precipitate in δ-ferrite particles. Therefore, the precipitation of the σ phase is via the eutectoid decomposition of $\delta \rightarrow \sigma + \gamma_2$. The σ phase has a similar precipitation mechanism when aged at 650°C for 31000 h, but the eutectoid decomposition of $\delta \rightarrow \sigma + \gamma_2$ is more evident in this condition.

This Section introduces the morphologies of the σ phase in austenitic, ferritic, and duplex stainless steels.

2.3.1. Austenitic Stainless Steels. Figure 5(a) shows the as-casting microstructure of the σ phase in AISI 309 LSi stainless steel, and then the σ phase precipitating in dendrite δ-ferrite [25]. This is the typical precipitation morphology of the σ phase. Figure 5(b) shows the microstructure of the σ phase in as-rolled AISI 316 stainless steel [26]. The σ phase is indicated by a lacy structure and precipitates with a fixed crystallographic direction because of the rolling function. Figure 5(c) shows the as-rolled microstructure of the σ phase with a 50% rolling ratio in AISI 304 stainless steel, and the σ phase exhibits dispersed globular morphologies [27]. This globular structure is a stable morphology.

2.3.2. Ferritic Stainless Steels. Generally, many secondary phases (e.g., χ phase, Lave phase, etc.) exist in ferritic stainless steels, so that a microstructural examination of the σ phase is necessary. The microstructures of the σ phase were studied in

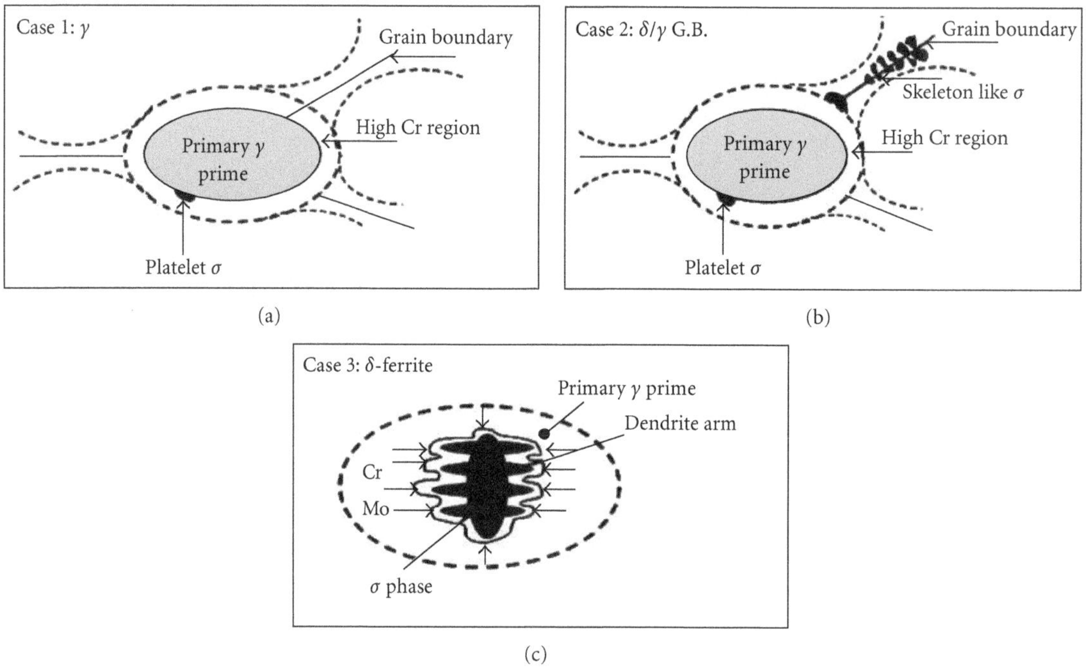

FIGURE 2: Schematic diagram of the σ phase precipitation [21].

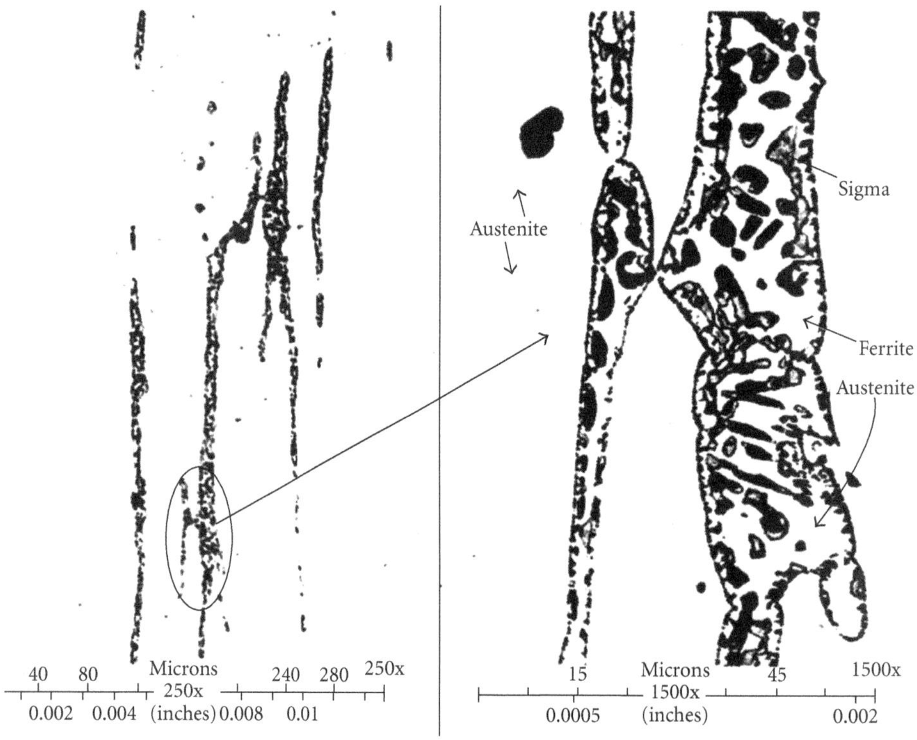

FIGURE 3: Microstructural observation of the $\delta \rightarrow \sigma + \gamma_2$ phase transformation in AISI 304 stainless steel (595°C, 31000 hr) [24].

Fe-29Cr-8W ferritic stainless steel aged at 850°C for different aging times [28]. The σ phase can precipitate obviously above 10 h. However, the precipitation mechanism of the σ phase accompanies the χ phase in the δ-ferrite matrixes as compared to austenitic stainless steels.

2.3.3. Duplex Stainless Steels. The microstructure of the σ phase was studied after heat treatment in A905 duplex stainless steel [26]. The σ phase (bright phase) precipitates in δ-ferrite grains when the aging condition is 30 min at 1100°C + 1 h at 900°C. However, $\sigma + \gamma_2$ cellular structures form in

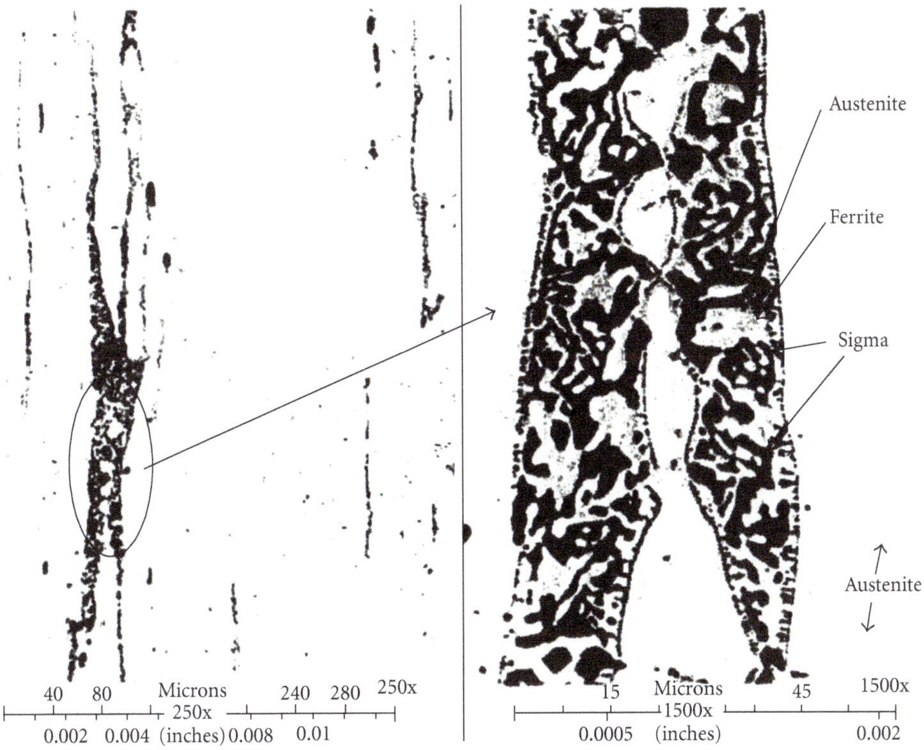

FIGURE 4: Microstructural observation of the $\delta \rightarrow \sigma + \gamma_2$ phase transformation in AISI 304 stainless steel (650°C, 31000 hr) [24].

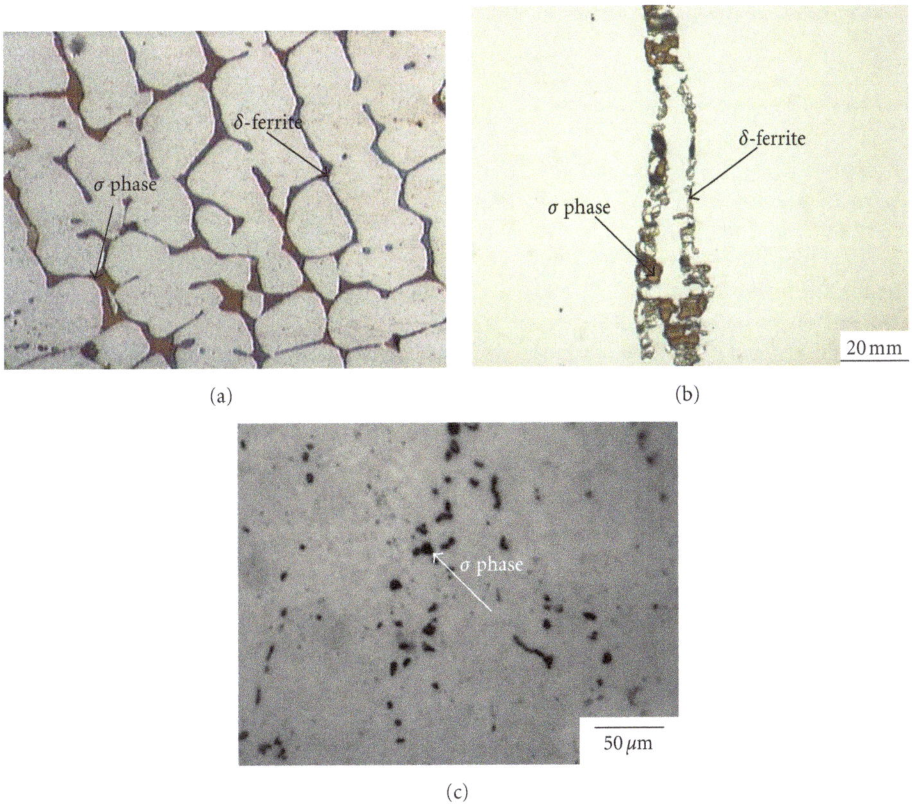

FIGURE 5: Microstructural observation of the σ phase with different working conditions in the austenitic stainless steels (a) as casting AISI 309 LSi stainless steel, (b) as rolled AISI 316 stainless steel, (c) with 50% rolling ratio in AISI stainless steel [25].

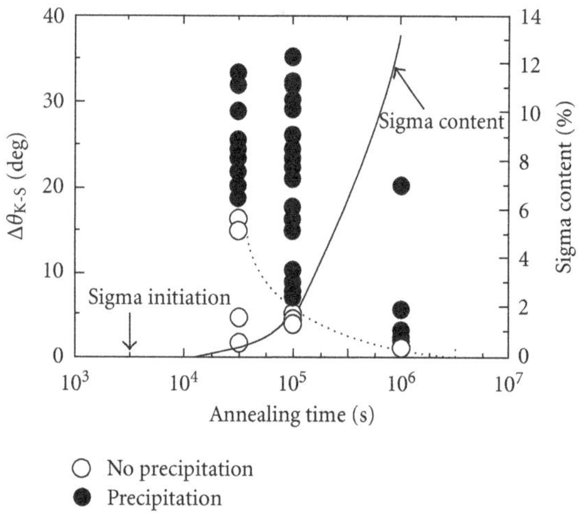

FIGURE 6: Effect of the deviation angle $\Delta\theta_{K\text{-}S}$ and the annealing time on the formation of σ phase at δ/γ interfaces in the weld metal [20].

δ-ferrite particles when the aging situation is 30 min at 1100°C + 10 h at 900°C. Finally, δ-ferrite is consumed by the $\sigma + \gamma_2$ cellular structures completely when the aging condition is 30 min at 1100°C + 10 h at 800°C.

2.4. Classification of Morphologies of the σ Phase. In stainless steels, classification of the morphology of the σ phase can be divided into four types: (1) grain boundary precipitation, (2) triple point precipitation, (3) corner precipitation, and (4) cellular precipitation [29]. A discussion of these four morphologies follows.

2.4.1. δ/γ Grain Boundary Precipitation. The σ phase is easy to precipitate at the δ/γ phase boundary, which is a high Cr region. When a high Cr region forms, the depleted Cr region is also formed and leads to a decrease in the corrosion resistance at the same time. Furthermore, the δ/γ interphase boundary is a high interface energy site, and, as many defects concentrate there, it is a beneficial site for the heterogeneous nucleation of the σ phase. When the σ phase nucleates at the δ/γ interphase boundary, some defects disappear, which releases the free energy of the materials. Consequently, this decreases the activation energy barrier to form a coherent interface. Sato and Kokawa [30] pointed out that the σ phase forms preferentially at one side of the δ/γ interfaces along intergranular austenite in duplex stainless steel weld metal annealed at 1100 K. The crystallographic orientation relationships at the δ/γ interfaces are the σ phase forms which have deviated largely from the K-S relationship ($\Delta\theta_{K\text{-}S}$), as shown in Figure 6. The minimum deviation angle for the sigma phase formation decreases with increasing annealing time. Furthermore, the formation of the σ phase is strongly affected by the coherency and interfacial energy of the δ/γ interface.

2.4.2. Triple Point Precipitation. The σ phase precipitates at the triple point of the δ-ferrite boundary and this precipitation type, called the "triple point σ phase." Barick [31] pointed out that the formation of the triple point σ phase must be aged below 600°C for 10000 h~15000 h. The triple point σ phase also forms at the incoherent twin boundaries and the intragranular inclusions.

2.4.3. Corner Precipitation. The corner precipitation of the σ phase means that the σ phase forms directly in corner δ-ferrite particles because the δ-ferrite is a high Cr content phase, and σ phase prefers to nucleate and precipitate at that point. When the σ phase precipitates at the corner δ-ferrite, it consumes the Cr content of the δ-ferrite particles.

2.4.4. Cellular Precipitation. Cellular precipitation means that the σ phase and secondary austenite ($\sigma + \gamma_2$) precipitate as laminar precipitation in the δ-ferrite particles. This reaction is called the eutectoid decomposition of $\delta \rightarrow \sigma + \gamma_2$. When the eutectoid decomposition of $\delta \rightarrow \sigma + \gamma_2$ is finished, the σ phase consumes the Cr, Mo, and Si contents of the δ-ferrite particles.

2.5. Precipitation of the σ Phase in Austenitic, Superferritic, and Duplex Stainless Steels. Villanueva et al. [32] discussed the precipitation mechanism of the σ phase in the as-rolled plate of AISI 316L austenitic (17Cr-12Ni-2.4Mo), DIN 1.4574 superferritic (28Cr-4Ni-2.4Mo-0.31Nb), and UNS S31803 duplex (22Cr-5.4Ni-2.4Mo) stainless steels with various aging temperatures. Three schematic diagrams (Figures 7–9) can be used to explain the precipitation mechanism of σ phase.

Figures 7(a)–7(d) show the precipitation mechanism of σ phase in AISI 316L stainless steel at different aging temperatures. Without heating (T_0), the δ-ferrite is indicated by a lacy structure and precipitates at the δ/γ interphase boundary and γ phase. When the aging temperature is increased to T_1, the σ and γ_2 phases precipitate in δ-ferrite particles. When the aging temperatures are increased to T_2, the cellular $\sigma + \gamma_2$ also forms in δ-ferrite particles, and the σ phase precipitates at the triple points and the δ/γ interphase boundaries. When the aging temperature is T_3, the precipitation of the laminar $\sigma + \gamma_2$ is more obvious than at the other aging temperatures.

Figures 8(a)–8(d) show the precipitation of σ phase in superferritic stainless steel. There is only δ-ferrite and no precipitation of the σ phase without heating (T_0) in superferritic stainless steel. When the aging temperature is T_1, the σ phase can occupy the triple points of the δ-ferrite grain boundaries. The σ phase grows at the δ-ferrite grain boundaries completely when the aging temperature increases to T_2. Finally, the σ phase precipitates and extends to the inner δ-ferrite grain when the aging temperature is raised to T_3.

Figures 9(a)–9(d) are the precipitation mechanism of σ phase in UNS S31803 duplex stainless steel. Here are δ and γ phases in UNS S31803 duplex stainless steel without heating (T_0). Some laminar cellular structures ($\sigma + \gamma_{New}$) are formed at the δ/γ interphase boundaries when the aging temperature is T_1. Furthermore, the laminar cellular structures continue to grow when the aging temperature is increased to T_2. Finally, the δ-ferrite is occupied completely.

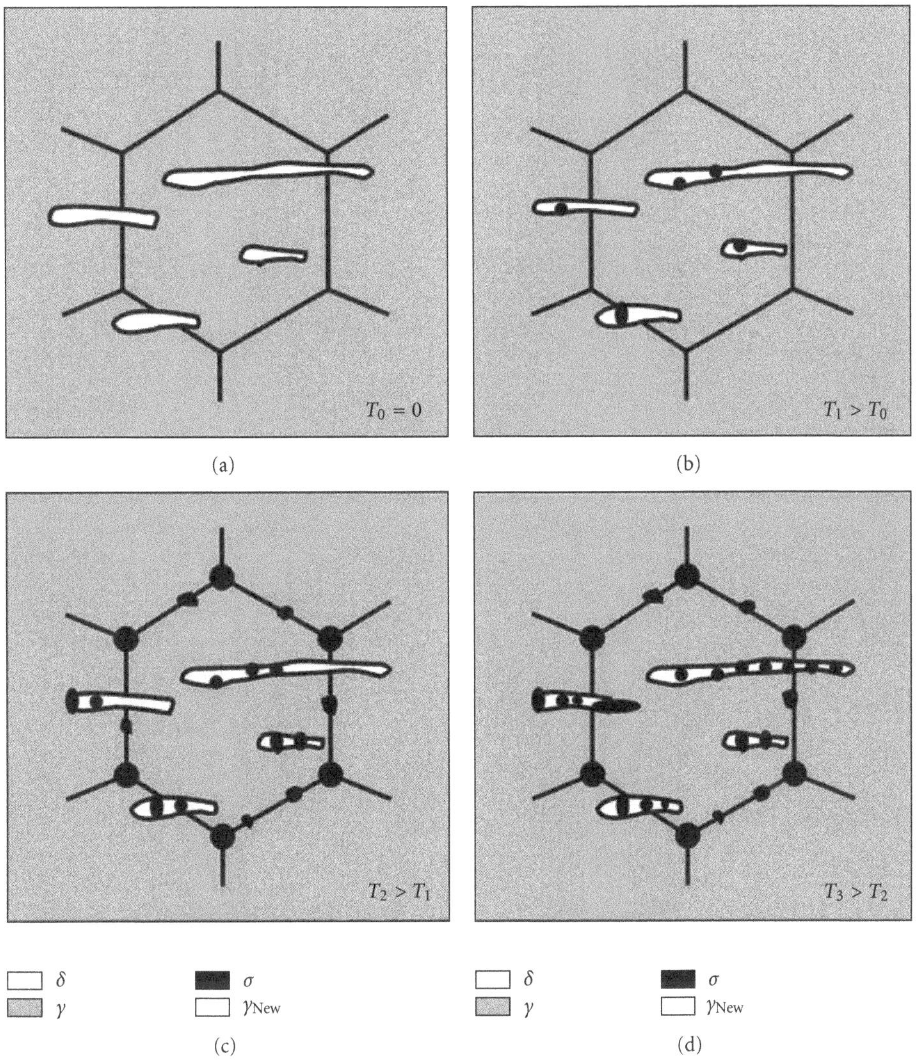

FIGURE 7: Precipitation mechanism of the σ phase in AISI 316L stainless steel [32].

Villanueva et al. [32], in describing the above results, pointed out the precipitation level of the σ phase: Duplex stainless steel > Superferritic stainless steel > Austenitic stainless steel.

2.6. Effect of the σ Phase on the Properties of Stainless Steels

2.6.1. Formation of Brittle Region. The embrittlement region of the σ phase forms in a temperature range of 650°C~900°C during the welding process in austenitic stainless steels. Hence, the welding energy and cooling rate must be controlled during welding in order to prevent the grain growth and precipitation of the brittle σ phase because the precipitation of the σ phase decreases the toughness and elongation. An embrittlement region of σ phase can be shown in Shaeffler diagram of Figure 10 [33]. Figure 10 of gray region is embrittlement region of σ phase, and this region is the widest in the mixed austenite plus ferrite phase region (A + F). On the other hand, the single austenite phase region (A) shows a small embrittlement range of the σ phase. Therefore, the possibility of the σ phase is lower in austenitic stainless steels. However, increasing the Cr content can promote the precipitation of the σ phase.

2.6.2. Extension of Hot Cracking during Welding. Konosu et al. [34] observed hot cracking in three stainless steel weld metals and the hot cracking extends along the γ/σ interface boundaries. The γ/σ interface boundaries are an initial fracture position of low elongation. This crack is a beneficial site and acts as a nucleated point of the σ phase. The growth of hot cracking cannot be neglected as it is very significant during the welding. Mataya and Carr [35] pointed out that hot cracking occurs because of the δ → σ phase transformation in the 21Cr-6Ni-9Mn austenitic stainless steel.

2.6.3. Formation of Depleted Cr Region and Reduction of Corrosion Resistance. The precipitation of the σ phase reduces the corrosion resistances to the intergranular, pitting, and crevice. Reis et al. [36] reported that pitting corrosion

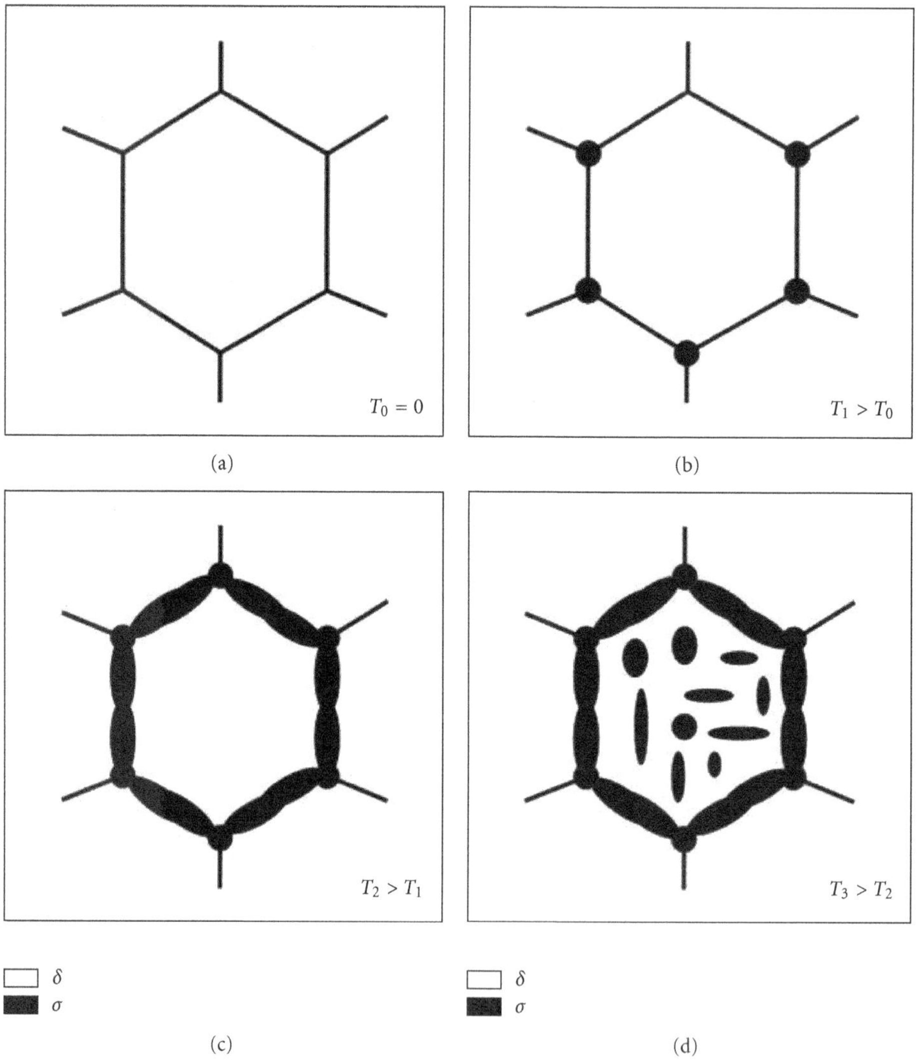

FIGURE 8: Precipitation mechanism of the σ phase in DIN 1.4575 stainless steel [32].

resistance has an important relationship with the Cr content. Hence, pitting corrosion always happens at low Cr content points. Generally, the austenitic boundary is a preferential site for pitting corrosion because the austenite has a low Cr content. However, the depleted Cr zone disappears gradually as a function of Cr diffusion, which issues from the austenite but not the δ-ferrite. The σ phase always forms at the δ/γ interface boundaries and causes the formation of a depleted zone [37]. The σ phase and γ-austenite induce the galvanic effect, and the γ-austenite is corroded preferentially. Consequently, the corrosion potent is decreased. The fine σ phase has a more obvious effect on intergranular corrosion resistance than does the coarser σ phase because the fine σ phase forms a networked structure at the interface, and the coarse σ phase precipitates independently at the interface. Ravindranath and Malhotra [38] pointed out that intergranular corrosion resistance has an important effect on the formation of the networked σ phase. Besides, an independent coarse σ phase has no influence on intergranular corrosion resistance.

2.7. Prediction Methods of the σ Phase. Chemical composition can be used to predict the precipitation of the σ phase. Gow and Harder [39] pointed out an empirical formula to examine the precipitation tendency of the σ phase, as expressed in(1)

$$\text{Ratio factor} = \frac{\%\text{Cr} - 16\%\text{C}}{\%\text{Ni}}. \tag{1}$$

If ratio factor >1.7, the σ phase can precipitate in stainless steels. When the C content is much higher, the ratio factor is much lower, and the precipitation tendency of the σ phase is decreased.

Woodyatt [40] also pointed out a predicted formula of the σ phase via an electron vacancy number (at. %), as represented in(2)

$$\begin{aligned} N_v &= 0.66\text{Ni} + 1.71\text{Co} + 2.66\text{Fe} + 4.66(\text{Cr} + \text{Mo} + \text{W}) \\ &\quad + 5.66\text{V} + 6.66\text{Zr} + 10.66\text{Cb}. \end{aligned} \tag{2}$$

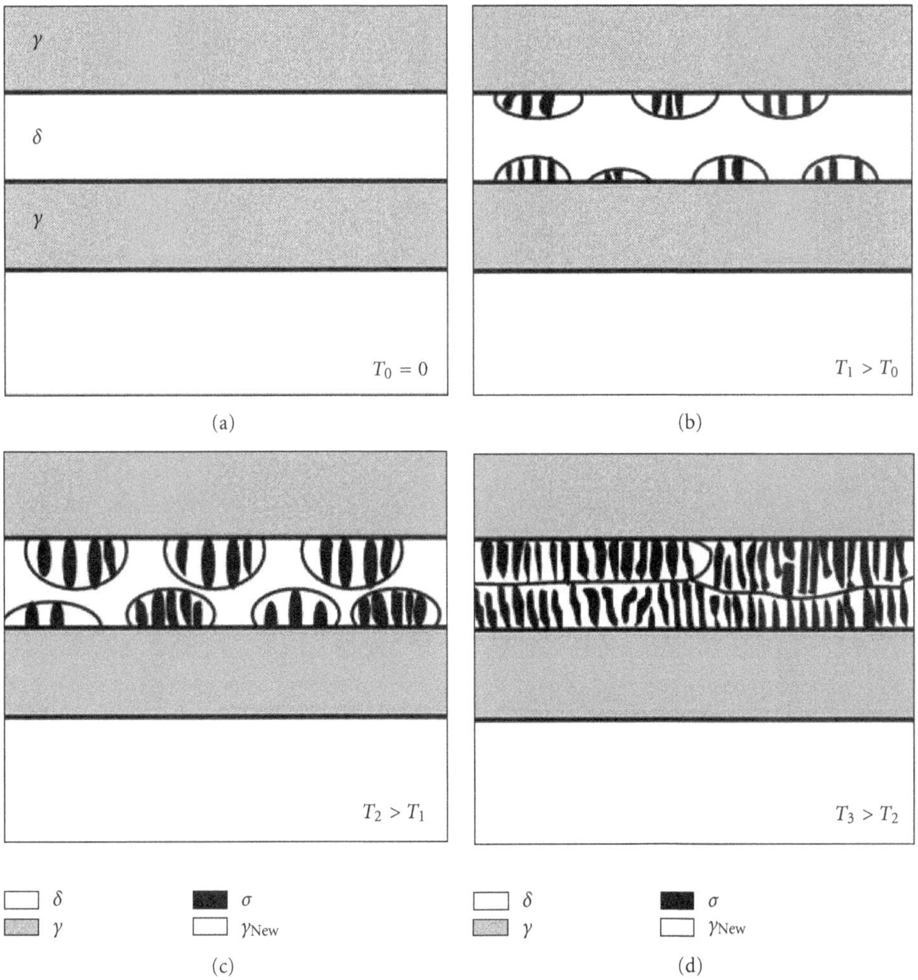

FIGURE 9: Precipitation mechanism of the σ phase in UNS S31803 stainless steel [32].

When the electron vacancy number ($\overline{N_v}$) > 2.52, the σ phase can be formed in stainless steels. According to Woodyatt's theory, the $\overline{N_v}$ in the AISI 310 stainless steel is equal to 2.88, so that this stainless steel indicates the precipitation of the σ phase. However, the interstitial elements (C and N) are neglected in this prediction formula.

Hull [41] pointed out an empirical formula to predict the σ phase precipitation, as exhibited in (3). This equation has been widely used because it takes into account many alloy elements.

$$\begin{aligned} \mathrm{Cr_{eq}} &= \mathrm{Cr} + 0.31\mathrm{Mn} + 1.76\mathrm{Mo} + 0.97\mathrm{W} + 2.02\mathrm{V} + 1.58\mathrm{Si} \\ &\quad + 2.44\mathrm{Ti} + 1.70\mathrm{Cb} + 1.22\mathrm{Ta} - 0.266\mathrm{Ni} - 0.177\mathrm{Co}. \end{aligned} \tag{3}$$

2.8. Retardation Methods of the σ Phase

2.8.1. Solid Solution Heat Treatment. The precipitation of the σ phase can be retarded by the use of the solid solution heat treatment. When stainless steels are heated to above 1050°C, the σ phase diffuses and dissolves into the γ-austenite matrix [42]. This dissolution process is called the "$\sigma \rightarrow \gamma$ phase transformation."

2.8.2. Subsequent Rolling Process. Gill et al. [43] reported that the dendrite σ phase is an unstable morphology and can lead to embrittlement of the stainless steels. The subsequent rolling process can be used to refine the σ phase particles from an unstable dendrite into stable globular morphologies. Hence, the embrittlement of the σ phase is decreased in the stainless steels by the subsequent rolling process.

2.8.3. Addition of Alloy Elements. In recent years, Ni has been used to replace N and decrease the N addition because the price of the Ni in steel continues to improve. Lin et al. [44] reported that N addition decreases the δ-ferrite content during the nitrogen atmospheric aging heat treatment and retards the phase transformation from δ to σ. However, C has a similar effect on the suppression of the σ phase, but its addition leads to the precipitation of the nitride and causes sensitization and embrittlement.

Brandis et al. [45] pointed out that Nitrogen had a better effect on the retardation of the σ phase than Carbon. Smuk et al. [46] studied the powder metallurgy of duplex stainless steels and found that Cu particles could pin the σ/γ_2 interface boundaries, as shown in Figure 11. This meant that the

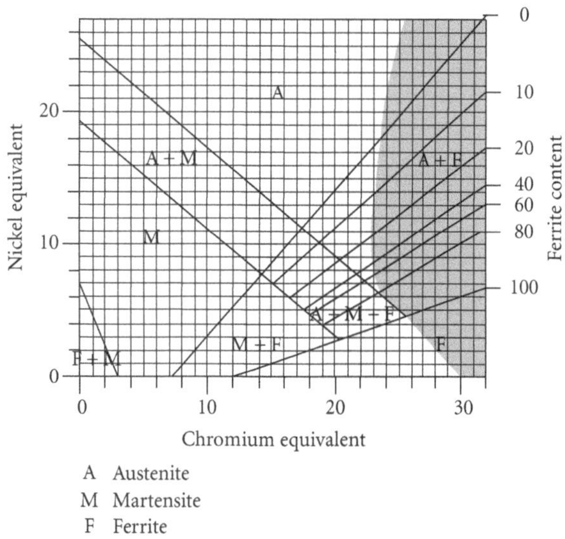

Figure 10: Shaeffler diagram showing the embrittlement region of the σ phase [33].

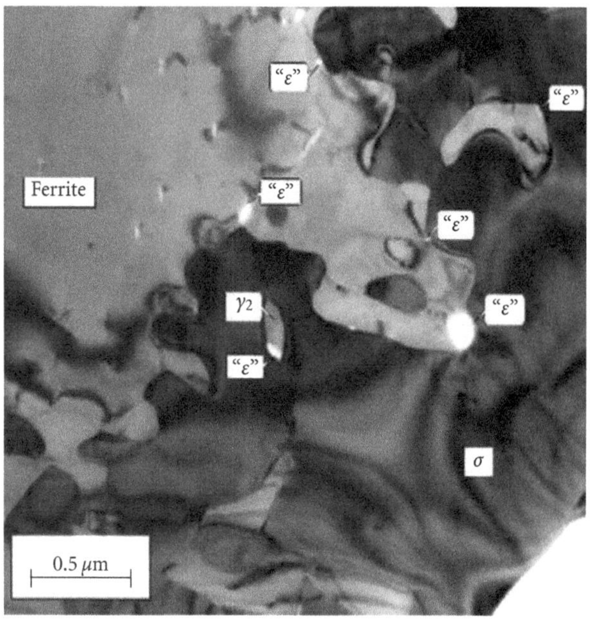

Figure 11: Cu particles can pin and prevent the motion of the σ/γ_2 interface [46].

Cu particles could prevent the motion of the σ/γ_2 interface boundaries and can change the morphologies of the σ phase. Consequently, the $\delta \rightarrow \sigma$ phase transformation could be suppressed efficiently.

2.9. Thermo-Calc Simulation of the σ Phase. Besides the metallographic technique and quantitative analysis, the precipitation examination of the σ phase can also be predicted using a Thermo-Calc software simulation. This thermodynamic simulation only needs the input of the element contents and experimental temperatures to produce the phase diagrams. This simulation method is called Calculation of Phase Diagrams (CALPHAD).

The simulation methods of the Thermo-Calc software in the σ phase have three databases: Steel & Fe-Alloys Database (TCFE), SGTE Solution Database (SSOL) [47] + Lindholm [48], and Modified SSOL [49]. Erneman et al. [49] used the three Thermo-Calc databases (TCFE, SSOL + Lindholm, Modified SSOL) to calculate the σ phase fraction in AISI 347 stainless steel at 700°C. The σ phase has a lesser volume fraction, which means a lower grain growth according to the TCFE simulation result. The simulation result of the Modified SSOL indicates a higher volume fraction of the σ phase, which means a higher grain growth. The simulation data of SSOL was in between the TCFE and SSOL databases. Saunders and Miodownik [50] used the Thermo-Calc software to simulate the precipitation quality of the σ phase with Mo, Mn, So, N, and C in Fe-22Cr-5.5Ni stainless steel between 670°C and 850°C. The series of simulation results were shown in Figures 12(a)–12(f). However, by adding 3% Mo into the original stainless steel, this steel becomes Fe-22Cr-5.5Ni-3Mo stainless steel, as shown in Figure 12(b). This CAPLAD result indicates that the σ phase has the highest content of all temperatures at 800°C. Moreover, Mo addition can extend the precipitation range to 900°C.

By adding 1.7% Mn into the Fe-22Cr-5.5Ni-3Mo stainless steel, this steel becomes Fe-22Cr-5.5Ni-3Mo-1.7Mn stainless steel. The CALPHAD result shows that Mn has a competitive effect with Cr and Mo, so it can promote the precipitation of the σ phase, as shown in Figure 12(c). By adding 0.4% Si into the Fe-22Cr-5.5Ni-3Mo-1.7Mn, this steel becomes Fe-22Cr-5.5Ni-3Mo-1.7Mn-0.4Si stainless steel. The CALPHAD result shows that Si is a strong ferrite stabilizer which can raise the σ phase content, as indicated in Figure 12(d). Hence, it can be concluded that Si promotes the $\delta \rightarrow \sigma$ phase transformation. By continuing to add 0.14N into the Fe-22Cr-5.5Ni-3Mo-1.7Mn-0.4Si, this steel becomes Fe-22Cr-5.5Ni-3Mo-1.7Mn-0.4Si-0.14N stainless steel. When N is added, the nitride (M_2N) forms, and the σ phase content decreases at the same time, as shown in Figure 12(e). This is because N addition promotes the $\delta \rightarrow \gamma$ phase transformation and retards the $\delta \rightarrow \sigma$ phase transformation.

Similarly, by adding 0.24C to the Fe-22Cr-5.5Ni-3Mo-1.7Mn-0.4Si-0.14N, this steel becomes Fe-22Cr-5.5Ni-3Mo-1.7Mn-0.4Si-0.14N-0.24C stainless steel. The CALPHAD result shows that C addition leads to the formation of $M_{23}C_6$, as shown in Figure 12(f). The content of carbide is higher, and the content of σ phase is lower in stainless steels. Furthermore, the σ phase fraction with the C addition is lower than that with Mn, Si or N additions.

2.10. Effect of Various Processes on the σ Phase

2.10.1. Effect of Solid Solution-Treated Temperature and Cooling Rates on the σ Phase. Chen and Yang [51] studied 2205 duplex stainless steel as the solid solution treated at 1020°C with different cooling rates ($1°C\,s^{-1}$, $0.5°C\,s^{-1}$, $0.25°C\,s^{-1}$,

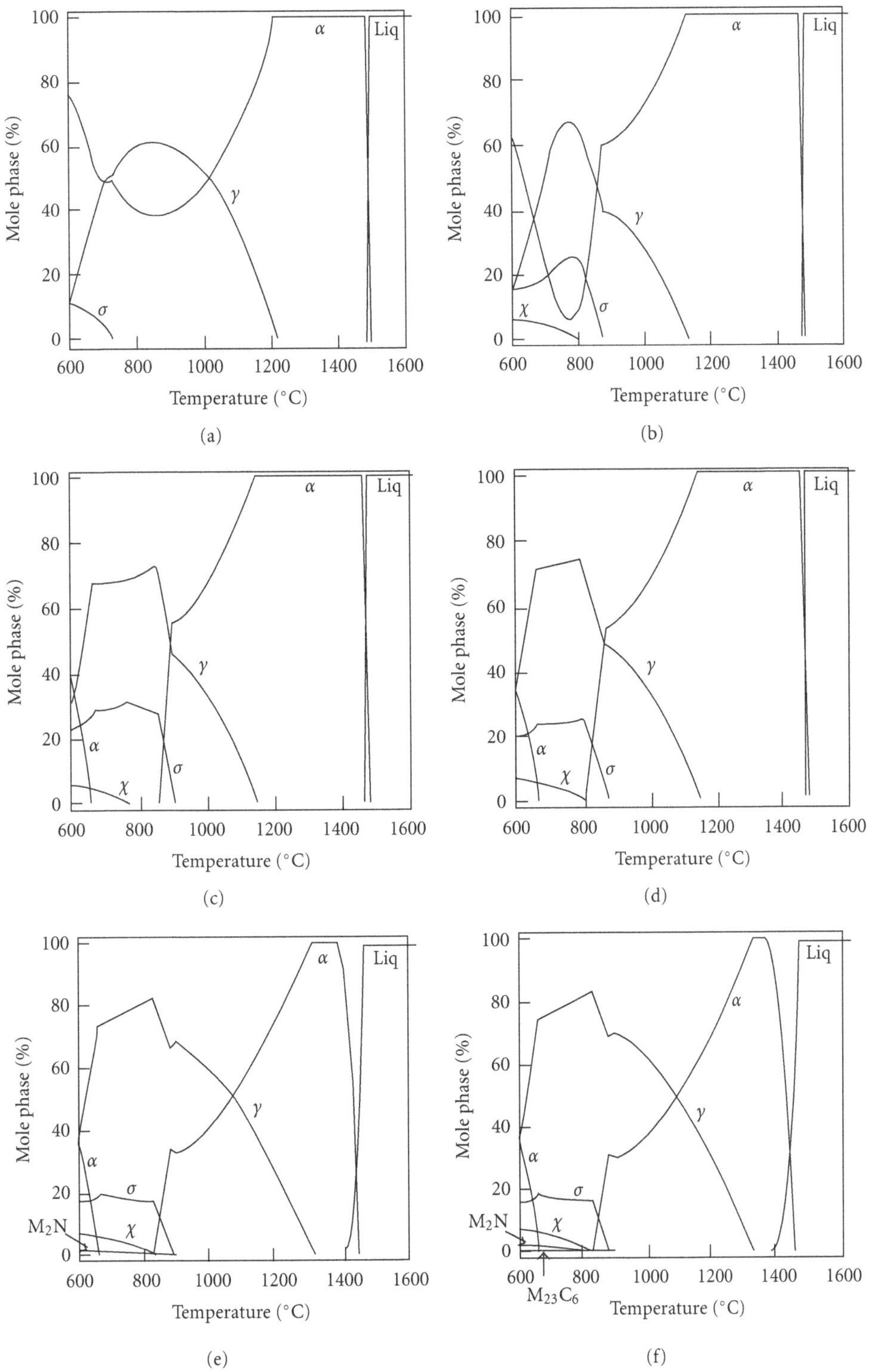

FIGURE 12: CALPHAD result of the phases (a) Original Fe-22Cr-5.5Ni, (b) Fe-22Cr-5.5Ni-3MoFe-22Cr-5.5Ni, (c) Fe-22Cr-5.5Ni-3Mo-1.7MnFe-22Cr-5.5Ni, (d) Fe-22Cr-5.5Ni-3Mo-1.7Mn-0.4Si, (e) Fe-22Cr-5.5Ni-3Mo-1.7Mn-0.4Si-0.14N, and (f) Fe-22Cr-5.5Ni-3Mo-1.7Mn-0.4Si-0.14N-0.24C stainless steels [46].

and 0.1°C s^{-1}). The results showed that there was no σ phase when cooling rates were 1°C s^{-1} and 0.5°C s^{-1}. The cooling rates of 0.25°C s^{-1} and 0.1°C s^{-1} indicated a slight σ phase and a lot of σ phases, respectively. On the other hand, the precipitation of the σ phase was concentrated mainly at the δ/γ interface boundaries and δ-ferrite. However, Chen and Yang [51] also observed the precipitation of the σ phase at two solid solution temperatures, 1020°C and 1080°C. Experimental result indicated that the content of the σ phase was higher with the cooling rate at 1020°C than at

Table 2: EDS analysis of δ, σ, and γ phases at various solid solution temperatures and a fixed cooling rate of $0.1^{\circ}C\,s^{-1}$ [51].

Element	Fe	Cr	Ni	Mo	Mn	Si
Solution treated at 1020°C and cooled at $0.1^{\circ}C\,s^{-1}$						
σ	61.7 ± 0.5	25.7 ± 0.6	3.46 ± 0.06	6.84 ± 0.08	1.15 ± 0.04	1.15 ± 0.04
δ	66.6 ± 0.7	24.1 ± 0.4	3.60 ± 0.06	4.09 ± 0.06	1.15 ± 0.04	0.44 ± 0.02
γ	68.0 ± 0.7	20.8 ± 0.4	6.45 ± 0.08	2.42 ± 0.06	1.45 ± 0.04	0.32 ± 0.02
Solution treated at 1080°C and cooled at $0.1^{\circ}C\,s^{-1}$						
σ	62.7 ± 0.5	25.2 ± 0.6	3.92 ± 0.06	5.90 ± 0.08	1.14 ± 0.04	1.13 ± 0.04
δ	66.7 ± 0.7	24.0 ± 0.4	3.68 ± 0.06	4.05 ± 0.06	1.15 ± 0.04	0.43 ± 0.02
γ	68.3 ± 0.7	20.9 ± 0.4	6.39 ± 0.08	2.45 ± 0.06	1.44 ± 0.04	0.33 ± 0.02
Solution treated at 1200°C and cooled at $0.1^{\circ}C\,s^{-1}$						
δ	67.0 ± 0.7	23.6 ± 0.4	3.83 ± 0.08	3.96 ± 0.06	1.16 ± 0.04	0.41 ± 0.02
γ	68.2 ± 0.7	21.3 ± 0.4	6.25 ± 0.08	2.54 ± 0.06	1.33 ± 0.04	0.35 ± 0.02

1080°C. Finally, Chen and Yang [51] used a solid solution temperature (1200°C) at a fixed cooling rate ($0.1^{\circ}C\,s^{-1}$) to discuss the precipitation of the σ phase. The experimental result showed that the Cr, Mo, and Si elements at 1200°C were higher than at 1080°C or 1200°C. This meant that the elemental diffusion of the Cr, Mo, and Si had an increasing effect on the precipitation of the σ phase at 1200°C at a cooling rate of $0.1^{\circ}C\,s^{-1}$, as represented in Table 2 [51].

2.10.2. Effect of Tensile Strain Rate and Time on the σ Phase. Hong and Han [52] conducted a high-temperature tensile test in superduplex stainless steel (Fe-24Cr-7Ni-3Mo-0.14N) at 850°C with a strain rate of $3.16\times10^{-3}\,s^{-1}$. The grain size of the σ phase increased in a deformed sample and maintained a constant value in the undeformed sample. This was because the controlled grain boundaries slip-strain induced the grain growth [53, 54] during the super-plasticity. Chandra and Kuchlmayr [55] reported that the σ phase content decreased with an increase in strain rate when the aging temperature was 900°C in Fe-19Cr-5Ni-2.7Mo stainless steel.

2.10.3. Effect of Cold Rolling Ratios on the σ Phase. Abe et al. [56] reported on the effect of various cold rolling ratios on the precipitation of the σ phase in Fe-10Cr-30Mn stainless steel. Experimental result indicated the LAXRD (Low-angle X-Ray Diffraction) analysis at 923 K under various cold rolling ratios and aging times (30% CW-10 h, 30% CW-100 h, 30% CW-1000 h, and 80% CW-1000 h). The CW is an abbreviation of the "Cold Working." The diffraction angle (2θ) of the LAXRD analysis is smaller than 85°, and this analytical method can show a clear diffraction signal. Many σ diffracted peaks can be detected, and there is no diffracted peak in the second phases. The diffracted peak of the σ phase is more obvious with 30% CW with an increase in aging time, which means that the precipitation of the σ phase does not decrease. Hence, cold rolling promotes the precipitation of the σ phase. Other diffracted peaks ($\sigma(410)$, $\sigma(212)$) of the σ phase were examined when the cold rolling ratio increased to 80% CW, and the $\sigma(312)$ peak was higher.

Figure 13 shows the TEM image in Fe-10Cr-30Mn stainless steel heat treated at 923 K for 3600 ks after cold rolling with various cold rolling ratios. The result shows that the grain size of the σ phase becomes smaller with increases in the cold rolling ratio, but that the distribution density of the σ phase increases. However, recrystallization only occurs above 30% CW, and there is no recrystallization with 10% CW. The σ phase only precipitates at the grain boundaries and triple points, as shown in Figure 13(a). A complete recrystallization happens with 60% CW and 80% CW, and then partial recrystallization occurs with 30% CW, as indicated in Figures 13(b)–13(d).

Huang and Chen [57] pointed out that cold working promoted the formation of the σ phase because the recrystallized and cold working boundaries were nucleated sites for the σ phase at low temperatures. At high temperatures, fine recrystallized grains formed and increased the area of the grain boundary so that the σ phase nucleated easily. Duhaj et al. [7] reported that the formation of the σ phase accelerated when the cold working degree was increased and that the σ phase content increased at the same time. It was concluded that the defects of stacking fault and twin acted as nucleation sites of the σ phase.

Figures 14(a)–14(c) show the relationship between recrystallization and the σ phase precipitation [56]. Initially, the σ phase nucleates at the recrystallized boundaries when crystallization occurs during cold rolling, as shown in Figure 14(a). This precipitation of the σ phase occurs because of the minimum strain energy during cold rolling. When the σ phase is formed, the recrystallized boundaries move to the deformed grains as a function of the recrystallized driving force (f_R). A recrystallized reacting force ($-f_p$) is also applied to the front of the recrystallized boundaries, as indicated in Figure 14(b). This recrystallized reacting force is due to the difference of the dislocation density from two sides of the recrystallized boundaries. Finally, the σ phase finishes the nucleation and grain growth and then escapes the recrystallized boundaries, as shown in Figure 14(c).

Koutsoukis et al. [58] studied the precipitation of the σ phase in S32654 super austenite stainless steel, cold rolled and aged at 850°C during cold rolling. The experimental result showed that the dispersion of the σ phase was more obvious when the cold rolling ratios were increased from 20% CW to 60% CW. The precipitation of the σ phase

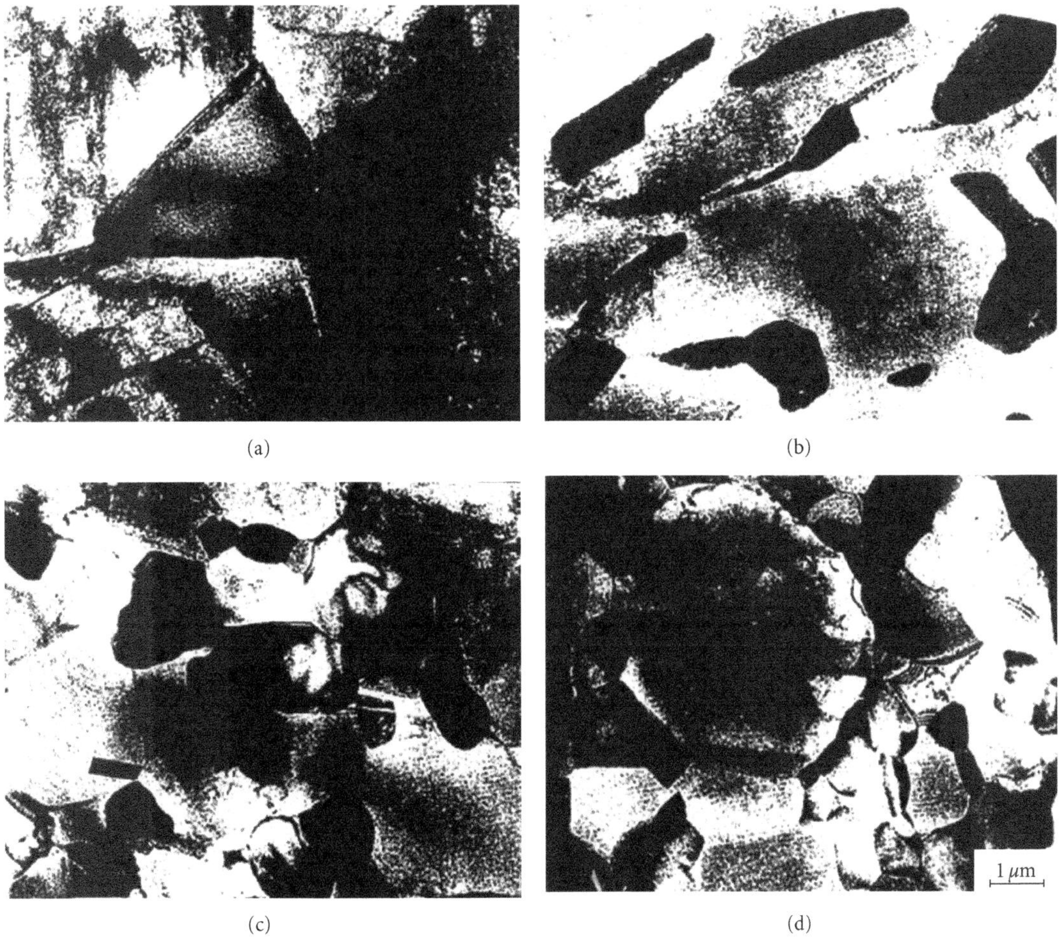

FIGURE 13: The effect of various cold rolling ratios on distribution and grain size of the σ phase (a) 10% CW, (b) 30% CW, (c) 60% CW, and (d) 80% CW [56].

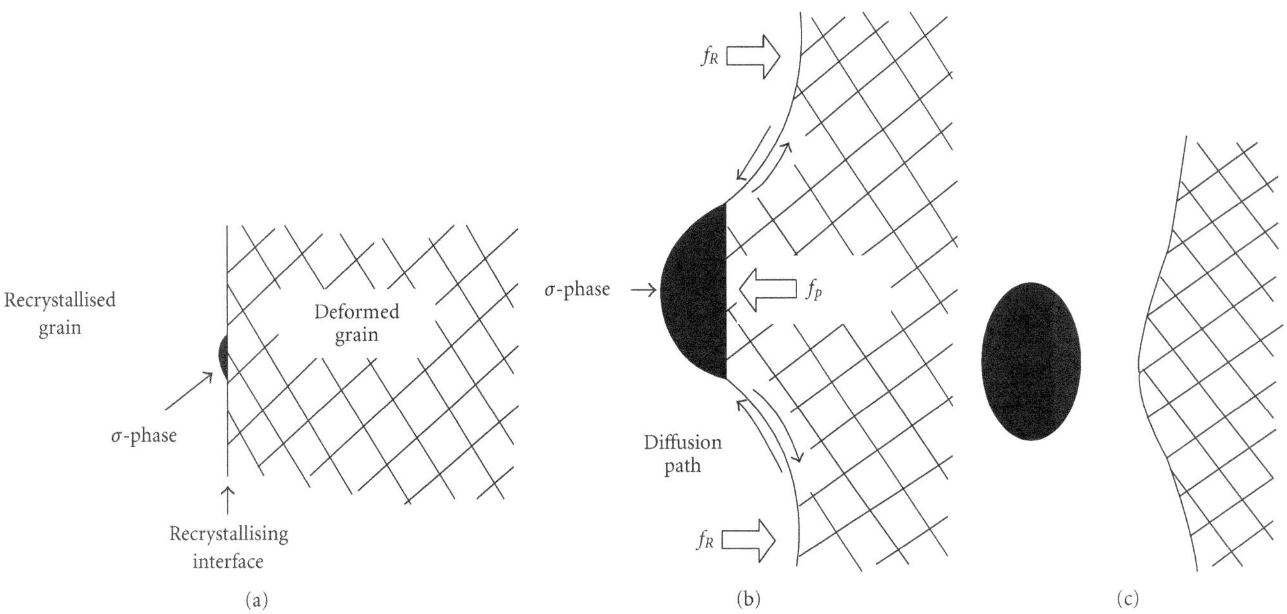

FIGURE 14: Nucleation and grain growth of σ phase at the recrystallized boundaries (f_R: recrystallized driving force, $-f_p$: recrystallized reacting force) [56].

increased the hardness value about 500 HV with 60% CW when aged at 650°C.

2.10.4. Effect of Hot Rolling Ratios on the σ Phase. Hsieh et al. [59] studied the precipitation mechanism of the σ phase in 19Cr-9Ni-2Mn austenitic and 18Cr-0.75Si ferritic stainless steels, hot rolled at 800°C under hot rolling ratios of 0%, 50%, and 75%. The experimental results indicated that the precipitation tendencies of the σ phase increased gradually with increase of the hot rolling ratio.

Experimental results show the morphological variations of the σ phase with hot rolling ratios of 0%, 50%, and 75% in 19Cr-9Ni-2Mn stainless steel. The morphologies of the σ phase changed from complete dendrites into fine globular structures when the hot rolling ratio was increased from 0% to 75%, with similar results in 18Cr-0.7Si stainless steel.

Hsieh et al. [60] pointed out that the dendrite σ phase was an unstable morphology and that the globular σ phase was a stable morphology. Gill et al. [61] also reported that the globular σ phase resulted from a refinement of the unstable σ phase. The globular σ phase consumed the Cr content of the δ-ferrite completely. Hence, the globular σ phase was maintained in the rolled state samples. Furthermore, the dendrite σ phase occurred because of the partial consumption of the Cr content of the δ-ferrite.

3. Summary

The above review has presented a clear overview of the precipitation of the σ phase in stainless steels. This review paper has provided the metallographic characteristics, precipitation mechanisms, properties effects, and prediction methods of the σ phase in stainless steels. The precipitation of the σ phase is the main reason for the degradation of stainless steels. Understanding the precipitation of the σ phase will allow for some control over the properties of stainless steels.

Acknowledgments

The authors would like to thank the National Science Council of the Taiwan, R.O.C. for financial support under Projects numbered NSC 100-ET-E-005-001-ET, NSC 101-2623-E-005-002-ET, NSC 100-2811-E-005-001, and NSC 101-2811-E-005-001. They are also obligated to thank Prof. Weite Wu and Prof. Dong-Yih Lin for their scientific writing guidance about this review paper.

References

[1] W. Treitschke and G. Tammann, "Enthalpy of formation for σ-phase solid solutions at 1060 K," *Anorganische Chemie*, vol. 55, p. 707, 1907.

[2] E. C. Bain and W. E. Griffiths, "An introduction to the Iron-Chromium Nickel alloys," *Transactions American Institute of Mining, Metallurgical and Petroleum Engineers*, vol. 75, pp. 166–213, 1927.

[3] E. R. Jett and F. Foote, "The Fe-Cr alloy system," *Metals and Alloys*, vol. 7, pp. 207–210, 1936.

[4] K. Yano and K. Abiko, "Role of carbon and nitrogen on the transformation of the σ phase in highly purified Fe-50 mass% Cr alloys," *Materials Transactions*, vol. 41, no. 1, pp. 122–129, 2000.

[5] B. Hattersley and W. Hume-Rothery, "Constitution of certain austenitic steels," *The Journal of the Iron and Steel Institute*, vol. 204, pp. 683–701, 1966.

[6] E. O. Hall and S. H. Algie, "The sigma phase," *International Materials Reviews*, vol. 11, pp. 61–88, 1966.

[7] P. Duhaj, J. Ivan, and E. Makovicky, "Sigma phase precipitation in austenitic steels," *The Journal of the Iron and Steel Institute*, vol. 206, pp. 1245–1252, 1968.

[8] M. E. Wilms, V. J. Gadgil, J. M. Krougman, and F. P. Ijsseling, "The effect of σ-phase precipitation at 800°C on the corrosion resistance in sea-water of a high alloyed duplex stainless steel," *Corrosion Science*, vol. 36, no. 5, pp. 871–881, 1994.

[9] C. M. Souza, H. F. G. Abreu, S. S. M. Tavares, and J. M. A. Rebello, "The σ phase formation in annealed UNS S31803 duplex stainless steel: texture aspects," *Materials Characterization*, vol. 59, no. 9, pp. 1301–1306, 2008.

[10] J. Lee, I. Kim, and A. Kimura, "Application of small punch test to evaluate sigma-phase embrittlement of pressure vessel cladding material," *Journal of Nuclear Science and Technology*, vol. 40, no. 9, pp. 664–671, 2003.

[11] D. Peckner and I. M. Bernstein, *Handbook of Stainless Steels*, McGraw-Hill, New York, NY, USA, 1st edition, 1977.

[12] M. H. Lewis, "Precipitation of (Fe, Cr) sigma phase from austenite," *Acta Metallurgica*, vol. 14, no. 11, pp. 1421–1428, 1966.

[13] F. B. Waanders, S. W. Vorster, and H. Pollak, "The influence of temperature on σ-phase formation and the resulting hardening of Fe-Cr-Mo-alloys," *Hyperfine Interactions*, vol. 120-121, no. 1–8, pp. 751–755, 1999.

[14] K. Shinohara, T. Seo, and K. Kumada, "Recrystallization and sigma phase formation as concurrent and interacting phenomena in 25%Cr-20%Ni steel," *Materials Transactions*, vol. 20, no. 12, pp. 713–723, 1979.

[15] E. Baerlecken and H. Fabritius, "Umwandlungskinetik der sigmaphase in einer Eisen-Chrom-Legierung mit 48% Chrom," *Arch Eisenhüttenwes*, vol. 26, pp. 679–686, 1955.

[16] C. M. Garzón and A. J. Ramirez, "Growth kinetics of secondary austenite in the welding microstructure of a UNS S32304 duplex stainless steel," *Acta Materialia*, vol. 54, no. 12, pp. 3321–3331, 2006.

[17] E. Folkhard, *Welding Metallurgy of Stainless Steels*, Springer, New York, NY, USA, 1st edition, 1988.

[18] J. Barcik, "Mechanism of σ-phase precipitation in Cr-Ni austenitic steels," *Materials Science and Technology*, vol. 4, no. 1, pp. 5–15, 1988.

[19] V. K. Sikka, M. G. Cowgill, and B. W. Roberts, "Creep properties of modified 9Cr-1Mo Steel," in *Proceedings of Topical Conference on Ferritic alloys for Use in Nuclear Energy Technologies*, ASM International, 1983.

[20] Y. S. Sato, H. Kokawa, K. Okamoto, S. Hirano, and M. Inagaki, "Rapid formation of the sigma phase in 304 stainless steel during friction stir welding," *Scripta Materialia*, vol. 49, no. 12, pp. 1175–1180, 1999.

[21] Y. S. Na, N. K. Park, and R. C. Reed, "Sigma morphology and precipitation mechanism in Udimet 720Li," *Scripta Materialia*, vol. 43, no. 7, pp. 585–590, 2000.

[22] A. F. Padilha and P. R. Rios, "Decomposition of austenite in austenitic stainless steels," *ISIJ International*, vol. 42, no. 4, pp. 325–337, 2002.

[23] B. Weiss and R. Stickler, "Phase instabilities during high temperature exposure of 316 austenitic stainless steel," *Acta Metallurgica*, vol. 3, no. 4, pp. 851–866, 1972.

[24] R. J. Gray, V. K. Sikka, and R. T. King, "Detecting transformation of delta-ferrite to sigma phase in stainless steels by advanced metallographic techniques," *Journal of Metals*, vol. 30, no. 11, pp. 18–26, 1978.

[25] C. C. Hsieh, *The study of $\delta/\sigma/\gamma$ phase transformation in 309LSi stainless steels after aging under nitrogen atmospheres*, M.S. thesis, I-Shou University, Taiwan, 2004.

[26] X. Tang, "Sigma phase characterization in AISI 316 stainless steel," *Microscopy and Microanalysis*, vol. 11, no. 2, pp. 78–79, 2005.

[27] C. C. Hsieh, D. Y. Lin, and W. Wu, "Dispersion strengthening behavior of σ phase in 304 modified stainless steels during 1073 K hot rolling," *Metals and Materials International*, vol. 13, no. 5, pp. 359–363, 2007.

[28] C. J. Park, M. K. Ahn, and H. S. Kwon, "Influences of Mo substitution by W on the precipitation kinetics of secondary phases and the associated localized corrosion and embrittlement in 29% Cr ferritic stainless steels," *Materials Science and Engineering A*, vol. 418, no. 1-2, pp. 211–217, 2006.

[29] C. C. Hsieh, D. Y. Lin, and T. C. Chang, "Microstructural evolution during the $\delta/\sigma/\gamma$ phase transformation of the SUS 309LSi stainless steel after aging under various nitrogen atmospheric ratios," *Materials Science and Engineering A*, vol. 475, no. 1-2, pp. 128–135, 2008.

[30] Y. S. Sato and H. Kokawa, "Preferential precipitation site of sigma phase in duplex stainless steel weld metal," *Scripta Materialia*, vol. 40, no. 6, pp. 659–663, 1999.

[31] J. Barcik, "The kinetics of σ-phase precipitation in AISI 310 and AISI 316 steels," *Metallurgical Transactions*, vol. 14, no. 4, pp. 635–641, 1983.

[32] D. M. E. Villanueva, F. C. P. Junior, R. L. Plaut, and A. F. Padilha, "Comparative study on sigma phase precipitation of three types of stainless steels: austenitic, superferritic and duplex," *Materials Science and Technology*, vol. 22, no. 9, pp. 1098–1104, 2006.

[33] A. L. Shaeffler, "Constitution diagram for stainless steel weld metal," *Metal Progress*, vol. 56, pp. 680–681, 1949.

[34] S. Konosu, H. Mashiba, M. Takeshima, and T. Ohtsuka, "Effects of pretest aging on creep crack growth properties of type 308 austenitic stainless steel weld metals," *Engineering Failure Analysis*, vol. 8, no. 1, pp. 75–85, 2001.

[35] M. C. Mataya and Martin J. Carr, "Characterization of inhomogeneities in complex austenitic stainless steel forgings," in *Deformation, Processing, and Structure*, ASM Materials Science Seminar, St. Louis, Mo, USA, 1982.

[36] G. S. Reis, A. M. Jorge, and O. Balancin, "Influence of the microstructure of duplex stainless steels on their failure characteristics during hot deformation," *Materials Research*, vol. 3, pp. 31–35, 2000.

[37] N. Lopez, M. Cid, and M. Puiggali, "Influence of σ-phase on mechanical properties and corrosion resistance of duplex stainless steels," *Corrosion Science*, vol. 41, no. 8, pp. 1615–1631, 1999.

[38] K. Ravindranath and S. N. Malhotra, "Influence of aging on intergranular corrosion of a 25% chromium-5% nickel duplex stainless steel," *Corrosion*, vol. 50, no. 4, pp. 318–328, 1994.

[39] J. T. Gow and O. E. Harder, "Balancing the composition of cast 25 per cent chromium-12 per cent nickel type alloys," *Transactions of the American Society for Metals*, vol. 30, pp. 855–935, 1942.

[40] L. R. Woodyatt, *Austenitic Stainless Steels: Microstructure and Mechanical Properties*, Springer, London, UK, 1st edition, 1984.

[41] F. C. Hull, "The effect of δ-ferrite on the hot cracking of stainless steel," *Welding Journal*, vol. 46, pp. 399s–409s, 1980.

[42] R. Badji, M. Bouabdallah, B. Bacroix, C. Kahloun, K. Bettahar, and N. Kherrouba, "Effect of solution treatment temperature on the precipitation kinetic of σ-phase in 2205 duplex stainless steel welds," *Materials Science and Engineering A*, vol. 496, no. 1-2, pp. 447–454, 2008.

[43] T. P. S. Gill, M. Vijayalakshmi, J. B. Gnanamoorthy, and K. A. Padmanabhan, "Transformation of delta-ferrite during the postweld heat treatment of type 316L stainless steel weld metal," *Welding Journal*, vol. 65, no. 5, pp. 122–128, 1986.

[44] D. Y. Lin, G. L. Liu, T. C. Chang, and H. C. Hsieh, "Microstructure development in 24Cr-14Ni-2Mn stainless steel after aging under various nitrogen/air ratios," *Journal of Alloys and Compounds*, vol. 377, no. 1-2, pp. 150–154, 2004.

[45] H. Brandis, W. Heimann, and E. Schmidtmann, "Effect of nitrogen on the precipitation behavior of steel X3CrNiMoNbN23 17," *TEW-Technische Berichte*, vol. 2, no. 2, pp. 150–166, 1976.

[46] O. Smuk, P. Nenonen, H. Hänninen, and J. Liimatainen, "Microstructures of a powder metallurgy-hot-isostatically pressed super duplex stainless steel forming in industrial heat treatments," *Metallurgical and Materials Transactions A*, vol. 35, no. 7, pp. 2103–2109, 2004.

[47] SGTE Solution Database, Department of Materials Science and Engineering, Royal Institute of Technology, Stockholm, Sweden.

[48] M. Lindholm, "A description of the thermodynamics of some Si containing systems," Licentiate thesis, Royal Institute of Technology, Stockholm, Sweden, 1995.

[49] J. Erneman, M. Schwind, L. Nylöf, J.-O. Nilsson, H.-O. Andrén, and J. Ågren, "Comparison between quantitative metallography and modeling of σ-phase particle growth in AISI 347 stainless steel," *Metallurgical and Materials Transactions A*, vol. 36, no. 10, pp. 2595–2600, 2005.

[50] N. Saunders and A. P. Miodownik, *CALPHAD (Calculation of Phase Diagrams) A Comprehensive Guide*, Elsevier Science, 1st edition, 1998.

[51] T. H. Chen and J. R. Yang, "Effects of solution treatment and continuous cooling on σ-phase precipitation in a 2205 duplex stainless steel," *Materials Science and Engineering A*, vol. 311, no. 1-2, pp. 28–41, 2001.

[52] S. H. Hong and Y. S. Han, "Phenomena and mechanism on superplasticity of duplex stainless steels," *Metals and Materials International*, vol. 6, no. 2, pp. 161–167, 2000.

[53] K. Holm, J. D. Embury, and G. R. Purdy, "The structure and properties of microduplex Zr-Nb alloys," *Acta Metallurgica*, vol. 25, no. 10, pp. 1191–1200, 1977.

[54] J. R. Spingarn and W. D. Nix, "Diffusional creep and diffusionally accommodated grain rearrangement," *Acta Metallurgica*, vol. 26, no. 9, pp. 1389–1398, 1978.

[55] T. Chandra and R. Kuchlmayr, "Effect of strain rate on sigma formation in ferrite-austenite stainless steel at high temperatures," *Journal of Materials Science*, vol. 23, no. 2, pp. 723–728, 1988.

[56] F. Abe, H. Arki, and T. Noda, "Discontinuous precipitation of σ-phase during recrystallization in cold rolled Fe-10Cr-30 Mn austenite," *Materials Sciences and Technology*, vol. 4, pp. 885–893, 1988.

[57] C. S. Huang and H. C. Chen, "A discussion on stainless steels welding (II)," *Welding and Cutting*, vol. 8, pp. 48–54, 1998.

[58] T. Koutsoukis, E. Papadopoulou, P. Kokkonidis, and G. Fourlaris, "Effect of ageing in cold rolled superaustenitic stainless steels," in *EMC 2008 14th European Microscopy Congress*, S. Zormalia, Ed., pp. 451–452, Springer, Berlin, Germany, 2008.

[59] C. C. Hsieh, D. Y. Lin, and W. Wu, "Precipitation behavior of σ phase in 19Cr-9Ni-2Mn and 18Cr-0.75Si stainless steels hot-rolled at 800°C with various reduction ratios," *Materials Science and Engineering A*, vol. 467, no. 1-2, pp. 181–189, 2007.

[60] C.C. Hsieh, D.Y. Lin, and W. Wu, "Precipitation behavior of sigma phase in 304 and 430 stainless steels as hot-rolled at 800°C," in *Proceedings of the American Physical Society (APS '07)*, Colorado Convention Center, March 2007.

[61] T. P. S. Gill, M. Vijayalkshmi, P. Rodriguez, and K. A. Padmanabhan, "On microstructure-property correlation of thermally aged type 316L stainless steel weld metal," *Metallurgical Transactions A*, vol. 20, no. 6, pp. 1115–1124, 1989.

Thermal Expansion Behaviour of Ternary Nickel-Based, Cobalt-Based, and Iron-Based Alloys Containing Very High Fractions of Carbides

Patrice Berthod and Lionel Aranda

Team 206 "Surface and Interface, Chemical Reactivity of Materials", Department of Chemistry and Physics of Solids and Surface, Institut Jean Lamour (UMR CNRS 7198), Faculty of Sciences and Technologies, University of Lorraine, Boulevard des Aiguillettes, BP 70239, 54506 Vandoeuvre-lès-Nancy, France

Correspondence should be addressed to Patrice Berthod, patrice.berthod@ijl.nancy-universite.fr

Academic Editors: F. Hori and M.-C. Wang

Some of the wear-resistant pieces or coatings, constituted of a metallic matrix and of carbides present in high fractions, are a mix, in similar quantities, of two materials displaying greatly different levels of hardness but also of thermal expansion coefficient. When temperature increases, the second difference of property may lead to particular geometrical behaviours. To study these differences, nine nickel-based, cobalt-based, and iron-based alloys containing very high quantities of carbides were elaborated by foundry. In their as-cast conditions, the microstructures of these alloys were characterized; their hardness and thermal expansion until 1200°C were measured and analysed, with regard to the evolution of the structures predicted by thermodynamic calculations. The hardness of the alloys is high (nickel alloys) or very high (cobalt and iron alloys, 600 Hv and more) while the thermal expansion is greatly influenced by carbides, notably when temperature has become very high. Some of the variations of thickness at the end of heating or during an isothermal stage at 1200°C, essentially contraction, directly result from the mechanical interaction between matrix and carbides which was accumulated during the heating.

1. Introduction

Refractory metallic alloys containing significant quantities of carbides are of a great importance for high temperature applications [1, 2], even if they are less common than the γ/γ'-type superalloys, solid-solution strengthened cobalt-based superalloys, or heat-resistant steels. Many of the carbides-strengthened alloys are based on nickel, cobalt, or iron and elaborated by foundry. They generally contain great quantities (several tens percents in weight) of chromium, which is a carbide-former element promoting, in presence of carbon, the formation of Cr_7C_3 or $Cr_{23}C_6$ carbides.

The chromium-rich cast nickel-based alloys are used for numerous applications, notably for turbine blades in aeronautics, for which chromium brings to the alloys a good resistance against high temperature oxidation [3, 4]. Carbon can be added to the (Ni, Cr-) based alloys in order to obtain high levels of mechanical resistance and an example of hardness [5–8]. Cobalt-based alloys rich in chromium are also considered for hot parts in aeronautic turbines and power generation machines, as well as for other applications like some of the fiberizing tools used in the glass industry [9]. Generally several tens of weight percents of chromium are added to cobalt to allow the alloy sufficiently resisting hot corrosion by various molten substances (salts, glasses, etc.), as well as high temperature oxidation. Its role can be also to develop carbides with carbon for improving the high temperature mechanical resistance against creep. High levels of hardness can be expected but they are generally achieved using other types of carbides, as for cutting tools [10], made of a cobalt matrix containing high amounts of dispersed WC carbides, or for coatings [5] consisting in Co-W_2C deposited by thermal spray on steels. It is well known that the carbon-containing iron-based alloys are able to offer great volume fractions of hard phases (as cementite, pearlite, or martensite) which allow leading to high hardness values. In their cases too, chromium can be added to the chemical composition to enhance mechanical resistance or hardness of bulks [11] or of hardfacing coatings [12] by forming numerous interdendritic chromium carbides.

Thus, chromium carbides can be met in nickel-, cobalt- or iron-based alloys to mechanically strengthen them at high temperature (moderate amounts of carbon and fractions of carbides) or for wear-resistance purpose (high-carbon contents and carbides fractions). In all cases, in service, the temperature of such alloys can increase up to high levels (several hundreds of Celsius degrees), either because of a high temperature in use (e.g., superalloy in turbine blades exposed to hot gases, near 1000–1300°C) or because of a heating due to friction/abrasion (e.g., cutting tool, lower temperatures). The difference of thermal expansion coefficient existing between matrix (15–20 $\times$ 10^{-6}°C^{-1}) and carbides (10 $\times$ 10^{-6}°C^{-1}) [13] may then induce a particular global thermal expansion behaviour for the whole alloy.

The aim of this work is to study the possible matrix-carbides mechanical interactions during heating and cooling and their consequences on the thermal expansion behaviours of several model nickel-based alloys, cobalt-based alloys, and iron-based alloys, all containing the same high quantity of chromium and, for each family, three different high levels in carbon to obtain three different high carbide fractions.

2. Details of the Experiments

2.1. Elaboration of the Studied Alloys. Nine high-carbon 30 wt.% Cr-containing ternary alloys were synthesized by foundry from pure elements. They are M-30Cr-xC alloys, with M = Ni, Co or Fe, and x = 2.5, 3.0 or 3.5 wt.%. The ingots, of a mass equal to about 30 g, were cast in the water-cooled copper crucible of a High Frequency induction furnace (CELES), under an inert atmosphere of 300 millibars of pure argon, by melting pure elements together (purity higher than 99% in mass): Co, Ni, Fe, and Cr coming from Alfa Aesar, and C added as pure graphite.

2.2. Metallographic Preparation and Characterization: Hardness Measurements. The ingots were thereafter cut using a Buehler Isomet 5000 precision saw. A sample per ingot was embedded in a cold resin (Escil CY230 + HY956), and the mounted samples were polished by using SiC paper from 120 to 1200 grit under water, then with a 1 μm-diamond paste. A Philips XL30 Scanning Electron Microscope was used for metallographic observations, essentially in the Back Scattered Electrons mode (BSE) under an acceleration voltage of 20 kV. The chemical composition of each alloy was controlled in the bulk, by using a Cameca SX microprobe in a delocalized mode (magnification: ×400) on several locations, while a Philips XPERT'Pro diffractometer was also used to complete the metallographic characterization.

The hardness of each mounted and polished alloy was measured by performing Vickers indentations under a 30 kg load, using a Testwell Wolpert device.

2.3. Dilatometry Runs and Thermodynamic Calculations. The thermal expansion was measured using a TMA 92-16.18 Setaram apparatus on parallelepipedic samples cut in each ingot the dimensions of which were about 4 mm $\times$ 4 mm (square main faces) $\times$ 2 mm (direction of the characterized expansion). The heating was performed from ambient temperature up to 1200°C (rate of 10°C min^{-1}), followed by a dwell of 1 hour and then a cooling down to ambient temperature (rate of -10°C min^{-1}). Three parts of the evolution of the sample's thickness were studied: the thermal expansion during the heating, the thermal contraction during the cooling, and the isothermal dimensional evolution during the 1-hour dwell at 1200°C.

In order to interpret the phenomena occurring during the dilatometry runs calculations of thermodynamic, stable states were carried out using the Thermo-Calc software [14] and a data base containing the descriptions of the Cr-C system [15], of the Ni-C, Ni-Cr and Ni-Cr-C [16–18] systems for the nickel alloys, of the Co-C, Co-Cr, and Co-Cr-C systems for the cobalt alloys [15, 17–19], and of the Fe-C, Fe-Cr, and Fe-Cr-C systems [20–23] for the iron alloys.

3. Results

3.1. Microstructures and Room Temperature Hardness of the Alloys. All the alloys in the as-cast conditions were examined using the Scanning Electron Microscope (SEM) in Back Scattered Electrons (BSE) mode. Nine micrographs illustrating their microstructures are given in Figure 1.

The three as-cast nickel alloys display microstructures constituted in both matrix and carbides. In all cases, the alloy's matrix is a nickel-FCC solid solution, and the carbides are both Cr_7C_3 and Cr_3C_2, as revealed by X-ray diffraction. This can be also seen on the BSE micrographs with the coarse proeutectic carbides; the outer part of which is darker than the middle (the Cr_7C_3 carbides were seemingly transforming into C-richer Cr_3C_2 carbides during cooling). All microstructures are effectively of a hypereutectic type, with presence of these acicular proeutectic carbides, which are discrete in the Ni-30Cr-2.5C alloy but which become more present and coarse for the two highest carbon contents, 3.0 and 3.5 wt.%. The hardness values increase with the carbon content, from 341 ± 10 for 2.5C to 386 ± 15 for 3.5C ($Hv_{30\,kg}$).

For the three as-cast cobalt alloys, the microstructure are, on the contrary, hypoeutectic for 2.5C and 3.0C and seemingly eutectic for 3.5C. The matrix is a mix of HCP and FCC cobalt solid solutions and carbides are essentially Cr_7C_3. These alloys are significantly harder than the nickel alloys since their hardness increase from 588 ± 15 for 2.5C to 646 ± 10 for 3.5C ($Hv_{30\,kg}$).

With the three iron-based alloys, the three microstructures characters, hypoeutectic, eutectic and hypereutectic, are represented with the Fe-30Cr-2.5C, Fe-30Cr-3.0C, and Fe-30Cr-3.5C alloys, respectively. As shown by X-ray diffraction, the matrix is Centred Cubic, and the carbides are Cr_7C_3 in all cases. The hardness increases from 584 ± 4 for 2.5C to 619 ± 12 for 3.5C ($Hv_{30\,kg}$).

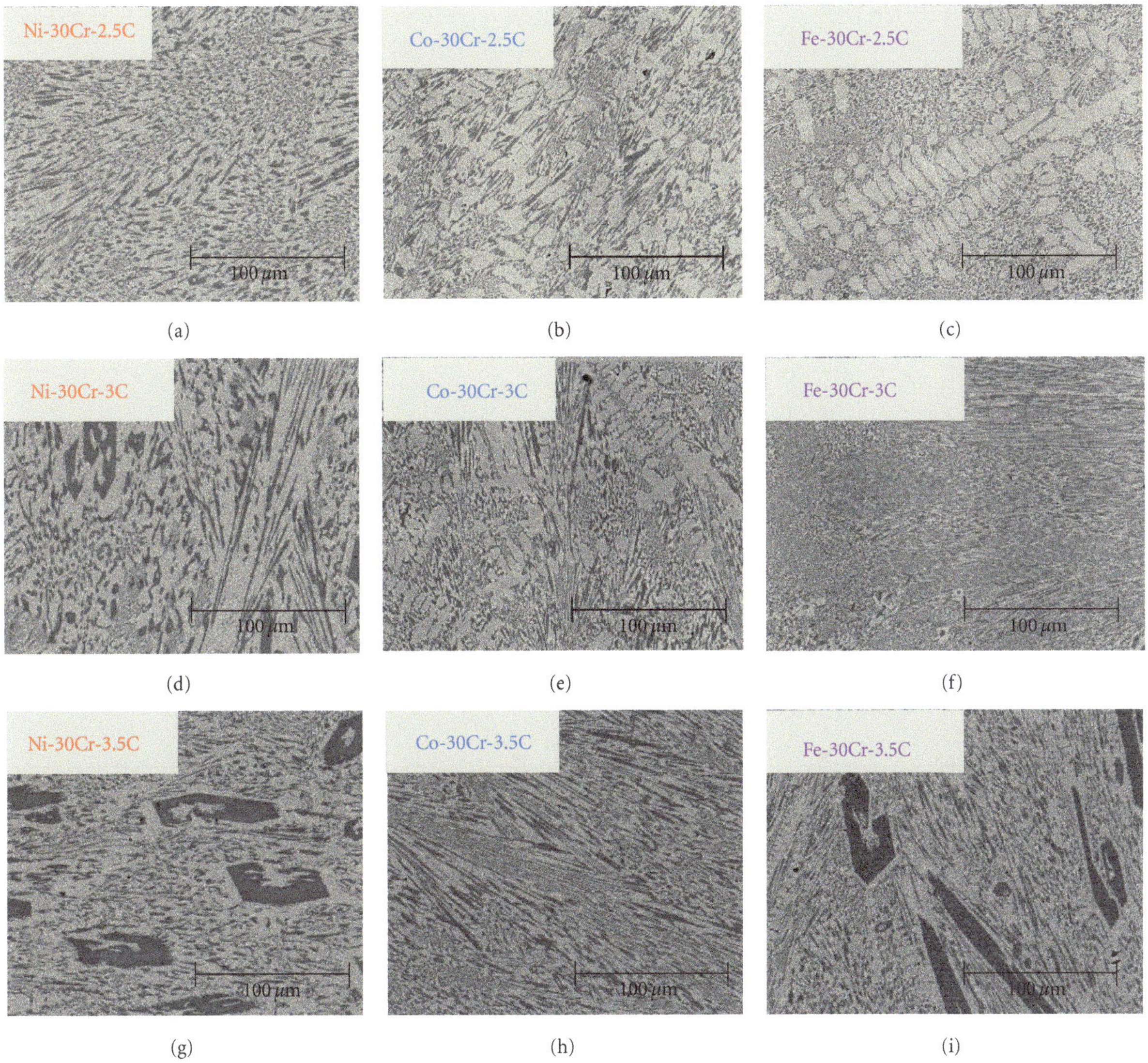

FIGURE 1: Microstructures of the studied alloys (SEM/BSE micrographs).

3.2. Thermal Expansion Behaviours. The obtained dilatometry curves are given in Figure 2 for the nickel alloys, Figure 3 for the cobalt alloys, and Figure 4 for the iron alloys. Only their parts above 100°C (and even sometimes 200°C) were taken under consideration because perturbations caused at very low temperature by the thermal equilibrium establishment of the apparatus.

The thermal expansion of the nickel alloys (Figure 2) is often almost linear from 100°C to 1200°C, with a thermal expansion coefficient of around $17\text{-}18 \times 10^{-6\circ}C^{-1}$. However, when approaching the maximal temperature, one can notice a deceleration of expansion in the case of the Ni-30Cr-3.5C alloy, over about 900°C. During the isothermal stage a contraction happens. It is revealed by a position of the 1200°C-extremity of the heating part of the dilatometry curve which is higher than the 1200°C-extremity of its cooling part (Figure 2, graphs on the left side). These contractions are more visible on the graphs presented in the right side of the same figure, with curves plotted versus time instead temperature. One can see an isothermal contraction initially rapid and slower thereafter, which is more pronounced for the highest carbon alloy than for the others: the absolute value of the total isothermal contraction increased from 0.09% to 0.24% (Table 1) for carbon content in the alloy increasing from 2.5 to 3.5 wt.% C. This generally results in a residual negative deformation after return at 100°C, the absolute value of which increases from 0.04% to 0.18% for a carbon content in the alloy increasing from 2.5 to 3.5 wt.% C.

The cobalt alloys do not present so clear evolution of their thermal expansion behaviour versus the carbon content (Figure 3). Indeed, if the Co-30Cr-2.5C alloy behaves like the nickel alloy with the same carbon content, curious phenomena affect the thermal expansion curves of the two C-richest cobalt alloys. The Co-30Cr-3.0C curve suddenly jumps near 1000°C just before knowing a contraction (sudden too) when approaching 1200°C. This contraction goes on during the stage over a particularly high amplitude (−0.52%, Table 2).

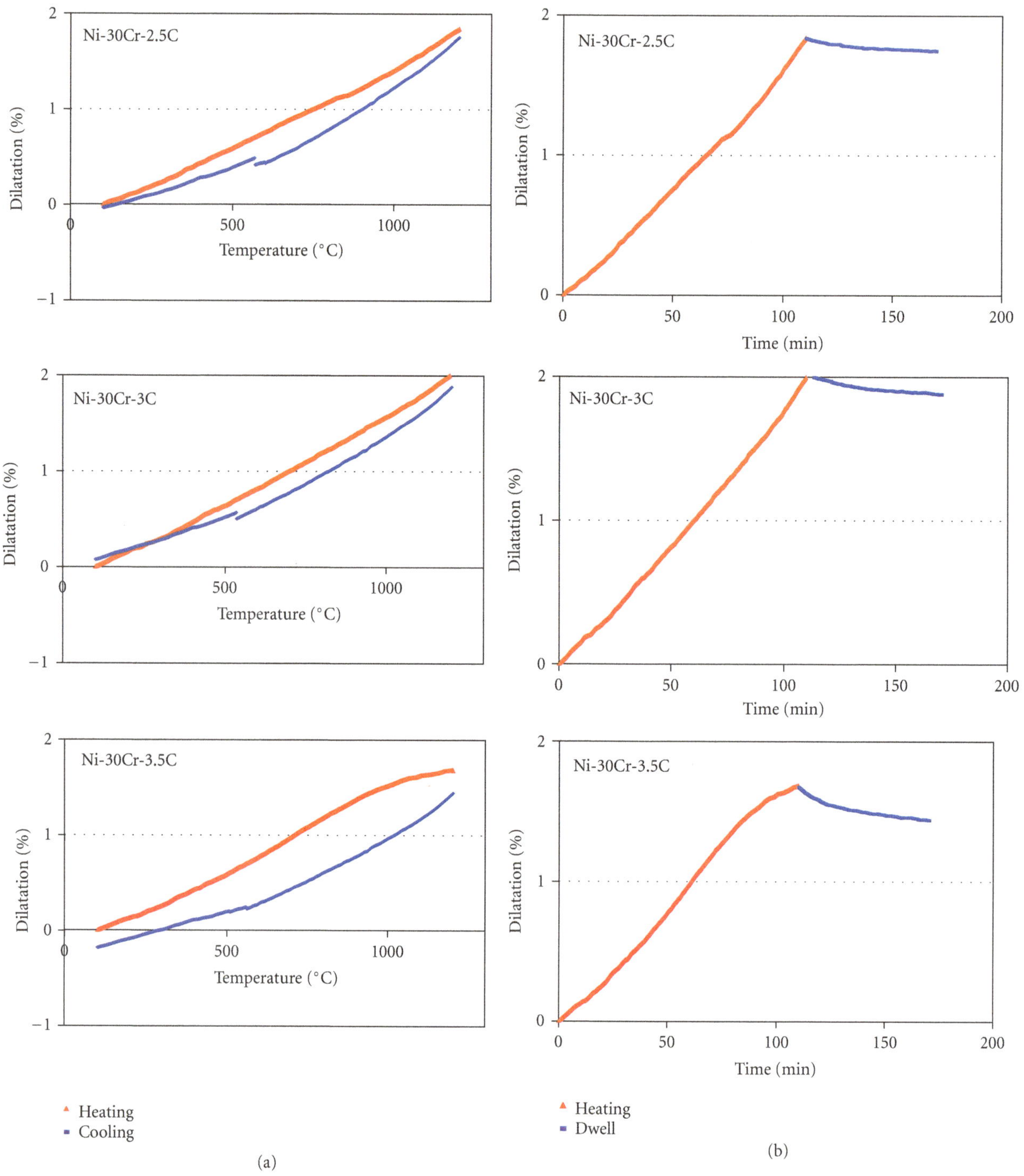

FIGURE 2: The whole thermal expansion curves of the nickel-based alloys versus temperature (left) and the (heating + isothermal) parts versus time (right).

A similar phenomenon seems affecting the high temperature part of the heating dilatometry curve of the Co-30Cr-3.5C alloy, but with amplitudes much lower. The cooling part of all curves presents a small dilatation near 700°C. Concerning the average thermal expansion coefficient (determined on the linear part of the heating dilatometry curve), it is of about 16 $\times$ 10^{-6}°C^{-1} for the three alloys, as is to say a little smaller than for the nickel alloys.

The iron alloys present dilatometry curves; the heating part of which can be divided into two parts. The average thermal coefficient (Table 3) on the low temperature part (typically <800°C) is generally lower than the one of the high temperature part (>800°C), and these two α coefficients both decrease when the carbon content increases (from 18 to less than 16 $\times$ 10^{-6}°C^{-1} for the low temperature part, and from 25 $\times$ 10^{-6}°C^{-1} to 14 $\times$ 10^{-6}°C^{-1} for the high

TABLE 1: Thermal expansion coefficients measured on the dilatometry curves, deformation at T reaching 1200°C, deformation and total deformation values after the 1200°C-isothermal stage ($\Delta\varepsilon_{1200\%}$) and after return to 100°C ($\Delta\varepsilon_{100\%}$)—nickel-based alloys.

Temperature (°C)	Average linear thermal expansion coefficient ($\times 10^{-6}$°C^{-1})		
	Ni-30Cr-2.5C	Ni-30Cr-3.0C	Ni-30Cr-3.5C
Average α (10^{-6} K^{-1})	**16.6**	**18.2**	**17.8**
T range for α	100–1200°C	100–1200°C	100–900°C
$\varepsilon_{1200\%}$ end heating	1.843	2.012	1.674
$\varepsilon_{1200\%}$ start cooling	1.751	1.881	1.439
Total $\Delta\varepsilon_{1200\%}$	−0.092	−0.131	−0.235
Total $\Delta\varepsilon_{100\%}$	−0.035	+0.077	−0.180

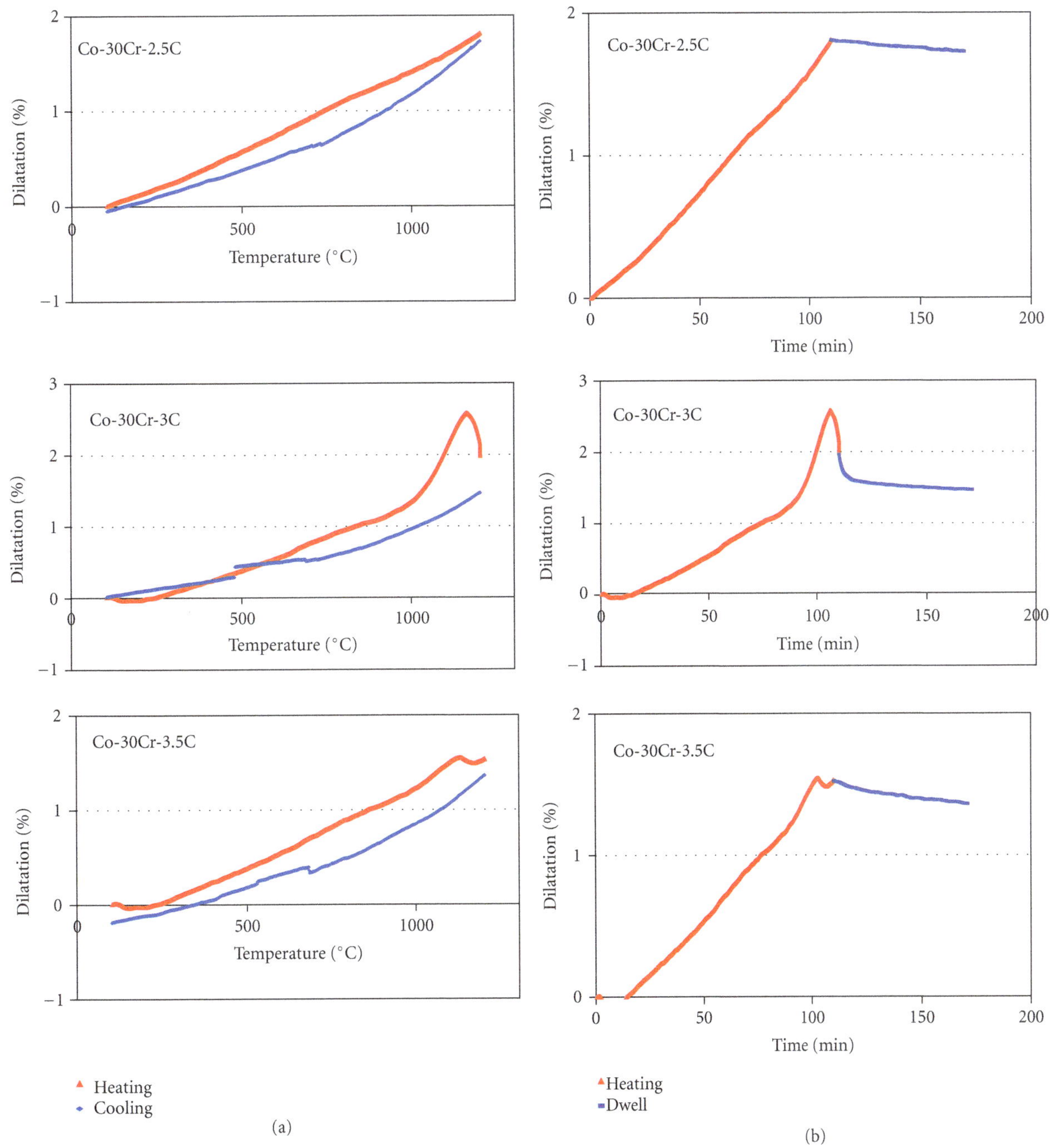

FIGURE 3: The whole thermal expansion curves of the cobalt-based alloys versus temperature (left) and the (heating + isothermal) parts versus time (right).

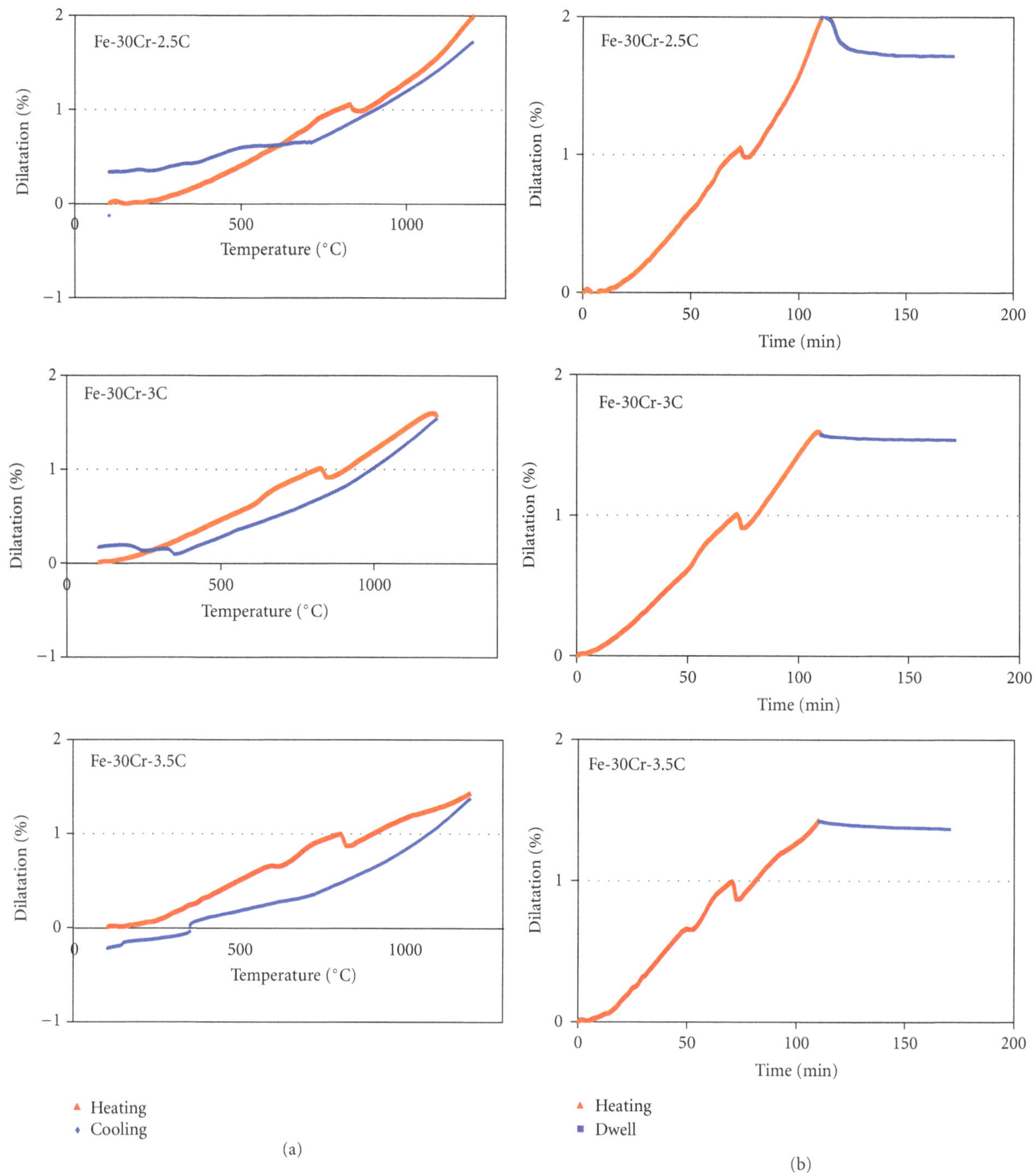

Figure 4: The whole thermal expansion curves of the iron-based alloys versus temperature (left) and the (heating + isothermal) parts versus time (right).

temperature part). At the end of heating and during the isothermal stage, there is also a contraction but which is curiously less pronounced when the carbon content of the alloy is higher, as is to say contrarily to what was observed for the nickel alloys. Indeed, the absolute value of the isothermal contraction decreases from 0.28 to 0.06% when the carbon content increases from 2.5 to 3.5 wt.%. However, the residual deformation after return to 100°C after cooling decreases (from +0.34 to −0.22%) when the carbon content in the alloy increases, as observed for the nickel alloys and for the cobalt alloys.

4. Discussion

Thus, with such high carbon contents, the Ni-30Cr, Co-30Cr, and Fe-30Cr systems present high densities of carbides, Cr_7C_3 exclusively (cases of the cobalt or iron alloys) or together with the C-richer Cr_3C_2 carbides for the nickel

TABLE 2: Thermal expansion coefficients measured on the dilatometry curves, deformation at T reaching 1200°C, deformation and total deformation values after the 1200°C-isothermal stage ($\Delta\varepsilon_{1200\%}$) and after return to 100°C ($\Delta\varepsilon_{100\%}$)—cobalt-based alloys.

Temperature (°C)	Average linear thermal expansion coefficient ($\times 10^{-6}$°C^{-1})		
	Co-30Cr-2.5C	Co-30Cr-3.0C	Co-30Cr-3.5C
Average α (10^{-6} K^{-1})	**16.5**	**15.8**	**16.1**
T range for α	100–1200°C	200–900°C	200–1000°C
$\varepsilon_{1200\%}$ end heating	1.812	1.977	1.527
$\varepsilon_{1200\%}$ start cooling	1.728	1.462	1.363
Total $\Delta\varepsilon_{1200\%}$	−0.084	−0.515	−0.164
Total $\Delta\varepsilon_{100\%}$	−0.049	+0.013	−0.190

TABLE 3: Thermal expansion coefficients measured on the dilatometry curves, deformation at T reaching 1200°C, deformation and total deformation values after the 1200°C-isothermal stage ($\Delta\varepsilon_{1200\%}$) and after return to 100°C ($\Delta\varepsilon_{100\%}$)—iron-based alloys.

Temperature (°C)	Average linear thermal expansion coefficient ($\times 10^{-6}$°C^{-1})		
	Fe-30Cr-2.5C	Fe-30Cr-3.0C	Fe-30Cr-3.5C
Average α 1 (10^{-6} K^{-1})	**17.8**	**15.2**	**15.7**
T range for α 1	230–725°C	200–800°C	200–800°C
Contraction T range	808–886°C	798–837°C	780–821°C
Average α 2 (10^{-6} K^{-1})	**24.5**	**21.5**	**13.9**
T range for α 2	870–995°C	870–1170°C	820–1185°C
$\varepsilon_{1200\%}$ end heating	1.996	1.578	1.422
$\varepsilon_{1200\%}$ start cooling	1.720	1.537	1.366
Total $\Delta\varepsilon_{1200\%}$	−0.276	−0.041	−0.056
Total $\Delta\varepsilon_{100\%}$	+0.336	+0.166	−0.223

alloys. Unsurprisingly too, the obtained hardness are very high and increase with the carbon content in the three systems, showing that such alloys are very hard and suitable for wear-resistance applications (especially the cobalt alloys and the iron alloys with their 600 Hv and more). This can be easily explained by the great hardness of the present carbides: 1336 $Hv_{50\,g}$ for Cr_7C_3 and 1350 $Hv_{50\,g}$ for Cr_3C_2 [24].

If the temperature of these alloys increases, these ones logically know thermal expansion. Concerning the nickel alloys, this expansion is globally linear, at least until reaching a very high level of temperature (1200°C for the Ni-30Cr-2.5C and Ni-30Cr-3.0C alloys, 900°C for the Ni-30Cr-3.5C alloy). This is not surprising since the matrix remains FCC over the whole temperature range, as verified with Thermo-Calc calculations (which also confirms the possibility to get both Cr_7C_3 and Cr_3C_2 carbides for the three alloys). The heating parts of the dilatometry curves of the cobalt alloys are also globally linear, despite the theoretical change of matrix from HCP to FCC as predicted by Thermo-Calc near 500°C. It is true that the matrix was not initially wholly HCP as shown by the XRD patterns performed on the as-cast cobalt alloys. However, one can think that the alloy has become wholly austenitic at the end of heating, without accompanying dilatometry phenomenon. The new appearance of a new HCP part of matrix during cooling seems being revealed by small discontinuities in the dilatometry curve at 700°C and in two cases (Co-30Cr-3.0C and Co-30Cr-3.5C) at 500°C, which is the temperature of the FCC → HCP change according to Thermo-Calc calculations. The iron alloys also know a linear thermal expansion at heating, but which is linear on two parts. The low-temperature one corresponds to a ferritic matric (revealed by the XRD patterns) and the high-temperature one corresponds to an austenitic matrix. This allotropic change, occurring over about 800–900°C, is also predicted by Thermo-Calc for these three alloys. One can notice that the dilatometry runs show that the temperature range of $\alpha \rightarrow \gamma$ transformation slightly decreases when the carbon content increases, from about 810–890°C for the Fe-30Cr-2.5C alloy to 780–820°C for the Fe-30Cr-3.5C alloy, which is slightly lower than for pure iron for which this transformation occurs near 910°C with the same contraction consequence.

Almost all the alloys have also encountered, either already before the end of heating or at least during the whole isothermal stage, a contraction which often results in a residual negative deformation after return to 100°C in our case. Such curious behaviour was previously encountered, and explained, for (Ni, Co, or Fe)-30Cr-0.8C alloys [25]: the difference of thermal expansion coefficient between the matrix (e.g., 18–20 × 10^{-6}°C^{-1} for a Ni-30Cr binary alloy) and the chromium carbides (about 10 × 10^{-6}°C^{-1}) leads to an increasing tensile elastic deformation of the carbides due to matrix expansion, accompanied by an increased compression stress applied by carbides to the matrix; when temperature is high enough to weaken the matrix, this one can be visco-plastically deformed in compression under the action of the carbides which reduces their accumulated tensile stresses. Such phenomena may appear if carbides are sufficiently present in the metallic alloys, but it can be reduced, and even eliminated, by fragmenting carbides by applying an adequate heat treatment if the carbides are not too coarse and not too numerous [26].

This contraction affects the Ni-30Cr-2.5C and Ni-30Cr-3.0C alloys only during the isothermal stage, and during this hour at 1200°C, the deformation in compression is greater for a higher carbon content, that is, for a higher carbides fraction. Due to a higher carbide fraction, the contraction begins before the end of the heating for the C-richest nickel alloy, and goes on, with a higher rate than for the two other alloys, during the isothermal stage. The same phenomenon affects the cobalt alloys, but with different forms. The Co-30Cr-2.5C alloy globally behaves like the nickel alloy containing the same carbon. Because of their higher carbide fractions, the linear parts of the thermal expansion of the Co-30Cr-3.0C and Co-30Cr-3.5C alloys display rates of deformation versus temperature which are lower. But when temperature is high enough, it seems that the compressed matrix becomes more (Co-30Cr-3.0C) or less (Co-30Cr-3.5C) free from the grip of carbides, but these ones finish to compress matrix at least during the isothermal stage. At the end of this one, just before cooling, the total deformation of the alloy is anyway lower for the alloys containing more carbides. Concerning the iron alloys, similar observations can be made, and the deformation at the end of heating as well as the deformation at the end of isothermal stage, is also lower if the carbon content or carbide fraction is higher in the alloy.

Thus there are significant consequences for the dimensional evolution of carbides-rich alloys in temperature cycling conditions. To avoid such effects and hysteresis of deformation between heating and cooling, it may be first possible to fragment carbides by applying a specific heat treatment, as already successfully obtained for alloys less rich in carbon. Another possible mean is to reinforce the matrix, for example, by solid solution strengthening, to allow it to deform reversibly by remaining in the elastic deformation domain and then avoiding visco-plastic deformation.

5. Conclusions

The alloys of this study, which are for most of them simplified versions of real alloys used to resist wear, for example, as coatings deposited on mechanical pieces, present logically high levels of hardness—especially when based on cobalt or iron—and also particular thermal expansion behaviours. In service, if frictions lead to very high temperatures, such behaviours may induce small but real deformation of these pieces, a lack of geometrical stability which can become a problem. But some means for diminishing this effect potentially may exist, by structural hardening allowing matrix to better resist visco-plastic deformation and then recovering a reversible character of deformation.

Acknowledgments

The authors wish to thank E. Souaillat, O. Hestin, M. Ba, and A. Dia, master students, for their contributions, as well as T. Schweitzer for his technical assistance.

References

[1] E. F. Bradley, *Superalloys: A Technical Guide*, ASM International, Materials Park, Ohio, USA, 1988.

[2] C. T. Sims and W. C. Hagel, *The Superalloys*, John Wiley & Sons, New York, NY, USA, 1972.

[3] P. Kofstad, *High Temperature Corrosion*, Elsevier Applied Science, London, UK, 1988.

[4] D. Young, *High Temperature Oxidation and Corrosion of Metals*, Elsevier, Amsterdam, The Netherlands, 2008.

[5] A. Klimpel, L. A. Dobrzański, A. Lisiecki, and D. Janicki, "The study of properties of Ni-W_2C and Co-W_2C powders thermal sprayed deposits," *Journal of Materials Processing Technology*, vol. 164-165, pp. 1068–1073, 2005.

[6] Z.-T. Wang and H.-H. Chen, "Microstructure and friction and wear behavior of induction clad Ni-based composite coating reinforced with micro- and nano-WC particulates," *Mocaxue Xuebao/Tribology*, vol. 25, no. 3, pp. 203–206, 2005.

[7] H. Han, S. Baba, H. Kitagawa et al., "Plasma-carburization of nickel-based self-fluxing alloy," *Vacuum*, vol. 78, no. 1, pp. 27–32, 2005.

[8] D. Zhang and X. Zhang, "Laser cladding of stainless steel with Ni-Cr_3C_2 and Ni-WC for improving erosive-corrosive wear performance," *Surface and Coatings Technology*, vol. 190, no. 2-3, pp. 212–217, 2005.

[9] P. Berthod, J. L. Bernard, and C. Liébaut, Cobalt Alloy and Fabrication of Articles from the Alloy, Patent WO99/16919.

[10] B. Roebuck and E. A. Almond, "Deformation and fracture processes and the physical metallurgy of WC-Co hardmetals," *International Materials Reviews*, vol. 33, no. 2, pp. 90–110, 1988.

[11] H. E. N. Stone, "Hardness, oxidation and constitution in chromium-iron alloys," *Journal of Materials Science*, vol. 14, no. 12, pp. 2787–2790, 1979.

[12] B. V. Cockeram, "Some observations of the influence of δ-ferrite content on the hardness, galling resistance, and fracture toughness of selected commercially available iron-based hardfacing alloys," *Metallurgical and Materials Transactions A*, vol. 33, no. 11, pp. 3403–3419, 2002.

[13] T. B. Shafffer, *Handbooks of High-Temperature Materials No.1. Materials Index*, Plenum Press, New York, NY, USA, 1964.

[14] Thermo-Calc version N: "Foundation for Computational Thermodynamics" Stockholm, Sweden, Copyright (1993, 2000).

[15] J. O. Andersson, "Thermodynamic properties of Cr-C," *Calphad*, vol. 11, no. 3, pp. 271–276, 1987.

[16] A. Gabriel, C. Chatillon, and I. Ansara, "Thermochemical and phase diagram analysis of the Ni-C, Co-C, and Co-Ni-C systems," *High Temperature Science*, vol. 25, no. 1, pp. 17–54, 1988.

[17] A. Fernández Guillermet, "Critical evaluation of the thermodynamic properties of cobalt," *International Journal of Thermophysics*, vol. 8, no. 4, pp. 481–510, 1987.

[18] A. F. Guillermet, "Thermodynamic analysis of the Co-Al-C and Ni-Al-C systems," *Zeitschrift für Metallkunde*, vol. 78, no. 10, pp. 700–709, 1987.

[19] A. F. Guillermet, "Thermodynamic properties of the iron-cobalt-carbon system," *Zeitschrift für Metallkunde*, vol. 79, no. 5, pp. 317–329, 1988.

[20] A. Fernandez Guillermet and P. Gustafson, "Assessment of the thermodynamic properties and the (p,T) phase diagram of iron," *High Temperatures—High Pressures*, vol. 16, no. 6, pp. 591–610, 1984.

[21] J. O. Andersson and B. Sundman, "Thermodynamic properties of the CrFe system," *Calphad*, vol. 11, no. 1, pp. 83–92, 1987.

[22] P. Gustafson, "A thermodynamic evaluation of the C-Fe system," *Scandinavian Journal of Metallurgy*, vol. 14, no. 5, pp. 259–267, 1985.

[23] J.-O. Andersson, "A thermodynamic evaluation of the Fe-Cr-C system," *Metallurgical Transactions A*, vol. 19, no. 3, pp. 627–636, 1988.

[24] G. V. Samsonov, *Handbooks of High-Temperature Materials No. 2. Properties Index*, Plenum Press, New York, NY, USA, 1964.

[25] P. Berthod, "Influence of carbides and of the dendritic orientation on the thermal expansion of Ni-base, Co-base and Fe-base simple cast alloys," *International Journal of Materials Research*, vol. 99, no. 3, pp. 265–272, 2008.

[26] P. Berthod, C. Heil, and L. Aranda, "Influence of the morphologic evolution of the eutectic carbides at high temperature on the thermal expansion behavior of refractory cast alloys," *Journal of Alloys and Compounds*, vol. 504, no. 1, pp. 243–250, 2010.

Effect of Artificial Aging on Plane Anisotropy of 6063 Aluminium Alloy

S. O. Adeosun, O. I. Sekunowo, M. A. Bodude, A. A. Agbeleye, S. A. Balogun, and H. O. Onovo

Department of Metallurgical & Materials Engineering, University of Lagos, Akoka, Lagos, Nigeria

Correspondence should be addressed to S. O. Adeosun, samsonoluropo@yahoo.com

Academic Editor: P. Lukac

Most aluminum profiles' production by deep-drawing and extrusion processes require certain degree of structural homogeneity because of the segregated second-phase particles in the as-cast structure. Rolled texture and directionality in properties often give rise to excessive earring, breakout, and tears. This study investigates the effect of heat treatment (artificial aging) on the anisotropic behavior of AA6063 alloy between rolling direction (0°) through 90° directions. The results show significant reduction in property variability in the aged samples along the rolling direction 0°, and 90° directions compared with the as-cast samples. This gave rise to improved % elongation, impact toughness, and substantial reduction (33.3%) in hardness. These results are capable of achieving huge savings in die conditioning and replacement with improved quality and sale of deep-drawn AA6063 alloy profiles for sustained profitability.

1. Introduction

In metal forming, texture gradients often ensued due to nonhomogeneous flow. Texture heterogeneities can also occur in other deformation modes, such as sheet rolling, wire drawing and tube extrusion. Thus, in many industrial forming processes for aluminum, mechanical loading is usually combined with some form of heat treatment such as annealing between deformation steps. This serves the purpose of mitigating substantially strain-hardening phenomenon during deformation. A particular example is the stretching of aluminum parts in a number of stages with intermediate annealing treatments. However, this heating and cooling cycle often results in time wastage and could be minimized.

The cause of work hardening during mechanical working varies in different aluminium alloy compositions.

Hardening of nonheat treatable Al-Mg alloys is due mainly to the presence of solute atoms in solid solution. In heat treatable Al-Mg-Si and Al-Cu alloys, strengthening is determined by precipitates formed during aging treatment. For room-temperature forming, the material behaviour of aluminum sheet is completely determined by work hardening and almost independent of the strain rate [1].

Previous study showed that profiles extruded from both homogenized and unhomogenized billets did exhibit the same mechanical properties and metallurgical features [2]. Hence, the whole homogenizing process could be eliminated without compromising any of the mechanical properties. It is established that the factors that determine behaviour of aluminium alloy component are the type, amount, and distribution of second-phase particles such as Al-Fe, Al- Fe-Si, and Al-Fe-Mn-Si dissolved in solution [3]. Furthermore, the phases formed depend mainly on the cooling rate and the Fe/Si ratio in the alloy [4]. Manganese as a common addition in 6XXX alloys increases strength as finely precipitated intermetallics modifies the shape of plate-like iron phases which reduces their embrittling effect [5]. The combination of manganese with Fe, Si, and Al also forms α-$Al_x(Fe,Mn)_ySi_z$ phase that acts as nucleation sites for Mg_2Si crystals, which eventually influences the alloys behavior [6].

Hence, precipitation-hardened aluminium alloys are commercially important group of materials because their mechanical properties can be modified by heat treatment.

Table 1: Chemical composition of Al 6063 alloy.

Element	Al	Mg	Si	Fe	Mn	Ti	Zn	Ni	Cu	Sn
% Composition	98.65	0.499	0.466	0.343	0.015	0.011	0.007	0.006	0.002	0.001

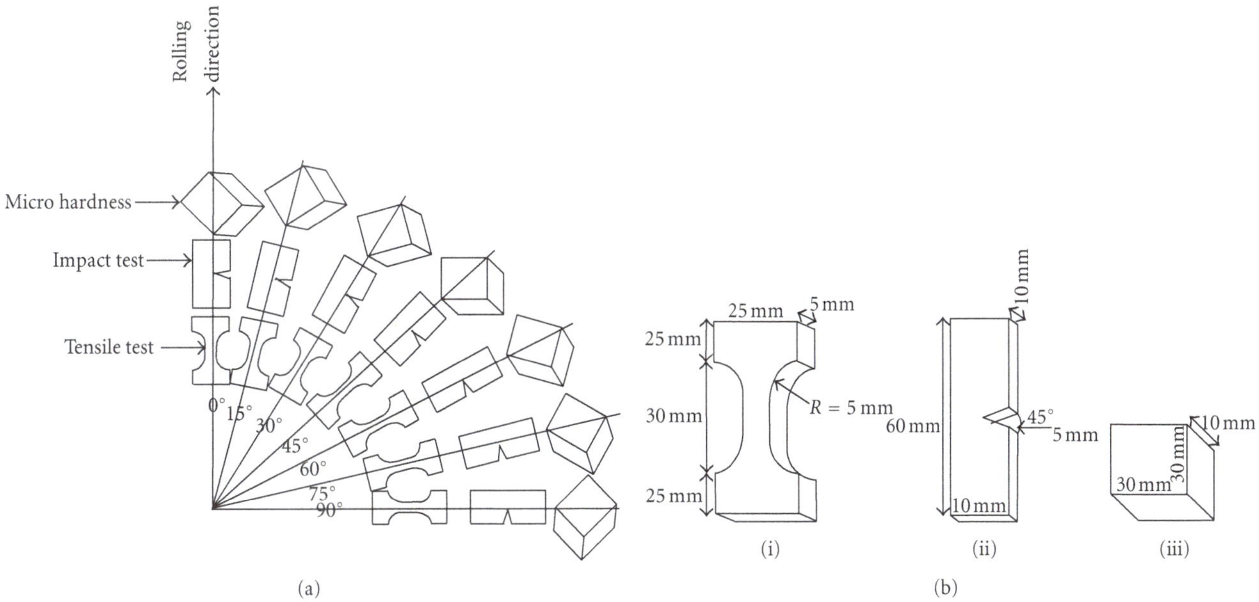

Figure 1: (a) Test pieces cut at different angles to the rolling direction (b) standard test samples (i) tensile, (ii) impact, and (iii) hardness.

However, the imperative of even distribution of the precipitated phases to achieve structural homogeneity is a major challenge. This paper studies the impact of ambient temperature rolling and artificial aging on the normal anisotropy, tensile strength, hardness, and impact resistance of 6063 aluminum alloy.

2. Experimental Methodology

The composition of AA6063 alloy used for this study is given in Table 1. Moulds for the castings were made using hardwood pattern of dimension 350 mm ×150 mm ×15 mm. In the aluminum alloy, 15 kg was charged into a crucible and placed in a pit furnace until molten at 670°C. The molten alloy was poured into the sand moulds and allowed to solidify.

The cleaned casts were divided into three groups, namely, control samples (CS), cold rolled sample (RS), and cold rolled and precipitation hardened sample (RAS) for further processing.

RS and RAS samples were homogenized in a Carbolite furnace at 515°C for 8 hours, air cooled and cold rolled (at 27.4°C surface temperature) to 451 mm × 163 mm ×10 mm samples in 13 passes (at 34.4°C surface temperature) (Figure 1). Standard tensile test pieces were machined (ASTM (E8)) from CS, RS, and RAS samples at 0°, 15°, 30°, 45°, 60°, 75°, and 90° directions using the rolling direction as datum line. Similarly standard, impact, and hardness tests pieces were machined (Figure 1).

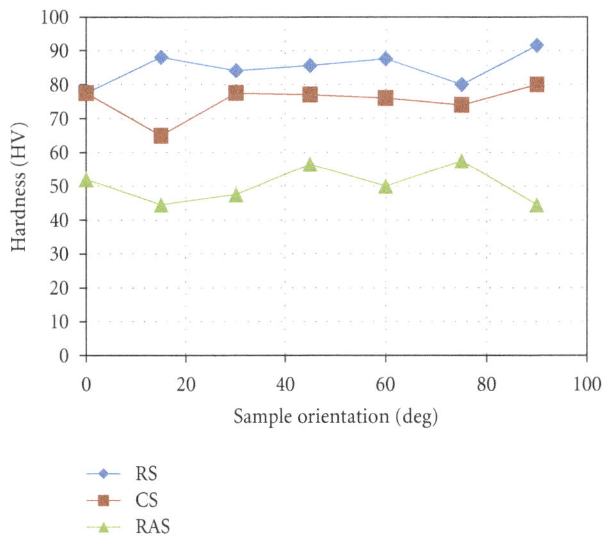

Figure 2: Hardness variations across sample surface.

RAS test pieces were subjected to solution heat treatment at 515°C for 8 hours and furnace cooled. These solutionized pieces were artificially aged at 190°C for 8 hours and air cooled to room temperature. Standard microstructural test pieces from CS, RS, and RAS were ground using emery paper with grit 220 to 600 microns in succession. The ground surfaces of the pieces were polished using a mixture of Al_2O_3

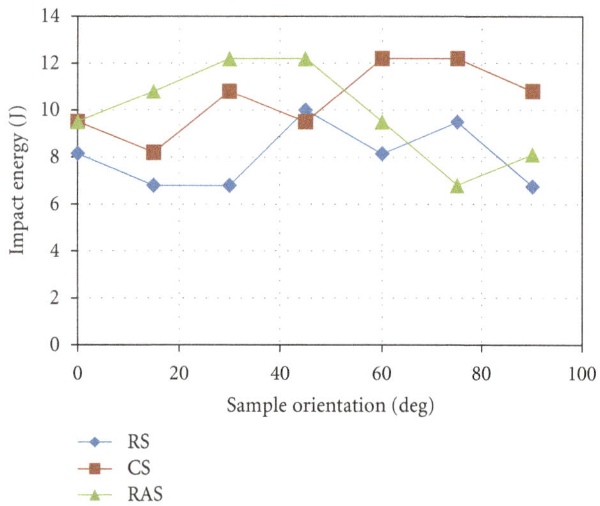

Figure 3: Impact energy variations across sample surface.

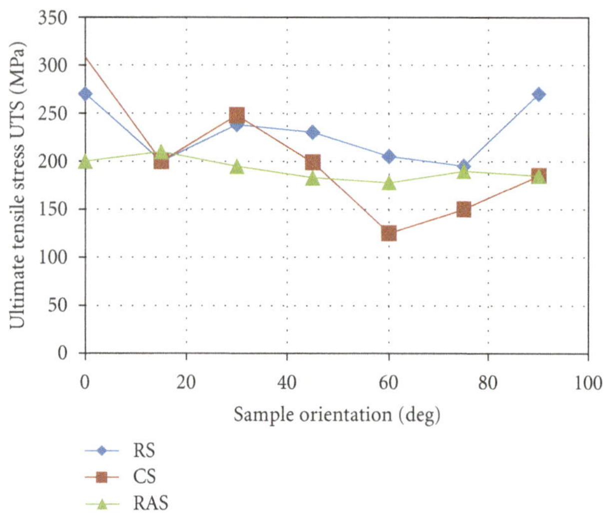

Figure 4: Ultimate tensile stress (UTS) variations across sample surface.

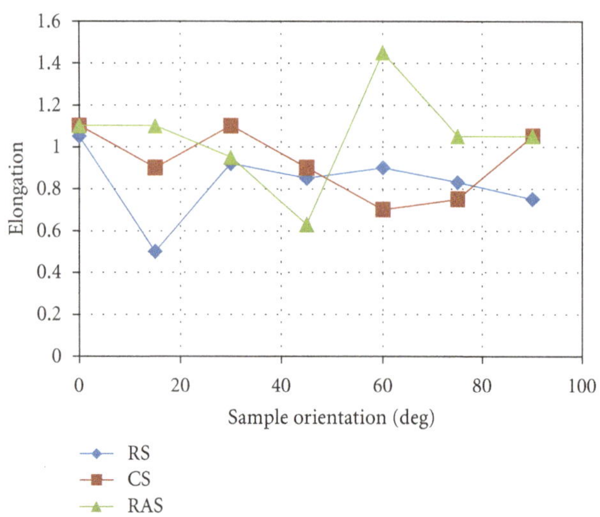

Figure 5: Elongation variations across sample surface.

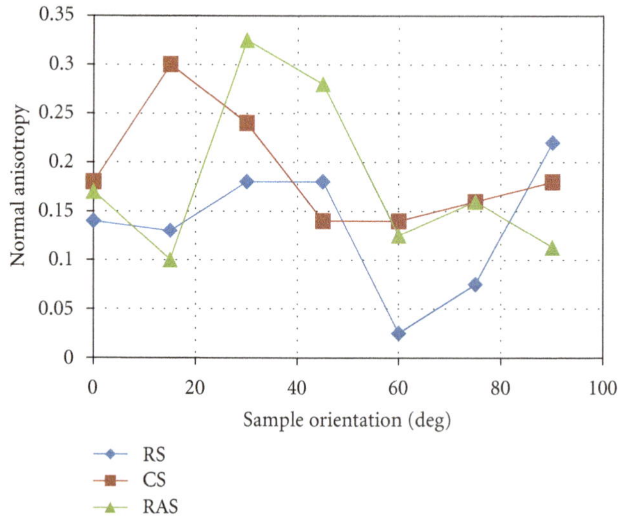

Figure 6: Normal anisotropy variations across sample surface.

and diamond paste before been etched a solution containing 5 grammes of sodium hydroxide (NaOH) dissolved in 100 mL of water, for 20 seconds and dried. The etched surfaces crystals morphology was carried out using a Digital Metallurgical Microscope at ×800 magnification, and the photomicrographs are shown in Figures 7(a)–7(i).

There were 21 samples each for tensile, hardness, and impact tests for the three categories of cast samples. A Vickers microhardness tester model "Deco" 2005 with a test load of 100 g and a dwell time of 10 s was used for the test pieces' hardness determination. Avery Impact Testing Machine (Charpy Tester) model type number 6703 and serial number E67424/4 with a striking velocity of 298.1 J/s was used for impact energy absorption capacity of the test pieces.

3. Results and Discussion

In the aged sample, the hardness in the rolling direction is 51.5 HV which is higher than at 15°, 30°, 60°, and 90° directions but lower to that at 45° (56 HV) and 75° (57.1 HV) directions. Hardness at 45° and 75° directions is similar in as-cast sample, but in the rolling direction it is 77.1 HV.

However, the as-cast hardness value declined at 15° (64.5 HV) but increased again at 30° (78.1 HV) and remain almost the same at 45° (76.2 HV), 60° (75.1 HV) and 75° (72.3 HV) respectively. Thus, peak hardness for as-cast sample is attained in the direction perpendicular to rolling direction. The rolled sample has 76.8 HV in the rolling direction and maximum of 91.1 HV in the 90° direction.

The hardness of rolled sample is superior to as-cast sample in all directions except in the rolling direction where both samples have similar and comparable hardness response (Figure 2). Hardness across the surface for as-cast and cold rolled samples varied considerably at 20% and 15%, respectively. The process of aging the as-cast structure of 6063 aluminum alloy considerably reduced its hardness by about 33.3%, and this paved way for improved formability.

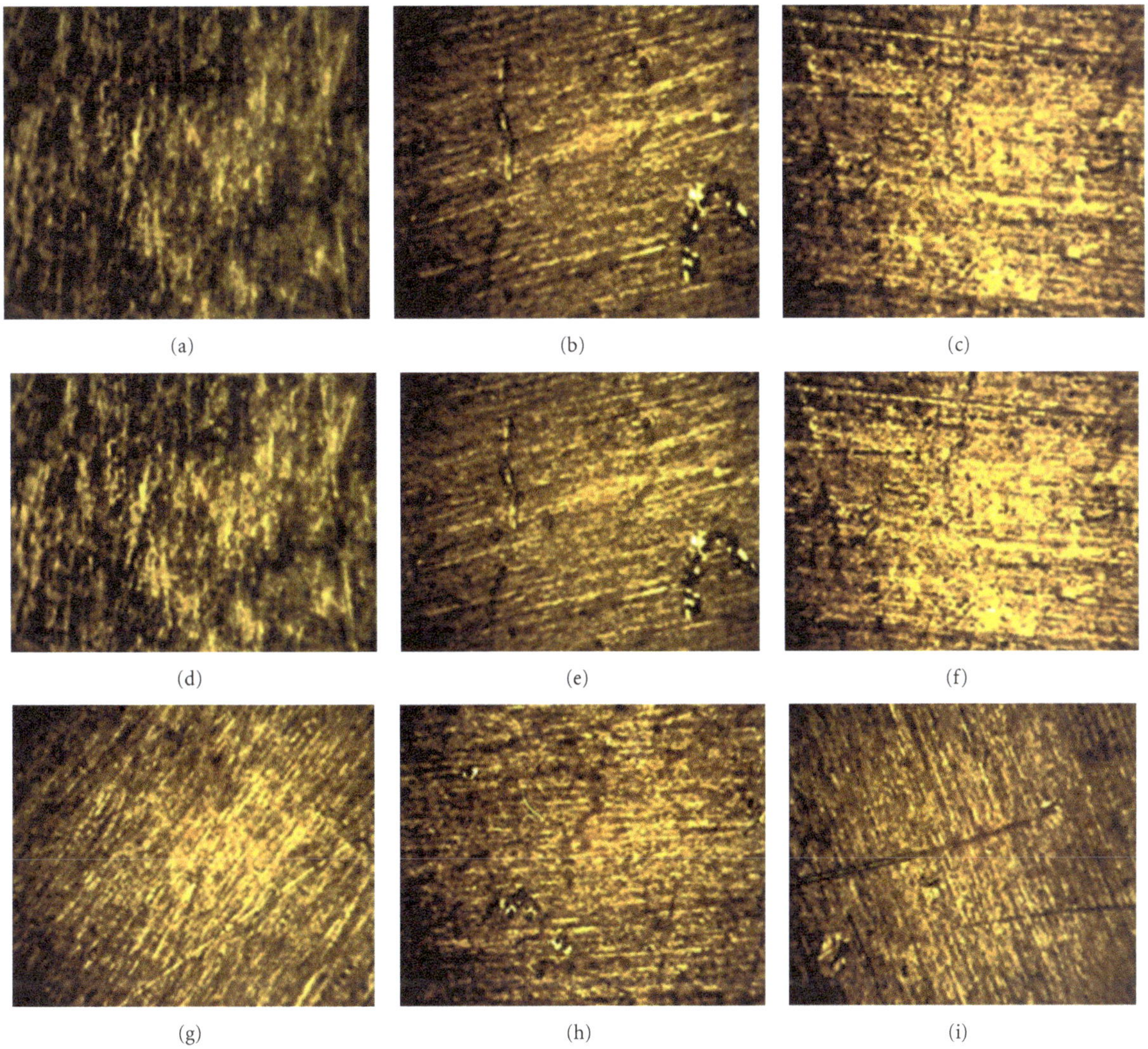

FIGURE 7: Micrographs of samples (a) as-cast at 0°, (b) rolled at 0°, (c) rolled and precipitation Hardened at 0°, (d) as-cast at 45°, (e) rolled at 45°, (f) rolled and precipitation Hardened at 45°, (g) as-cast at 90°, and (h) rolled at 90° (i) rolled and precipitation hardened at 90°.

The impact energy resistances of the samples show sinusoidal pattern (Figure 3). The as-cast sample has minimum impact energy absorption (8.1 J) at 15° and maximum of 12.3 J at 60° and 75° directions. Rolled sample has 6.8 J at 15°, 30°, and 90° directions which is less to that in rolling (8 J), 45°, and 75° directions (9.5 J). For the aged sample, a minimum energy absorption capacity of 6.8 J occurred at 75° direction and maximum of 12.2 J at 30° and 45° directions. Minimal variation in impact toughness behaviour of test samples is observed. This is because the preponderance and the texture of AlFeSi, and other toughness promoting intermetallic phases present do not differ appreciably across the sample surface.

The tensile strength of as-cast sample is maximum (305 MPa) in the rolling direction and 125 MPa minimum at 60° direction. Tensile strength of 270 MPa is obtained in the rolled sample both in the rolling and 90° directions. The tensile strength of the aged sample oscillates between 210 MPa and 180 MPa with peak and minimum strengths at 15° and 60° directions, respectively (Figure 4). The elongations of the samples are similar (1.15) in the rolling direction and are better to that in other directions in as-cast and rolled samples. In the 60° direction, the elongation of aged sample is 1.4 (Figure 5).

In the rolling direction, the normal anisotropy of the samples displayed close responses as the values lie between 0.14 and 0.17. The maximum normal anisotropy for as-cast sample is 0.3 at 15° direction, while the sample's minimum is 0.14 at 45° direction. Normal anisotropy is 0.21 in the 90° direction for rolled sample and 0.03 minimum at 60° direction. The aged sample has maximum normal anisotropy of 0.33 at 30° direction 0.29 in 45° direction (Figure 6).

In the rolling direction Mg_2Si crystals are formed at the grain boundaries in α-aluminum matrix of as-cast sample but are discontinuous threadlike feature in the rolled matrix caused by deformation process. In the aged matrix, the intermetallic crystals are precipitated alongside α- aluminum crystals in company of other intermetallics with high nominal volume fraction than in the as-cast and rolled matrixes (Figures 7(a), 7(b), and 7(c)).

At 45°, there is significant increase in the nominal volume fraction of Mg_2Si crystals which were precipitated in tandem with the α-aluminum crystal and other intermetallics in as-cast matrix. The cold rolled matrix showed decrease in volume of Mg_2Si crystals precipitated while its crystals point in the slip direction. However, there was an increase in nominal volume fraction of Mg_2Si crystals resident at the grain boundaries in aged-sample matrix than in rolled matrix (Figures 7(d), 7(e), and 7(f)).

In the as-cast matrix fairly fine crystals of Mg_2Si are precipitated side by side with crystals of α-aluminum and other intermetallics. As the as-cast sample is being rolled, there is an increase in volume clustering of Mg_2Si crystals in the matrix.

However, by aging the cold rolled sample, recrystallization and precipitation of fresh Mg_2Si crystals occurred in between crystals of α-aluminum and those of other intermetallics in directional pattern (Figures 7(g), 7(h), and 7(i)).

Nucleation of semicoherent precipitates, θ' and θ, often occurs at elevated aging temperature [7]. The θ' and θ precipitates were usually impenetrable obstacles to dislocations resulting in increased work hardening of the rolled sample over other samples. The precipitate spacing also changes with aging, as well as the heterogeneity of the precipitate distribution in the matrix. All these are responsible for the changes in mechanical properties observed with test samples.

4. Conclusion

The results of this study have shown that moderate thermal treatment and enough time allowed for recovery ensure redistribution of the Mg_2Si precipitates and other intermetallics within the matrix. This phenomenon is the precursor for the improved formability, better fracture toughness, and significant reduction in hardness coupled with minimal anisotropy.

Acknowledgment

The authors are grateful to the Management and Staff of Nigerian Aluminum Extrusion (NIGALEX) Company, Oshodi, Lagos Nigeria for the provision of 6063 aluminum alloy used in this study.

References

[1] S. Kurukuri, *Simulation of thermally assisted forming of aluminum sheet*, Ph.D. thesis, University of Twente, Tweante, The Netherlands, 2010.

[2] G. Al-Marahleh, "Effect of heat treatment parameters on distribution and volume fraction of Mg_2Si in the structural Al 6063 alloys," *American Journal of Applied Science*, vol. 3, no. 5, pp. 1819–1823, 2006.

[3] N. C. W. Kuijpers, W. H. Kool, P. T. G. Koenis, K. E. Nilsen, I. Todd, and S. Van der Zwaag, "Assessment of different techniques for quantification of α-Al(FeMn)Si and β-AlFeSi intermetallics in AA 6xxx alloys," *Materials Characterization*, vol. 49, no. 5, pp. 409–420, 2002.

[4] R. A. Siddiqui, H. A. Abdullah, and K. R. Al-Belushi, "Influence of aging parameters on the mechanical properties of 6063 aluminium alloy," *Journal of Materials Processing Technology*, vol. 102, no. 1, pp. 234–240, 2000.

[5] A. K. Gupta, D. J. Lloyd, and S. A. Court, "Precipitation hardening in Al—Mg—Si alloys with and without excess Si," *Materials Science and Engineering A*, vol. 316, no. 1-2, pp. 11–17, 2001.

[6] G. Sha, K. O'Reilly, B. Cantor, J. Worth, and R. Hamerton, "Growth related metastable phase selection in a 6xxx series wrought Al alloy," *Materials Science and Engineering A*, vol. 304—306, no. 1-2, pp. 612–616, 2001.

[7] H. Sehitoglu, T. Foglesong, and H. J. Maier, "Precipitate effects on the mechanical behavior of aluminum copper alloys," *Metallurgical and Materials Transactions A*, vol. 36, pp. 1–13, 2005.

Permissions

The contributors of this book come from diverse backgrounds, making this book a truly international effort. This book will bring forth new frontiers with its revolutionizing research information and detailed analysis of the nascent developments around the world.

We would like to thank all the contributing authors for lending their expertise to make the book truly unique. They have played a crucial role in the development of this book. Without their invaluable contributions this book wouldn't have been possible. They have made vital efforts to compile up to date information on the varied aspects of this subject to make this book a valuable addition to the collection of many professionals and students.

This book was conceptualized with the vision of imparting up-to-date information and advanced data in this field. To ensure the same, a matchless editorial board was set up. Every individual on the board went through rigorous rounds of assessment to prove their worth. After which they invested a large part of their time researching and compiling the most relevant data for our readers. Conferences and sessions were held from time to time between the editorial board and the contributing authors to present the data in the most comprehensible form. The editorial team has worked tirelessly to provide valuable and valid information to help people across the globe.

Every chapter published in this book has been scrutinized by our experts. Their significance has been extensively debated. The topics covered herein carry significant findings which will fuel the growth of the discipline. They may even be implemented as practical applications or may be referred to as a beginning point for another development.

The editorial board has been involved in producing this book since its inception. They have spent rigorous hours researching and exploring the diverse topics which have resulted in the successful publishing of this book. They have passed on their knowledge of decades through this book. To expedite this challenging task, the publisher supported the team at every step. A small team of assistant editors was also appointed to further simplify the editing procedure and attain best results for the readers.

Our editorial team has been hand-picked from every corner of the world. Their multi-ethnicity adds dynamic inputs to the discussions which result in innovative outcomes. These outcomes are then further discussed with the researchers and contributors who give their valuable feedback and opinion regarding the same. The feedback is then collaborated with the researches and they are edited in a comprehensive manner to aid the understanding of the subject.

Apart from the editorial board, the designing team has also invested a significant amount of their time in understanding the subject and creating the most relevant covers. They scrutinized every image to scout for the most suitable representation of the subject and create an appropriate cover for the book.

The publishing team has been involved in this book since its early stages. They were actively engaged in every process, be it collecting the data, connecting with the contributors or procuring relevant information. The team has been an ardent support to the editorial, designing and production team. Their endless efforts to recruit the best for this project, has resulted in the accomplishment of this book. They are a veteran in the field of academics and their pool of knowledge is as vast as their experience in printing. Their expertise and guidance has proved useful at every step. Their uncompromising quality standards have made this book an exceptional effort. Their encouragement from time to time has been an inspiration for everyone.

The publisher and the editorial board hope that this book will prove to be a valuable piece of knowledge for researchers, students, practitioners and scholars across the globe.

List of Contributors

Miguel A. Barron, Dulce Y. Medina and Isaias Hilerio
Departamento de Materiales, Universidad Autonoma Metropolitana Azcapotzalco, Avenida San Pablo 180, Colonia Reynosa-Tamaulipas, 02200 Mexico, DF, Mexico

Gabriel Plascencia
Centro de Investigacion e Innovacion Tecnologica-IPN, Cerrada Cecati S/N, Colonia Santa Catarina, 02250 Mexico, DF, Mexico

P. Koltun
CSIRO Process Science and Engineering, Gate 5, Normanby Road, Clayton, VIC 3168, Australia

A. Tharumarajah
Envi Solutions, 20 Galahad Crescent, Glen Waverley, Melbourne, VIC 3150, Australia

Udaya Bhat Kuruveri, Prashanth Huilgol and Jithin Joseph
Department of Metallurgical and Materials Engineering, NITK, Surathkal 575025, India

Fanke Meng and Huimin Lu
School of Materials Science and Engineering, Beihang University, No.37 Xueyuan Road, Haidian District, Beijing 100191, China

P. Sreeraj
Department of Mechanical Engineering, Valia Koonambai kulathamma College of Engineering and Technology, Kerala 692574, India

T. Kannan
Department of Mechanical Engineering, SVS College of Engineering, Coimbatore, Tamil nadu 642109, India

Subhashis Maji
Department of Mechanical Engineering, IGNOU, Delhi 110068, India

Fanke Meng and Huimin Lu
School of Materials Science and Engineering, Beihang University, No. 37 Xueyuan Road, Haidian District, Beijing 100191, China

Bijayalaxmi Panda
Amity School of Engineering and Technology, Amity University, Noida, Uttar Pradesh 201303, India

Hemant Upadhyay
Blast Furnace Department, Jindal Steel & Power Ltd., Raigarh, Chhattisgarh 496001, India

T. K. Kundu
Department of Metallurgical and Materials Engineering, Indian Institute of Technology Kharagpur, Kharagpur, West Bengal 721302, India

Debasish Chatterjee
Department of Metallurgy & Materials Engineering, Indian Institute of Technology, Roorkee 247667, Uttarakhand, India

Prantik Mukhopadhyay
Electron Microscope Group, Defence Metallurgical Research Laboratory, Kanchanbagh, Hyderabad 500058, India

Siddhartha Tiwari
Department of Materials Science and Metallurgical Engineering, University Institute of Engineering & Technology, CSJM University, Kanpur 208024, India
Department of Metallurgical and Materials Engineering, Indian Institute of Technology Kharagpur, Kharagpur 721302, India

Priyanka Rajput and Sanjay Srivastava
Department of Materials Science and Metallurgical Engineering, University Institute of Engineering & Technology, CSJM University, Kanpur 208024, India

J. A. Montemayor-Aldrete, R.Mendez-Olvera and M. del Castillo-Mussot
Departamento de Estado Solido, Instituto de Fısica, Universidad Nacional Autonoma de Mexico, Apartado Postal 20-364, 01000 Mexico City, DF, Mexico

P. Ugalde-Velez
Departamento de Materiales (CBI), Universidad Autonoma Metropolitana-Azcapotzalco, Avenida San Pablo 180, Col. Reynosa, 02200 Mexico City, DF, Mexico

G. Torres-Villasenor
Instituto de Investigacion en Materiales, Universidad Nacional Autonoma de Mexico, Apartado Postal 70-360, 04510 Mexico City, DF, Mexico

A. Ghorai and Arjun Das
Department of Physics, Chandernagore College, Chandernagore 712136, India

D. D. Radev
Institute of General and Inorganic Chemistry, BAS, "Acad. G. Bonchev" Street, Bl.11, Sofia 1113, Bulgaria

Subir Paul
Department of Metallurgical and Material Engineering, Jadavpur University, Kolkata 700032, India

S. Chenna Krishna, Abhay K. Jha, Bhanu Pant and S. C. Sharma
Materials and Metallurgy Group, Materials and Mechanical Entity, Vikram Sarabhai Space Centre, Trivandrum 695022, India

N. Supriya
Analytical, Spectroscopy and Ceramics Group, Propellants, Polymers, Chemicals and Materials Entity, Vikram Sarabhai Space Centre, Trivandrum 695022, India

Koshy M. George
Materials and Mechanical Entity, Vikram Sarabhai Space Centre, Trivandrum 695022, India

G. Antoni
LMA UPR 7051, CNRS, Aix-Marseille Universite, 13331 Marseille Cedex 03, France

Rupa Dasgupta
CSIR-Advanced Materials and Processes Research Institute (AMPRI), Hoshangabad Road, Madhya Pradesh, Bhopal 462064, India

Eva Fazakas and Lajos Karoly Varga
Wigner Research Centre for Physics, Institute for Solid State Physics and Optics, Hungarian Academy of Sciences, P.O. Box 49, Budapest 1525, Hungary

Bela Varga
Faculty of Materials Science and Engineering, Transylvania University, Bulevardul Eroilor, No. 29, 500036 Brasov, Romania

K. Boubaker
Unite de Physique des Dispositifs `a Semi-Conducteurs, Faculte des Sciences de Tunis, Tunis El Manar University, 2092 Tunis, Tunisia